Markus Christen

The Role of Spike Patterns in Neuronal Information Processing

Markus Christen

The Role of Spike Patterns in Neuronal Information Processing

A Historically Embedded Conceptual Clarification

Südwestdeutscher Verlag für Hochschulschriften

Impressum/Imprint (nur für Deutschland/only for Germany)
Bibliografische Information der Deutschen Nationalbibliothek: Die Deutsche Nationalbibliothek verzeichnet diese Publikation in der Deutschen Nationalbibliografie; detaillierte bibliografische Daten sind im Internet über http://dnb.d-nb.de abrufbar.
Alle in diesem Buch genannten Marken und Produktnamen unterliegen warenzeichen-, marken- oder patentrechtlichem Schutz bzw. sind Warenzeichen oder eingetragene Warenzeichen der jeweiligen Inhaber. Die Wiedergabe von Marken, Produktnamen, Gebrauchsnamen, Handelsnamen, Warenbezeichnungen u.s.w. in diesem Werk berechtigt auch ohne besondere Kennzeichnung nicht zu der Annahme, dass solche Namen im Sinne der Warenzeichen- und Markenschutzgesetzgebung als frei zu betrachten wären und daher von jedermann benutzt werden dürften.

Coverbild: www.ingimage.com

Verlag: Südwestdeutscher Verlag für Hochschulschriften GmbH & Co. KG
Heinrich-Böcking-Str. 6-8, 66121 Saarbrücken, Deutschland
Telefon +49 681 37 20 271-1, Telefax +49 681 37 20 271-0
Email: info@svh-verlag.de

Approved by: Zürich, ETH, Diss., 2006

Herstellung in Deutschland (siehe letzte Seite)
ISBN: 978-3-8381-3347-8

Imprint (only for USA, GB)
Bibliographic information published by the Deutsche Nationalbibliothek: The Deutsche Nationalbibliothek lists this publication in the Deutsche Nationalbibliografie; detailed bibliographic data are available in the Internet at http://dnb.d-nb.de.
Any brand names and product names mentioned in this book are subject to trademark, brand or patent protection and are trademarks or registered trademarks of their respective holders. The use of brand names, product names, common names, trade names, product descriptions etc. even without a particular marking in this works is in no way to be construed to mean that such names may be regarded as unrestricted in respect of trademark and brand protection legislation and could thus be used by anyone.

Cover image: www.ingimage.com

Publisher: Südwestdeutscher Verlag für Hochschulschriften GmbH & Co. KG
Heinrich-Böcking-Str. 6-8, 66121 Saarbrücken, Germany
Phone +49 681 37 20 271-1, Fax +49 681 37 20 271-0
Email: info@svh-verlag.de

Printed in the U.S.A.
Printed in the U.K. by (see last page)
ISBN: 978-3-8381-3347-8

Copyright © 2012 by the author and Südwestdeutscher Verlag für Hochschulschriften GmbH & Co. KG and licensors
All rights reserved. Saarbrücken 2012

Contents

    Zusammenfassung   5

    Summary   6

1 Introductory Remarks   9
    1.1 Goals of this Study 9
    1.2 Methodology 11
    1.3 Organization of the Thesis 11
        1.3.1 Main Line of Argumentation 11
        1.3.2 Typographic Structure and Layout 13
    1.4 Abbreviations and List of Symbols 14
        1.4.1 Abbreviations 14
        1.4.2 List of Symbols 15

I Historical Roots of the Information Processing Brain   17

2 The Brain on the Cusp of the Information Age   19
    2.1 Introduction 19
        2.1.1 The Information Processing Brain 19
        2.1.2 The Historical Context 21
    2.2 Cornerstones of the Information Age 23
        2.2.1 The Conceptualization of Information 23
        2.2.2 Information Theory 24
        2.2.3 Cybernetics 27
    2.3 Preconditions for the Information Processing Brain 29
        2.3.1 The Scheme for Analysis 29
        2.3.2 The Neuron Doctrine 31
        2.3.3 Spikes and Messages 33
        2.3.4 Measuring Single Fibres 36

3 The Birth of the Information Processing Brain   39
    3.1 1940-1970: Overview 39
    3.2 1940-1970: Detailed Analysis 41

		3.2.1	A new Role for the Neuron?	41
		3.2.2	Applying the Information Vocabulary	43
		3.2.3	The Emergence of a Toolbox for Spike Train Analysis	59
		3.2.4	The (Random) Network	61
		3.2.5	The Active and Reliable Brain	64
		3.2.6	The Modelling Approach	69

4 1940 – 1970: The Dynamics of Research   **73**
- 4.1 Journals and Publications 73
- 4.2 Conferences 77
- 4.3 Protagonists 83
- 4.4 Synopsis: Qualitative and Quantitative Analysis 95
  - 4.4.1 Up to 1940: Preconditions 95
  - 4.4.2 1940-1970: Main Developments 96

II Patterns in Neuronal Spike Trains   101

5 Neural Coding and Computation   **103**
- 5.1 Neural Coding 103
  - 5.1.1 What is a 'Neural Code'? 103
  - 5.1.2 Rate Coding vs. Temporal Coding 110
  - 5.1.3 Single Neurons vs. Neuronal Assemblies 112
- 5.2 Neural Computation 113

6 Defining Patterns   **115**
- 6.1 Characteristics of Patterns 115
  - 6.1.1 Defining 'Spike Pattern' 115
  - 6.1.2 Types of Events 118
  - 6.1.3 The Stability of Patterns 121
- 6.2 The Background of a Pattern 123
  - 6.2.1 The Significance of Patterns 123
  - 6.2.2 The POISSON Hypothesis 123
  - 6.2.3 Randomization Methods 126
- 6.3 Noise and Reliability 129
  - 6.3.1 What is Noise in Neuronal Systems? 129
  - 6.3.2 Noise Sources in Neurons 130
  - 6.3.3 The Reliability of Neurons 132
  - 6.3.4 Can Neurons Benefit from Noise? 135
- 6.4 Functional Aspects of Spike Patterns 136
  - 6.4.1 Causing Patterns: Network, Oscillation, Synchronization 136
  - 6.4.2 Using Patterns: LTP and Coincidence-Detection 142
- 6.5 Spike Patterns: Examples 144
- 6.6 The Stoop-Hypothesis 148
  - 6.6.1 Neurons as Limit Cycles 148
  - 6.6.2 Locking – The Coupling of Limit Cycles 150

		6.6.3	Emergence of a Coding Scheme	152
		6.6.4	The Role of Patterns	157
		6.6.5	The Integral Framework	159

7 Detecting Patterns — 163
7.1 The Problem of Pattern Detection 163
    7.1.1 Outlining the Problem 163
    7.1.2 Pattern Detection Methods: Overview 164
    7.1.3 Spike Train Distance Measures: Overview .. 165
7.2 Histogram-based Pattern Detection 167
    7.2.1 Classes of Histograms 167
    7.2.2 1D Interval Histograms 172
7.3 Correlation Integral based Pattern Discovery .. 175
    7.3.1 The Correlation Integral 175
    7.3.2 Smeared log-log Steps 176
    7.3.3 Pattern Length Estimation 178
7.4 Template-based Pattern Recognition 183
    7.4.1 The Template Approach 183
    7.4.2 Problems with Templates 185
7.5 Clustering-based Pattern Quantification 186
    7.5.1 The Clustering Algorithm 186
    7.5.2 Relating Patterns and Clusters 187
7.6 The LZ-Distance of Spike Trains 190
    7.6.1 The LZ-Distance Measure 191
    7.6.2 Assessment of the LZ-Distance 192
7.7 Pattern Detection: Summary 196

8 Investigating Patterns — 199
8.1 Description of the Data 199
    8.1.1 Rat Data: Olfactory System (mitral cells) .. 199
    8.1.2 Cat Data: Visual System (LGN, striate cortex) .. 200
    8.1.3 Monkey Data: Visual System (LGN, V1, MT) .. 200
8.2 First Prediction: Firing statistics 201
8.3 Second Prediction: Firing Reliability 203
8.4 Third Prediction: Neuronal Clustering 206
8.5 Fourth Prediction: Classes of Firing 209
8.6 Fifth Prediction: Pattern Stability 211

9 Conclusions — 213
9.1 Support for the Stoop-Hypothesis 213
9.2 The Stoop-Hypothesis in the General Context . 215
9.3 Outlook: Experimental Approaches 215

III Appendix	217
Figures	219
Tables	220
Bibliography	221
Index	255

Zusammenfassung

Es ist heutzutage ein Gemeinplatz, das Gehirn als informationsverarbeitendes System zu betrachten. Obgleich dieser Ausdruck sowohl innerhalb der Neurowissenschaft wie auch in einem weiteren Umfeld breite Verwendung findet, drückt er kein präzises Verständnis über die Vorgänge im Nervensystem aus. Klar ist zwar, dass der Begriff nicht (mehr) eine enge Analogie zwischen klassischen Computern und dem Gehirn zum Ausdruck bringen soll. Heute wird vielmehr ein prinzipieller Unterschied zwischen diesen beiden Arten der Informationsverarbeitung postuliert. Der Informationsbegriff der Neurowissenschaft wie auch die mit der Informationsverarbeitung verbundenen Prozesse, die oft als ein neuronaler Kode (*neural code*) oder eine neuronale Berechnung (*neural computation*) aufgefasst werden, bleiben damit weiterhin Gegenstand intensiver Forschungen. Diese Forschungen benötigen klare und definierte Begriffe. Die vorliegende Arbeit will zu dieser Klärung beitragen, indem der Begriff des 'Feuermusters' (*spike pattern*) definiert und hinsichtlich seiner Anwendung in Kontext des *neural coding* und der *neural computation* dargelegt werden soll.

Diese Arbeit verbindet eine naturwissenschaftliche Analyse – welche die Begriffsklärung, eine Hypothesenbildung und die Untersuchung experimenteller Daten umfasst – mit einer wissenschafthistorischen Untersuchung. Letztere will die historischen Wurzeln der heutigen Debatten um *neural coding* und *neural computation* freilegen. Der Schwerpunkt der historischen Analyse liegt in den 1940er bis 1960er Jahren – jenen Jahrzehnten also, in welchen eine Verwissenschaftlichung des Informationsbegriffs im Verbund mit der aufkommenden Informationstheorie und Kybernetik stattfand, die zu einem eigentlichen 'Informations-Vokabular' führten, mit zentralen Begriffen wie Kode (*code*), Berechnung (*computation*) und Rauschen (*noise*). Im historischen Teil werden zudem jene zuvor stattgefundenen Entwicklungen innerhalb der Neurowissenschaft skizziert, welche Voraussetzungen für die Anwendbarkeit des 'Informations-Vokabulars' schufen. Wichtige Beispiele sind die Entwicklung von Geräten für die zuverlässige Messung des neuronalen Feuerns und die Diskussion über *messages* in solchen Messungen des neuronalen Feuerverhaltens. Danach zeigen wir anhand eines detaillierten historischen Analyse-Schemas auf, welche Entwicklungen zum Begriff des 'informationsverarbeitenden Gehirns' beigetragen haben. Diese Untersuchung wird mit bibliometrischen und scientometrischen Analysen ergänzt, welche zur Identifikation wichtiger Konferenzen und zentraler Protagonisten des damaligen Diskurses dienen.

Auch der naturwissenschaftliche Teil beinhaltet eine Übersicht über einige der heute wichtigen Theorien über die neuronale Kodierung bzw. die neuronale Informationsverarbeitung. Danach folgt eine Schärfung des Begriffs 'Feuermuster' anhand seiner vielfältigen Verwendungsweisen in der wissenschaftlichen Literatur. Wir zeigen auf, dass dieser Begriff immer zusammen mit statistischen Hypothesen über das 'Nicht-Muster' verwendet werden muss, geben eine Übersicht der gängigen Randomisierungsverfahren und diskutieren die Rolle des *neuronal noise*. Danach präsentieren wir im Detail eine von RUEDI STOOP und Mitarbeitern entwickelte Hypothese, in welcher die verschiedenen diskutierten Aspekte – insbesondere die Feuermuster und *neuronal noise* – eine Verbindung eingehen. Diese STOOP-Hypothese basiert auf der Feststellung, dass sich kortikale Neuronen unter quasistationären Bedingungen als Grenzzyklen beschreiben lassen. Diese werden von den zahlreichen einkommenden synaptischen Impulsen (Hintergrund-Aktivität, oft als eine Form von *neuronal noise* aufgefasst), die sich unter der Bedingung der Quasi-Stationarität als konstanten Strom auffassen lassen, getrieben. Die Kopplung solcher Grenzzyklen führt zum generischen Phänomen der Frequenzeinrastung (*locking*). Damit lässt sich ein Schema neuronaler Kodierung erstellen, wonach sich die unterschiedliche Hintergrund-Aktivität zweier Neuronen, die wiederum Ausdruck spezifischer sensorischer Information sind, in ein spezifisches Feuermuster niederschlagen.

Aufbauend auf dieser Hypothese werden fünf Voraussagen abgeleitet, welche sich bei der Untersuchung neuronaler Daten zeigen lassen sollten. Demnach sollte erstens nur eine Minderheit von kortikalen Neuronen eine mit dem POISSON-Modell vereinbare Feuerstatistik zeigen. Zweitens erwarten wir Unterschiede in der Zuverlässigkeit des Feuerns hinsichtlich des *timings* von Nervenimpulsen und der Reproduzierbarkeit spezifischer Feuermuster je nach Ort der Messung entlang des Pfades der neuronalen Informationsverarbeitung. Drittens sollten sich in der parallelen Messung einer Vielzahl von Neuronen Gruppen von Neuronen mit ähnlichen Feuermustern finden. Viertens erwarten wir, dass eine bereits festgestellte Klassierung von Neuronen in drei Gruppen (solche mit unkorreliertem Feuern, solche mit instabilen Feuermustern und solche mit stabilen Feuermuster) auch im von uns untersuchten Datenset feststellbar ist. Fünftens erwarten wir, dass Feuermuster, die mutmasslich Ergebnis einer neuronalen Informationsverarbeitung sind, in einem von uns definierten Sinn stabiler sind als solche, die von Stimulus-Eigenschaften herrühren, die vermutlich für den Organismus keine Rolle spielen.

Ein zentraler Teil der Arbeit bildet eine Darstellung des Problems der Erkennung von Feuermustern. Wir bieten dazu eine Übersicht über gängige Histogramm und *Template* basierende Methoden und zeigen die damit verbundenen Schwierigkeiten auf. Danach präsentieren wir eine Reihe neuer Methoden für die Mustererkennung, basierend auf dem Korrelationsintegral, dem sequenziellen superparamagnetischen Clustering-Algorithmus und der LEMPEL-ZIV-Komplexität. Möglichkeiten und Grenzen dieser Methoden werden im Detail vorgeführt und in einem allgemeinen Schema zum Problem der Mustererkennung in neuronalen Daten – so genannten *spike trains* – integriert.

Im empirischen Teil zeigen wir einerseits die Anwendbarkeit der von uns entwickelten Methoden auf. Andererseits konnten wir, obwohl die uns zur Verfügung gestandenen Daten ursprünglich für anderweitige Zwecke erhoben wurden und damit nicht optimal waren, die genannten Hypothesen bestätigen. Dies sind wichtige Indizien für die Gültigkeit der STOOP-Hypothese zur neuronalen Informationsverarbeitung. Eine eigentliche experimentelle Prüfung der Hypothese war indes aufgrund externer Gründe nicht möglich. Wir zeigen deshalb auch, welche weiteren Schritte für die Prüfung der STOOP-Hypothese angezeigt sind. In ihrer Gesamtheit soll die vorliegende Arbeit auch als Einführung in die Problematik der neuronalen Informationsverarbeitung verstanden werden. Aus diesem Grund wurde auf gestalterische Fragen grossen Wert gelegt, so dass diese Arbeit dem Leser als Hilfsmittel und Nachschlagewerk dienen kann.

Summary

Today, understanding the brain as a 'information processing device' is a commonplace. Although this expression is widely used within and outside of neuroscience, it does not really express a precise understanding of the neuronal processes. However, it is clear that the 'information processing brain' does no longer express a close relationship between classical (digital) computers and the brain. Rather, a clear distinction between these two modes of information processing is postulated. Nevertheless, the concept of information in neuroscience, as well as the related concepts of neural coding and neural computation, are still a topic of intensive research within neuroscience. This also requires some clarification on the conceptual level. This PhD thesis intends to contribute to this clarification by defining the term 'spike pattern' and by evaluating the application of this concept within the framework of neural coding and neural computation.

This thesis unifies a scientific analysis – including a conceptual clarification, the formulation of a hypothesis and data analysis – with a historical investigation. The latter intends to analyze the historical roots of today's discussion on neural coding and neural computation. The focus of the historical analysis lies in the period of the 1940s to 1960s – the decades in which a scientific conceptualization of information in relation to the emerging information theory and cybernetics is observed. This led to the formation of a 'information vocabulary', whose central terms are 'code', 'computation' and 'noise'. Furthermore, in the historical part, those development within the history of neuroscience before this period that formed the preconditions for the application of this vocabulary will be sketched. Important examples are the construction of measurement devices that allowed the reliable recording of neuronal firing, and the discussion about 'messages' in neuronal data. Then we show, using a detailed scheme of analysis, which specific developments led to the formation of the notion of a 'information processing brain'. This investigation is accompanied by several bibliometric and scientometric studies, which lead to the identification of important conferences and protagonists during this period.

The scientific part also includes a review of some current theories of neural coding and neural information processing. This will be followed by a sharpening of the concept of 'spike pattern', referring to its current use in the neuroscience literature. We show that this concept always involves a clear statistical hypothesis of a 'non-pattern', we provide a overview of the common randomization methods for spike data and we discuss the role of neuronal noise. Then we present a hypothesis, developed by RUEDI STOOP and co-workers (STOOP-hypothesis), in which the discussed concept – notably spike patterns and noise – can be put in a common context. The STOOP-hypothesis is based on the finding that cortical neurons under quasi-stationary conditions can be described as limit cycles. They are driven by the numerous incoming synaptic impulses (background-activity, often understood as one aspect of neuronal noise) which result, assuming a GAUSSian central limit behavior, in a almost constant current. The coupling of limit cycles leads to the generic phenomenon of locking. In this way a coding scheme emerges such that the background-activity that is imposed on two coupled neurons reflecting specific sensory information, is encoded in a specific firing pattern. Based on this hypotheses, five predictions are derived: First, we expect that only a minority of neurons display a POISSON firing statistics. Second, we expect a different reliability of timing and pattern for neurons measured along the neuronal information processing pathway. Third, in multi-array recordings, we expect groups of neurons that show a similar firing behavior. Fourth, we expect to reproduce earlier findings of three classes of neuronal firing (uncorrelated firing, unstable and stable pattern firing) in our extended data set. Fifth, we expect that firing patterns that may reflect neuronal computation are more stable (in a precise sense defined by us) than patterns that reflect aspects of stimuli that are probably of no interest for the organism.

A central part of this work consists in providing a discussion of the pattern detection problem. We give an overview of current histogram and template based methods for pattern detection and discuss the difficulties that arise when these methods are applied. Then we provide several novel methods for pattern detection based on the correlation integral, the sequential superparamagnetic clustering algorithm, and the LEMPEL-ZIV-complexity. Possibilities and limitations of these methods are outlined in detail and integrated in a general scheme of the spike pattern detection problem.

In the empirical part, we demonstrate the application of the developed methods. We were furthermore able to verify our predictions although the data available was not optimal for us. These indicators support the STOOP-hypothesis. However, a detailed experimental test of the hypothesis was not possible due to external reasons. Therefore, we also discuss further experimental steps that may serve for testing the hypothesis. In total, this thesis can also be understood as an introduction into the problem of neuronal information processing. Therefore, care has been used on the structure and layout of this thesis, so that it can serve as a help and a reference book for the interested reader.

Chapter 1

Introductory Remarks

In the first chapter, the goals, the methods used, and the main line of argumentation are presented. The thesis is divided in two parts. The historical part discusses the history of the 'information processing brain' and uncovers the historical roots of central concepts – notably 'neural code', 'neural noise' and 'neural computation'. The scientific part provides a framework for defining the term 'spike pattern' within the current discussion on neural coding and computation, relates this concept with a hypothesis on neural information processing, discusses the methodological problems associated with spike pattern detection and demonstrates the application of our methods for a variety of neuronal data. The chapter concludes with a list of the abbreviations and symbols used.

1.1 Goals of this Study

Today, biological neural systems are viewed as an alternative information processing paradigm, that often proves far more efficient than conventional signal processing. Although the underlying structures (neurons and their connectivity) can be accurately modelled, the principles according to which they process information are not well understood. The search for general principles of neuronal information processing is a characteristic aspect of modern neuroscience, where concepts originating from a technological world – code, noise, computation – entered the biological world. As growing evidence suggests that neuronal information encoding differs largely from that of traditional signal processing, the brain became a model for new technology that, however, is still waiting to be developed.

The concept of 'neural coding' is a general term that unifies many different questions whose solutions may contribute to such a new technology. One basic question within this framework is, what 'information' in a neural context really means and how this 'information' is 'encoded' and 'processed' within the neural system. The number of proposals that address these and related questions is enormous, but they usually refer to some kind of 'pattern' within spike trains that reflects the information or its processing. Consequently, the detection of pattern occurrence can be considered a fundamental step in the analysis of neuronal coding. However, although spike patterns are generally believed to be a fundamental concept in neuroscience, their definition is dependent on circumstances and intentions: Pattern can be detected in single spike trains or multiple spike trains, others focus on higher-order patterns or suggest to allow time-rescaling of parts of the data. Thus, a clarification of the concept of patterns in spike trains, as well as an evaluation of their possible meaning for the neural system within a theoretical framework, is still lacking.

According to the preliminary concept (dated September 30th, 2001) and the research plan (dated May 06th, 2003) of this thesis, the work is embedded into a theoretical framework developed by RUEDI STOOP an co-workers, which is presented in detail in section 6.6. Briefly, the STOOP-hypothesis postulates that most neurons can be described as noise-driven limit cycles, whose coupling lead to the generic phenomenon of locking. In this way, a coding scheme emerges such that the changing background-activity, imposed on two coupled neurons and reflecting specific sensory information, is encoded in specific firing patterns. Latter are subject of a pattern detection problem, whose solution requires methods that are developed within this thesis.

This work was largely financed by the Swiss National Science Foundation (SNF) project No. 247076 *Locking als makroskopisches Antwortverhalten und Mittel der Informationskodierung von kortikalen Neuronen: Vergleich zwischen Messung und Simulation*. Unfortunately, the SNF did not concede the money for the biological part of the project. We were therefore not able to perform the necessary experiments in order to test the predictions that emerge within the progress of this study. We thus focussed on methodological questions by introducing and discussing several new methods for spike train analysis (see chapter 7). We are very thankful, that ADAM KOHN (*Center of Neural Science*, New York University), VALERIO MANTE and MATTEO CARANDINI (*Smith-Kettlewell Eye Research Institute*, San Francisco), KEVAN MARTIN (*Institute of Neuroinformatics*, ETH/University Zurich), and ALISTER NICOL (Cognitive & Behavioural Neuroscience Lab, *The Babraham Institute*, Cambridge), provided us with data that allowed the test of our methods and the finding of some empirical support for the framework developed. According to the research plan, the **scientific part** this work intends to...

... clarify the concept 'spike pattern',

... relate the pattern concept to the general framework of computation by locking,

... discuss the methodological problems associated with pattern detection,

... introduce novel, unbiased methods for spike pattern detection,

... test these methods for model and *in vivo* data.

This thesis intends to embed the specific methodological and empirical investigations related to spike patterns into a broader presentation of the main theoretical problems that are currently discussed within the neural coding debate. We believe, that such a discussion should not be 'historically blind' and should therefore outline how 'neural coding' became a relevant topic within neuroscience. In order to clarify the historical context and origins of the concepts, which are prominent in today's computational neuroscience, we provide a introduction to the historical roots of the information processing brain by focusing the time period 1940 to 1970 in the first part of this thesis. The **historical analysis** of this development leading to the 'information processing brain' should...

- ... provide a scheme for analysis in order to discuss the historical development,
- ... outline the characteristic elements of this development,
- ... show the emergence of main concepts of today's discussions,
- ... discuss the role of information theory and cybernetics,
- ... underlie the historical analysis with bibliometric and scientometric studies.

The historical analysis has been mainly performed from the end of December 2004 to the beginning of March 2005 at the *Max-Planck-Institute for the History of Science* in Berlin. The author intended to provide an interdisciplinary PhD thesis that tries to bridge the gap between 'science' and 'humanities', first stated by P.E. SNOW in his Rede lecture in 1959 [235]. This also incorporates the combination of two different writing styles. Therefore, special diligence has been used for formal aspects, so that the thesis can also serve as a guide of reference for readers interested in these topics.

1.2 Methodology

The historical part is based on an extensive literature search covering more than 500 publications, conference-proceedings, books and further sources. These sources have been obtained by systematically tracking all historical references in today's literature and the earlier literature, until (to some degree) the body of literature represented a 'closed network' of citations on the matters of interest in this thesis. Furthermore, several journals have been thoroughly analyzed for the time period 1940 to 1970 in order to receive a more complete picture on the publication activity of the period under investigation. The bibliometric and scientometric analysis are based on the *ISI Web of Knowledge*® databases, the *MedLine*® database and a self-created conference database. The methodological details of the individual investigations are outlined in sections 4.1, 4.2 and 4.3. Around 350 references, which have been actually used for the historical part, are listed in the first section of the bibliography. Secondary literature (historical and philosophical, around 60 references) is listed in a separate section. Due to time restrictions, we did not include an extensive archive search for persons and institutions of particular interest.

To embed the scientific part of this thesis in the actual discussion on neural coding, again an extensive literature search has been undertaken covering about 500 publications. The largest part of this survey is cited and listed in the third section of the bibliography. The models and programs used in this thesis have been developed in the *Mathematica*® software environment. For the biological experiments, we relied on external collaborators. The experimental procedures used by them are described in section 8.1.

1.3 Organization of the Thesis

1.3.1 Main Line of Argumentation

Since the mid 20th century, the 'information processing brain' became a central term within neuroscience for describing the functional processes within neural systems. We are interested in the origin of this term and the concepts that form the theoretical framework of

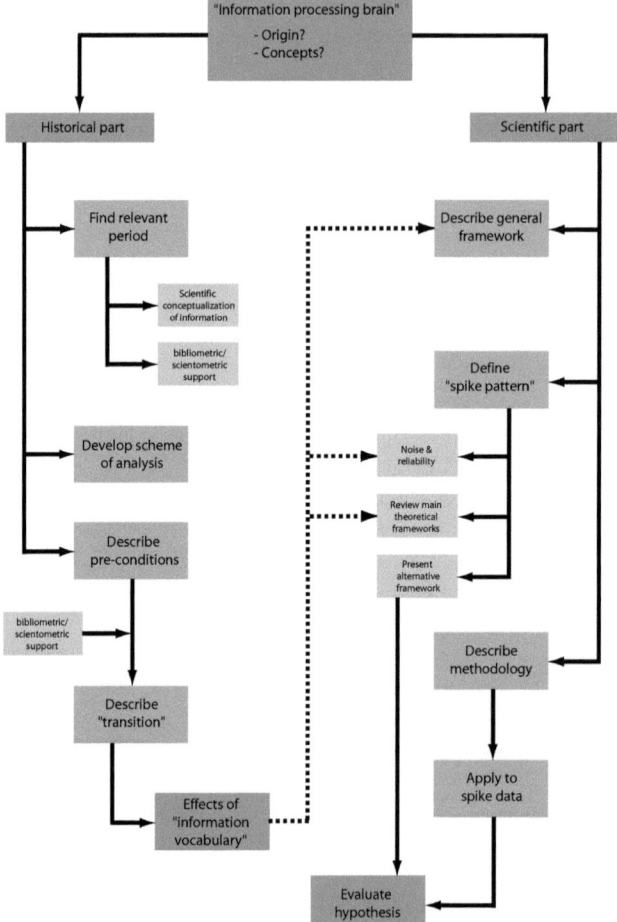

Figure 1.1: Scheme of argumentation (see text).

the information processing brain – notably neural coding and neural computation (for an overview of the argumentation, see Fig. 1.1). The origin of the term is analyzed in the historical part of the thesis. We first show (also using bibliometric tools, see chapter 4) that the 'information processing brain' became an important term in the decades after the Second World War (1940s to 1960s) – the same period when the scientific conceptualization of information (reflected in the emergence of information theory and cybernetics) can be observed (chapter 2). We then develop a scheme of analysis for analyzing the historical development in more detail. We sketch the situation in neuroscience in the various identified fields that set the preconditions for applying the 'information vocabulary' in neuroscience (chapter 2). We then get an overview of the historical roots of the concepts 'neural code', 'neural noise' and 'neural computation' that are now widely used in (computational) neuroscience. The historical analysis is supported by a quantitatively oriented bibliometric analysis focussing on journal publications and conference proceedings in the period of investigation (chapter 4). In this way, we identify the main scientific protagonists of that period and analyze their influence on a broader scientific audience.

For the scientific part, we outline today's neural coding discussion by presenting a definition of neural coding, which is based on a literature study, and by (shortly) discussing the concept of neural computation (chapter 5). Then we provide a definition of spike patterns including all different kinds of patterns that have been proposed in the literature. We show that any pattern definition needs the definition of a 'background' and we introduce the main statistical null hypothesis proposed in the literature that serve as tests for detecting patterns. We furthermore provide an overview on neuronal noise and the main theoretical frameworks, where spike patterns gain a functional role. This discussion is concluded by presenting the STOOP-hypothesis that combines the various concepts (coding, computation, pattern, noise) into a general theoretical framework by postulating, that neurons should be described as noise-driven limit cycles, where changes in the background-activity are encoded in specific firing patterns (chapter 6). This leads to a pattern detection problem, which is exemplified by five predictions that emerge out of the general framework provided by the STOOP-hypothesis. Therefore, we provide an in-depth discussion of the problem on pattern detection and introduce new methods for pattern discovery (chapter 7). These methods are then applied to spike data originating from various sources (chapter 8). The findings are used to evaluate the predictions of the hypothesis presented in chapter 6. Finally, we list the main open questions and possible further experiments in order to test the STOOP-hypothesis in more details (chapter 9).

1.3.2 Typographic Structure and Layout

We use several typographic and graphical means to guide the reader through the thesis, to show connections and links between the different topics we discuss, and to emphasize the main points. The general story is contained in the **body text**.[1] Each chapter starts with a **chapter abstract**, that outlines the main points that will be discussed in the chapter. **Names** of persons are printed in CAPITAL LETTERS, names of institutions or titles of papers and books are printed in *italic letters*, and web-resources are printed in `typewriter`

---

[1] **Footnotes** are used to outline minor side aspects or to explain a certain point in more detail. They are used mainly in the historical part.

letters. For the **citation form**, we chose the classical scientific form by using numbers referring to the bibliography at the end of the thesis. Citations are marked using double-quotes " " and a index number combined with a page reference (e.g. [123]:23-24 indicates a citation in reference number 123 spanning from page 23 to 24.

> "Within the historical part, important citations are accentuated in a separate paragraph."

In the historical part, authorship and year of publication of references are explicitly mentioned in the text, if they are necessary for understanding the argument. Single quotation marks ' ' are used for all other quotations.

> **In-depth texts:** They are used to discuss special aspects in more detail or to provide examples for some definitions. Examples are numbered according to their appearance in each chapter (e.g. example 2.1 would be the first example in the second chapter).

Definition 1.1 *Within the scientific part, concept we want to clarify are outlined in definitions. They are numbered according to their appearance in each chapter.*

In the **appendix**, we provide the list of figures and tables, the bibliography, the index (containing catchwords and names) and personal information (publications and a short biography) about the author of the thesis.

1.4 Abbreviations and List of Symbols

1.4.1 Abbreviations

BCL	Biological computer laboratory
C-distance	Correlation distance
EDVAC	Electronic discrete variable automatic computer
EEG	Electro-encephalogram
EPSP	Excitatory postsynaptic potential
fMRI	functional magnetic resonance imaging
GABA	Gamma-amino butyric acid
ISI	Interspike interval
ISP	Intensive Study Program
LGN	Lateral geniculate nucleus
LTD	Long term depression
LTP	Long term potentiation
LZ-distance	LEMPEL-ZIV-distance
MT	Middle temporal area
NMDA	N-methyl-D-asparate
NRP	Neurosciences Research Program
PET	Positron emission tomography
PSTH	Post (Peri) stimulus time histogram
SfN	Society for Neuroscience
V1	Primary visual cortex

1.4.2 List of Symbols

$\mathbb{A} = \{a_1, a_2, \ldots\}$	Alphabet of symbols a_i
α	Fitting parameter (exponential $e^{\alpha t}$ or power-law $t^{-\alpha}$ fit)
$[-b, b]$	Matching interval (of a template)
c_v	Coefficient of variation
$c(\mathsf{X}_n)$	Size of $\mathsf{P}_{\mathsf{X}_n}$
$C_N^{(m)}$	Correlation integral
$\mathcal{C}$	Set of codewords
$d(\mathsf{X}_n, \mathsf{Y}_n)$	LEMPEL-ZIV distance between the bitstrings X_n and Y_n
$E_{i,j}$	The i-th event in a sequence originating from the j-th neuron
f_c	Code relation
$f_{\Omega K}(\phi)$	Circle map
F	FANO factor
$\mathcal{F}(\kappa)$	Fit-coefficient
$g_K(\phi)$	Phase-response function
I	Current
$\mathcal{I}$	Set of code input words
K	Coupling parameter
$K_x(t)$	Template kernel
$K(\mathsf{X}_n)$	LEMPEL-ZIV complexity of X_n
κ	Parameter determining the cut-off interval of a distribution
l	Length of a sequence / periodicity of an oscillation
$l_{P_l(m)}$	The length of $P_l(m)$
L	Length of a spike train
m	Embedding dimension
$M(t^*)$	Matching function
M_{tresh}	Matching threshold
$\mathcal{M}(t^*)$	Matching
$\mathsf{M}(t^*)$	Matching string
$\mathsf{M}(y)$	Measurement of $y(t)$
N	Number of (embedded) points
$N(X_i, \Delta X_i)$	The number of appearance of intervals $X_i \pm \Delta X_i$ within a spike train
$N(P_l(m))$	The number of appearance of $P_l(m)$ within a spike train
$N^{\text{exp}}(P_l(m))$	The expected number of appearance of $P_l(m)$ within a spike train
ω	Frequency (of an oscillation)
Ω	Winding number (for $g_K(\phi_1) \equiv 0$)
$p_{i,j}$	The probability of occurrence of $E_{i,j}$
$\tilde{p}_i$	The relative frequency of intervals $X_i \pm \Delta X_i$ within a spike train
$\tilde{p}_{i,j}$	The probability of occurrence of $X_{i,j} \pm \Delta X_{i,j}$
$\mathsf{P}(t)$	POISSON process
$\mathcal{P}_l$	Pattern group of a sequence of length l
$P_l(m)$	The m-th element of $\mathcal{P}_l$
$\mathsf{P}_{\mathsf{X}_n}$	Set of phrases
$\Pi = \{\pi_i\}$	Partition of a state space

π_i	The i-th part of a partition, labelled by a symbol a_i
ϕ	Phase
$\mathbf{r} = \{r_1, \ldots r_n\}$	Spike train, local rate representation
ρ	Rate of a POISSON process
s	Cluster stability
$s_\mathcal{P}$	Pattern stability
S	Number of spike trains within a data set
$\sigma(\mathbf{y})$	Variance of the time series y
$\mathbf{t} = (t_1, \ldots, t_L)$	Spike train of length L, timing representation
t_c	Code transformation
δt	Sampling rate of a measurement $\mathtt{M}(y)$
T	Duration of a spike train
$T_{i,j}$	Timing of $E_{i,j}$
$\Delta T_{i,j}$	Jitter of $T_{i,j}$
T_{pop}	Time scale of a population code
T_{rate}	Time scale of a rate code
T_{spike}	Time scale of a single spike
T_{time}	Time scale of a temporal code
T	Temperature
$\mathcal{T}_l(t)$	Template
τ_i	The i-th bin
$\Delta \tau$	Bin width
$\theta(\cdot)$	Heaviside function
$\mathbf{x} = \{x_1, \ldots x_L\}$	Spike train of length L, ISI representation
$\bar{x} = \{\bar{x}_1, \ldots \bar{x}_l\}$	Template sequence
$X_{i,j}$	Time interval between $E_{i,j}$ and $E_{i+1,j'}$
$\Delta X_{i,j}$	Variation of $X_{i,j}$
$\mathsf{X}_n = \{\mathsf{x}_1, \ldots, \mathsf{x}_n\}$	Bitstring of length n, $\mathsf{x}_i \in \{0,1\}$
$\mathsf{X}_n(i,j)$	Phrase
$\xi_k^{(m)}$	Point embedded in dimension m
$y(t)$	Real-valued function of t
$\mathbf{y} = \{y_1, \ldots y_n\}$	Time series of length n, resulting from a measurement of $y(t)$
$\langle \mathbf{y} \rangle$	Mean of y
$\langle \mathbf{y}_1, \mathbf{y}_2 \rangle$	Dot product of two time series (or vectors) of equal length

Part I

Historical Roots of the Information Processing Brain

Chapter 2

The Brain on the Cusp of the Information Age

This chapter outlines the context in which we place our historical analysis. We set our main focus of analysis on the scientific activities in the period from 1940 to 1970. We explain our notion of the concept 'information age' as a period of the scientific conceptualization of information by including a short overview of the genesis of information theory and cybernetics. After introducing a general scheme for analysis that splits the historical development in six strands, we conclude by outlining the main developments that set the precondition for applying the information-vocabulary in brain research.

2.1 Introduction

2.1.1 The Information Processing Brain

The report of the *Neurosciences Research Program* work session on 'neural coding' in January 1968 starts with the sentence:

> "The nervous system is a communication machine and deals with information. Whereas the heart pumps blood and the lungs effect gas exchange, whereas the liver processes and stores chemicals and the kidney removes substances from the blood, the nervous system processes information" [177]:227.

This sentence outlines a development that led to today's widely used notion of the 'information processing brain' [434]. The basic unit of the brain, the neuron, is considered to be an entity that "transmits information along its axon to other neurons, using a neural code" [646]:R542. Finally, the 'neural coding problem' is described as "the way information (in the syntactic, semantic and pragmatic sense) is represented in the activity of neurons" [468]:358. However, the terms 'information', its 'processing' or the 'neural code' are often just vaguely defined and used in a rather general sense – also within neuroscience.[1] This indicates, that the scientific problems associated with these terms are not solved.

The term 'information processing brain' must be distinguished from the much more narrow metaphor of the 'brain as a computer'. The latter makes stronger claims about the way the processes in the brain should be understood, for example by suggesting ALAN TURING's concept of computation [248] as an useful framework for understanding the processes within neuronal networks. Today, the 'brain-computer analogy' is a historical metaphor in the sense that it is no longer considered as a guiding principle that may help to understand the functioning of brains. Rather, it is emphasized that the brain implements an 'alternative way' of signal processing which is distinct from today's computers [434]. In a broader context, the term 'information processing brain' reflects several important changes in the way biology in general and brain research in particular have been performed during the last few decades. Three observations demonstrate this change:

- 'Information' has become a central concept in the biological sciences. Processes in molecular biology, developmental biology and neuroscience are often considered as processes where 'information' is 'read', 'transformed', 'computed' or 'stored'.[2]

- 'Neuroscience' is a modern term, introduced in the late 1950s.[3] Its introduction indicates that new disciplines, notably molecular biology, were considered to be important for understanding the brain.

- Today, the brain is not only an entity that can be explained or modelled using recent technological concepts. It also became an entity, whose analysis may help to improve, or find new, technology.[4]

These observations open a wide field of questions for historians and philosophers of science. Some of them have already been discussed, for example the introduction of the information vocabulary in molecular biology and developmental biology.[5] Also the history of brain research in general up to the 20th century has been well analyzed.[6] The history of

---

[1] A conceptual remark: We use the term 'brain research' to indicate any scientific activity that has the brain of animals or humans as its subject. The term 'neuroscience' refers to research focussing on the functional organization and the resulting behavior of the brain [298]:7. In a standard textbook of today, *Principles of Neuroscience*, ERIC KANDEL, JAMES SCHWARTZ and THOMAS JESSELL define the task of neuroscience "to explain behavior in terms of the activities of the brain" [535]:5 As the 'information processing brain' is generally analyzed in the context of explaining behavior, we use 'neuroscience' to denote the research activity that interests us since the 1940s and 'brain research' as a more general term that covers also earlier developments.

[2] As an example consider the review-article *Protein molecules as computational elements in living cells* by DENNIS BRAY in 1995 [399], where he describes proteins in cells as functional elements of biochemical 'circuits' that perform computational tasks like amplification, integration and information storage.

[3] The term 'neuroscience' was probably first used in its modern sense by RALPH GERARD in the late 1950s (see [293]:Preface).

[4] The mission statement of the Institute of Neuroinformatics of the University/ETH Zurich may serve as an example: it claims to "discover the key principles by which brains work and to implement these in artificial systems that interact intelligently with the real world" (see http://www.ini.ethz.ch/).

[5] For developmental biology see for example SUSAN OYAMA [330]. For molecular biology see for example LILY KAY [323].

[6] As examples, consider the monograph of OLAF BREIDBACH [299] and the references therein and the monograph of MICHAEL HAGNER [317].

neuroscience since the mid 20th century, however, has been less well analyzed, although this period is characterized by an institutionalization of neuroscience, and an enormous growth of both the number of scientists and the number of publications in the field (see section 4.1). The introduction of the 'information vocabulary' (for a precise definition of this term see section 2.2.2) in neuroscience also falls in this period. This development is the main focus of the historical part of this thesis. We want to know:

- What were the preconditions for applying the 'information vocabulary'?
- What was the motivation for its introduction?
- What was the effect of this new terminology within neuroscience?

These questions are addressed mainly in chapters 2 and 3, whereas chapter 4 provides a quantitative support for some of our conclusions.

2.1.2 The Historical Context

The turn of 18th to the 19th century is often considered as the beginning of 'modern' brain research in the sense that the brain was no longer considered as the 'organ of the soul'. Rather the question emerged, whether one can understand a human being by understanding his (material) brain.[7] In the period between the beginning of the 19th century and the middle of the 20th century, major concepts of modern brain research were developed: the neuron-doctrine, the reflex-arc theory, the localization of functional units (e.g. BROCA's area) in cortex and the development of experimental techniques for stimulating the brain *in vivo*. The last decades of this period (from 1900) will be outlined in section 2.3.

The Second World War is often considered as a turning point in the development of science in the western world.[8] Several different aspects mark this transition. The first aspect concerns the involvement of scientists in war-related research programs. Although scientists already gained a big influence in the First World War, in the Second World War scientists were organized in very large and interdisciplinary scientific projects – the most famous one is the *Manhattan project*. The second aspect concerns the destruction of a whole scientific culture in Germany (starting from before the war due to the persecution of Jewish scientist), from which immigration countries, especially the United States, could profit to a substantial degree.[9] This aspect and its influence on the development of neuroscience will be analyzed in further detail in section 4.3. A third, directly war-related aspect is the creation of a new tool for scientific work, the computer. Its development was largely driven by the need for computer power for ballistic calculations. Beside these more direct effects, several indirect effects are important (in the following, we focus on developments in the United States). First, new research topics emerged, also as a result of the interdisciplinary cooperations between scientist in order to attack war-related problems. WARREN WEAVER – at that time director of the scientific department of the *Rockefeller Foundation* – characterized this transition in

---

[7]See BREIDBACH: [299], [298]:8-9; CLARKE/JACYNA: [304]; and HAGNER: [318], [319]:11-12.

[8]Consider for example the contributions in MENDELSOHN et al. [327].

[9]THEODORE BULLOCK, one of the protagonists in the early neural coding debate (see chapter 4), mentioned in his autobiography, that up to the 1930s, it was the goal of privileged American physiologists to do a post-doc in Europe, especially Germany or Scandinavia [302]:119.

1948 as an orientation towards problems of 'organized complexity' [275]. Biological questions became attractive for basic (theoretical) sciences. Second, new research fields also emerged shortly after the War – most prominently cybernetics and information theory (see 2.2), but also general systems theory and operations research. A third important aspect is the change in financing structure. This development has been well studied for the United States, where the military (the *Department of Defense* and the military-controlled *Atomic Energy Commission*) became a major funding source for science and kept this position in the Cold War period. LILY KAY has expounded the influence of this transition on the development of molecular biology [323]. It would be of great interest to analyze, how the emerging neuroscience were affected by this shift in financing. In so far as the funding sources were mentioned in the papers we investigated, military support was quite frequently evident, although we did not quantify this aspect.[10]

We did not analyze to what extent brain research was directly involved in war-related research activities. We found, however, no indications of a directly, war-induced change in research topics or institutional structures in brain research. Concerning the indirect aspects, the opening of the disciplinary boundaries is a central element that influenced neuroscience since the mid 20th century – at least in the United States, as the neuroscientists MAXWELL COWAN, DONALD HARTER and ERIC KANDEL recently noted [305]:345-347. They identified three major events in this respect: The activities of DAVID MCKENZIE RIOCH, who brought together scientists from behavioral research and brain research at the *Walter Reed Army Institute* in the mid 1950s; the establishment of the *Neurosciences Research Program* (NRP) in 1962 by FRANCIS SCHMITT and collaborators; and the creation of the first department of neurobiology in the United States at the *Harvard Medical School* by STEPHEN KUFFLER in 1967. Note that all three events happened quite some time after the war. This indicates, that the disciplinary boundaries of brain research remained stable for a longer period compared to genetics and the emerging molecular biology.

The founding of the NRP is directly related to the success of molecular biology in breaking the 'genetic code' and the new tools that were provided by molecular biology [338, 341]. This aspect is demonstrated by the following statement of FRANCIS SCHMITT in a NRP progress report of 1963: "This 'new synthesis' [is] an approach to understanding the mechanisms and phenomena of the human mind that applies and adapts the revolutionary advances in molecular biology achieved during the postwar period. The breakthrough to precise knowledge in molecular genetics and immunology – 'breaking the molecular code' – resulted from the productive interaction of physical and chemical sciences with the life sciences. It now seems possible to achieve similar revolutionary advances in understanding the human mind." (quoted from [341]:530). The NRP program was established in 1962 at the *Massachusetts Institute of Technology* (MIT) by FRANCIS SCHMITT and other collaborators. It intended to integrate the classical neurophysiological studies (top down approach) with the new methods provided by molecular biology (bottom-up approach). This interdisciplinary approach was decisive for SCHMITT. In a handwritten memo, dated September 1961, he listed nine 'basic disciplines' for his – then named – 'mental biophysics project': solid-state physics, quantum chemistry, chemical physics, biochemistry, ultrastructure (electron microscopy and

---

[10]Today, the increasing influence of military funding on neuroscience – especially on technology-oriented aspects like the development of brain-machine interfaces (in 2003 and 2004, the *Defence Advanced Research Projects Agency* invested almost 10% of its whole research budget, 24 Million Dollars, into projects of this field) – has recently led to some criticism [321].

x-ray diffraction), molecular electronics [!], computer science, biomathematics and literature research [341]:532. One month later, he expanded the list to 25 fields, many of which fall into the 'top-down approach'. Since its beginning, the NRP sponsored meetings, work sessions (usually with 10-25 participants), and *Intensive Study Programs* (~150 participants) for senior and junior scholars. It published timely summaries of these deliberations in the form of the *Neurosciences Research Program Bulletins*, which were distributed worldwide with no charge (up to January 1, 1971; later, a charge was added) to individual scientists, laboratories and libraries. A large part of these texts has been published in *Neurosciences Research Symposium Summaries*, of which seven volumes were published from 1966 to 1973. The goal of these instruments was to "facilitate and promote rapid communication within the field of neuroscience" [215]:vii. In 1982, the NRP moved from MIT to the *Rockefeller University*. The various NRP activities were very important for the emergence of a neuroscience community in the United States in the early 1960s [341]:546.

Since the 1960s, a large increase in both the number of publications and the number of journals within neuroscience can be observed (see chapter 4, Figs. 4.1 and 4.2). However, it is striking that the history of modern neuroscience (since the mid 20th century) does not contain scientific breakthroughs comparable to those that happened in molecular biology, as HAGNER and BORCK pointed out [316]. The introduction of new theoretical concepts (like the HEBBian rule, long-term potentiation) and technologies (for example brain imaging technologies like PET and fMRI) happened rather gradually. Furthermore, 'modern' concepts like the neural net and the HEBBian rule have clearly identifiable forerunners in the 19th century – an aspect that has been analyzed in detail by OLAF BREIDBACH [299, 300]. We can therefore expect, that also the integration of the information vocabulary (which was formulated in detail just after the Second World War, see next section) into neuroscience and the emergence of the 'information processing brain' happened gradually and is not related to a single scientific breakthrough.

2.2 Cornerstones of the Information Age

2.2.1 The Conceptualization of Information

The phrase 'information age' has become one of several labels for the current era. We are not interested in the questions of which criteria should be used to identify the 'information age' and which historical developments can be considered as its forerunners. We rather propose to relate the term 'information age' to the scientific conceptualization of the term 'information' – a historical process that has been analyzed in detail by WILLIAM ASPRAY in 1985 [295]. We identify the 'information age' with the period in which models to explain computation, new scientific fields (information theory and cybernetics) and – most importantly – the computer, have been developed and obtained relevance for analyzing scientific problems in many different fields. Thus, the beginning of the information age can be located in the 1930s and its duration covers our period of interest (1940 to 1970).

The scientific conceptualization of information occurred during the decade that followed the Second World War [295]:117. In that period, a small group of mathematically oriented scientists – identified as WARREN MCCULLOCH, WALTER PITTS, CLAUDE SHANNON, ALAN TURING, JOHN VON NEUMANN and NORBERT WIENER [295] – developed a theoretical basis for conceptualizing 'information', 'information processing' (or 'computation') and 'coding'. They created or specified the vocabulary for today's widespread discussion about information processing in natural (and even social) systems. The involvement of these persons in the debate on neuronal information processing is one aspect we consider in our historical analysis.

In parallel to this development, an increased importance of mathematics and 'mathematical thinking' in engineering and (later) in biology can be observed.[11] The *Bell Laboratories* illustrate this development ([328]: Chapter 1):Early industrial mathematicians were hired as consultants for individual engineering groups. In May 1922, the Engineering Department at *Western Electric* (which later founded the *Bell Laboratories*) created a small, separate mathematical section, that mainly consisted of one mathematician – T.C. FRY. In 1925, FRY's group was integrated in the newly formed *Bell Laboratories*. This group prospered and, in the 1930s, acquired direct control of a fund of its own and no longer operated as a mere consulting unit. Mathematics became a fundamental science in building up a nationwide (U.S.A.) communication infrastructure (e.g. the use of graph theory for constructing efficient telephone-networks). In 1964, FRY estimated the growth of the number of industrial mathematicians in the United States by counting the members of the *American Mathematical Society* employed by industry or government. By that standard, there was only one industrial mathematician in 1888, 15 in 1913, 150 in 1938 and 1800 in 1963 – approximately an exponential growth rate (quoted from [328]:77). The conceptualization of information in combination with a growing importance of 'mathematical thinking' enhanced the development of new, mathematically oriented disciplines such as information theory and cybernetics, which soon had a considerable influence on biology.

2.2.2 Information Theory

Information theory studies information[12] in signalling systems from a very general point of view in order to derive theorems and limitations universally applicable to all systems that can be understood as 'signalling systems'. The famous 'schematic diagram of a general communication system' of CLAUDE SHANNON [227] (Fig. 2.1) identifies the involved entities as 'information source', 'message', 'transmitter' (or encoder), 'signal', 'channel', 'noise', 'receiver' (or decoder) and 'destination'. The main practical problem which the theory intends to solve, is the reliability of the communication process. The reliability is affected by channel noise and various kinds of source-channel mismatch. The semantics aspect, the 'meaning', of the message are considered as irrelevant for this engineering problem – the famous 'semantics disclaimer'. The necessity to detach semantics for being able to derive a measure for information in an engineering context had already been formulated in 1928

---

[11]In 1941, the mathematician T.C. FRY, the first 'pure' mathematician in the *Bell Laboratories*, characterized 'mathematical thinking' by four attributes: a preference for results obtained by reasoning as opposed to results obtained by experiments; a critical attitude towards the details of a demonstration ('hair-splitting' from an engineer's point-of-view); a bias to idealize any situation with which the mathematician is confronted ('ignoring the facts' from an engineer's point-of-view); and, finally, a desire for generality (quoted from [328]:4). MILLMAN noted that this definition "still agrees well with usage at *Bell Laboratories*". These attributes are also appropriate to describe the 'cultural divide' between the mathematical sciences and biology, as EVELYN FOX KELLER pointed out ([309]: Chapter 3).

[12]The term 'information' was used many years before the emergence of information theory. It was recorded in print in 1390 to mean "communication of the knowledge or 'news' of some fact or occurrence" (*Oxford English Dictionary*, quoted from [295]:117). The word derives from the Latin *informatio*. It is also used in the sense of 'instruction', 'accentuation' and 'shaping' [339].

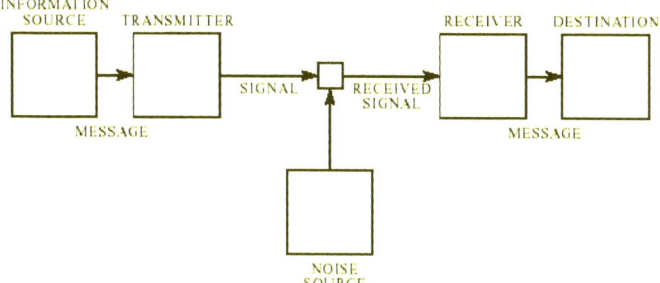

Fig. 1 — Schematic diagram of a general communication system.

Figure 2.1: SHANNON's scheme of a communication system, taken from [227]. The term 'channel', which is not mentioned in the diagram, denotes the small square on which the noise source acts.

by R. HARTLEY. His goal was to find a "quantitative measure [for information] whereby the capacities of various systems to transmit information may be compared" [116]:535. This requires an elimination of (what he called) the psychological factor in order "to establish a measure for information in terms of purely physical quantities" [116]:536. His proposal included a logarithmic law by defining $H = \log s^n$, where H is the information measure, s the number of symbols available and n the length of the symbol sequences. Furthermore, he also discussed the problem of (in today's terminology) the bandwidth of the channel.[13]

Although SHANNON's work had its precursors, he presented in 1948 the first general framework, where aspects like 'noise', 'channel capacity' and 'code' found a precise definition. His concept of an information measure operates with the probabilities p_i of symbols s_i, $i = 1, \ldots, s$, by defining $H = -\sum_{i=1}^{n} p_i \log_2 p_i$. The choice of the base of the logarithm corresponds to the choice of a unit for measuring information. The common base is 2 and the measure is called the 'bit' — a contraction of 'binary digit' proposed by J.W. TUCKEY. Maybe the most important contribution from a theoretical point of view was SHANNON's fundamental theorem for a noiseless channel, which set a limit on the information that can be transmitted over a channel. Furthermore, his work set a framework for further investigating several practical problems, e.g. error-correcting codes.

In the years after the publication of SHANNON's article *A mathematical theory of communication* in 1948, information theory became a 'hot topic' in many fields of science. This is surprising indeed, as the 'founding article' was technically written, appeared in a specialized journal and pertained to no other field than telecommunication. HENRY QUASTLER,

---

[13]Other precursors of information theory are LEO SZILARD [244] (who related the problem of information to entropy), R.A. FISHER [82] (who discussed information loss in a statistical context) and H. NYQUIST (who discussed information transmission in an engineering context, telegraph transmission theory).

a promoter of information theory in biology, noted in a well-written summary of information theory, that the article "did not look like an article designed to reach wide popularity among psychologists, linguists, mathematicians, biologists, economists, estheticists, historians, physicists... yet this has happened" [188]:4. Three causes explain the fast-rising popularity of information theory. First, the coincidence of the publication of SHANNON's work with NORBERT WIENERS monograph on cybernetics (see below). In this framework, information played a crucial role for controlling systems. Also cybernetics became popular very rapidly and both new fields supported each other in this respect. The second reason is that WARREN WEAVER, an influential figure in science promotion at that time, acknowledged the work of SHANNON and published with him in 1949 the book *The mathematical theory of communication* that made the theory accessible to a broader public [226] (note the shift from *A mathematical theory...* to *The mathematical theory...*). The third reason is that the vocabulary of information theory, exemplified by the 'general scheme' (Fig. 2.1), was attractive for many other branches of science. We classify this spectrum of application along the dimension of how much 'meaning' (the forbidden term of information theory) is related to the problem under investigation. At the one end of the spectrum, where 'meaning' is irrelevant, information theory provided a framework to further develop coding theory, probability theory and statistical physics. This is the most 'accepted' extension of information theory for the exponents of this field. A review of developments in information theory in the 1950s, published in 1961 by H. GOLDSTINE in *Science*, did not discuss any application of information theory in other than exact sciences [104]. Also today, the standard textbooks *Elements of Information Theory*, published in 1991 and written by THOMAS COVER and JOY THOMAS, discusses application of information theory only to physics, mathematics, statistics, probability theory, computer and communication science – with economics (portfolio theory) as the only social science exception [437]:2. At the other end of the spectrum, information theory was used to discuss human communication processes. For example, the third symposium on information theory of the *Royal Institution* in 1955 focussed on theoretical and mathematical aspects of human communication [66]. Many psychologists used information theory to estimate human channel capacities, for example at the *Bell Laboratories* [328]:454-464. These studies were related to the problem of sensory coding, as we will discuss in section 3.2.2.

Any application of information theory to fields where the problem of 'meaning' appeared, was accompanied by persistent warnings. For example SHANNON made this point explicit several times when applications of information theory to other fields were discussed at the *Macy conferences*. In the discussion of the paper *Communication patterns in problem-solving groups*, presented by the social scientist ALEX BAVELAS in 1951, he noted in the discussion: "Well, I don't see too close a connection between the notion of information as we use it in communication engineering and what you are doing here. (...) I don't see quite how you measure any of these things in terms of channel capacity, bits and so one. I think you are in somewhat higher levels semantically than we are dealing with the straight communication problem" [260]:22. Also the leading British information theoretician, COLIN CHERRY, criticized the extrapolation of information theory to other fields (see [323]:146). An advocate against an uncritical use of information theory was the logician YEHOSHUA BAR-HILLEL. In 1953, he noted: "Unfortunately, however, it often turned out that impatient scientists in various fields applied the terminology and the theorems of Communication Theory to fields in which the term 'information' was used, presystematically, in a semantic sense, that is, one involving contents or designata of symbols, or even in a pragmatic sense, that is, one involving the users of these symbols" [22]:147-148. Although there were several attempts to clarify the notion of 'meaning' in information theory, a satisfactory solution has not yet been provided.[14] For the developing neuroscience, however, information theory provided a vocabulary of new terms that should become important. This *information vocabulary* consists of the terms '(neural) code' or 'coding', '(neural) noise', '(neural) channel', and '(neural) information', respectively 'information processing' (computation).

2.2.3 Cybernetics

The emergence of cybernetics in the 1940s as a general framework to understand processes in different scientific disciplines is a major event in the modern history of science.[15] Several historical roots of cybernetics have been identified by historians. However, the question that can be considered as the 'founding problem' of cybernetics was related to the scientific needs of the Second World War. At that time, a problem arose in connection with anti-aircraft fire control. Targets performed haphazard evasive maneuvers to confuse gun directors. From a theoretical point of view, this can be interpreted as a 'noise problem', where the target coordinate $\vec{x}(t)$ had some resemblance to a random noise signal. If the time until an anti-aircraft projectile fired at time t reaches its target is t', a prediction of the future target coordinate $\vec{x}(t + t')$ is needed. NORBERT WIENER at the *Massachusetts Institute of Technology* and ANDREI KOLMOGOROV in the Soviet Union independently found a way to estimate the future coordinate [328]:43-45.[16] WIENER's treatment of the subject was, however, considered to be too mathematical by many engineers. The internal nickname of WIENER's original yellow-covered publication was the 'yellow peril' [328]:43 and the joke circulated that some copies should have been given to the enemy such that they would waste their time by trying to understand the paper (quoted after [323]:121). His paper was, after the war, sidestepped in a later investigation of the problem by R.B. BLACKMAN, H.W. BODE and C.E. SHANNON. Engineers preferred the latter, because it used familiar concepts about filters and noise. This disruption between a purely mathematical-abstract treatment and a more engineering-orientated treatment of the prediction problem might also be the reason why information theory is primarily connected with SHANNON, although WIENERS

---

[14] The earliest attempt traces back to DONALD MACKAY, a leading figure in the early neural coding debate, who noted as early as 1948 that the concept of information in communication theory should be supplemented by a concept of 'scientific information' (quoted after [22]:Footnote 2). In 1956, MACKAY presented a definition of 'meaning' as 'selective information' [148]. Another attempt was made in 1953 by YEHOSHUA BAR-HILL and RUDOLF CARNAP. Their proposal, however, was restricted to formal languages [22]. An extension of this approach has been undertaken by J. HINTIKKA [122] in 1968. A more modern approach is FRED DRETSKE's concept of semantic information [307], proposed in 1981.

[15] A recent overview can be found in the collection of essays in the second volume of the re-issued *Macy* conferences proceedings [333, 334]. Another informative essay, that especially focusses on the role of NORBERT WIENER is PETER GALISON's essay *The Ontology of the Enemy: Norbert Wiener and the Cybernetic Vision* [310]. An introduction into the history of cybernetics is also provided by NORBERT WIENER himself in his introduction of his monograph *Cybernetics: or Control and Communication in the Animal and the Machine* [280]:1-29.

[16] WIENER did not know about the Russian activities during the war, but he later appreciated the contribution of KOLMOGOROV. It is furthermore interesting to note that KOLMOGOROV's work was published in 1941 in a Russian journal, whereas the work of WIENER was only a secret, internal report, that had been distributed in 1943 within a few research groups in the US and Great Britain. WIENER's book on this subject appeared in 1949.

analysis of the prediction and feedback problem also involved a definition of 'information'.

WIENER's solution of the prediction problem did not have practical implication (a mechanized, usable version of his anti-aircraft predictor was finished after the war), but he was able to put the problem in a more general framework, applicable to biological systems.[17] Furthermore, WIENER was able to propagate this view. Famous stages in this process were the publication of the paper *Behavior, Purpose and Teleology* [208] in 1943,[18] the foundation of the *Teleological Society* (JOHN VON NEUMANN, WARREN MCCULLOCH and RAFAEL LORENTE DE NÓ were among the participants) [310]:248, the start of the *Macy*'s conference series in 1946, and – finally – the publication of the ground breaking monograph *Cybernetics: or Control and Communication in the Animal and the Machine* [280] in 1948. Soon after this rather short period, cybernetics became an integrating science that claimed to be applicable to many different fields in technical sciences, biology and social sciences. Cybernetics took much inspiration from neurophysiological knowledge, as nervous systems are a prime example of 'purposeful feedback systems'. The more important aspect is, however, that cybernetics promoted a principle, which claimed that the behavior and thought processes of brains are accessible by studying machines. This principle of the cybernetic research program was most stringently formulated in the answer of ARTURO ROSENBLUETH and NORBERT WIENER to the critique of TAYLOR:

> "We believe that men and other animals are like machines from the scientific standpoint because we believe that the only fruitful methods for the study of human and animal behavior are the methods applicable to the behavior of mechanical objects as well. Thus, our main reason for selecting the terms in question was to emphasize that, as objects of scientific enquiry, humans do not differ from machines" [207]:320.

Many scientists working in cybernetics referred explicitly to this principle. One example is ROSS ASHBY, who wrote in his monograph *Design for a Brain*, which appeared in 1952: "The work has as basis the fact that the nervous system behaves adaptively and the hypothesis that it is essentially mechanistic; it proceeds on the assumption that these two *data* are not irreconcilable."[19]:v. Furthermore, the principle was extended towards an epistemic ideal on the nature of explanation: Understanding the brain means, in its last consequence, building a brain. In the 1960s and early 1970s, this epistemic ideal can regularly be found in the cybernetic community. The biologist JOHN YOUNG, wrote in 1964: "if we understand the nervous system we should be able to take it to pieces, repair it, and even make others like it"[288]:10. The theoretician MICHAEL ARBIB noted in 1972: "If you understand something, you can 'build' a machine to imitate it."[14]:4. This idea of analogy and imitation was pushed even further in order to become the *credo* of neuromorphic engineering, which has been summarized by JAMES ANDERSON and EDWARD ROSENFELD in 1988 as "if you really understand something, you can usually make a machine do it" [294]:xiii.

---

[17]Interestingly, a similar development happened in Germany during the war. The engineer HERMANN SCHMIDT brought together a group of engineers and biologists in 1940 to discuss control mechanisms in technical and biological systems [118]:420-421. This meeting did, however, not have any consequences on the development of biology in Germany and the early German roots of cybernetics were rediscovered after 'American' cybernetics had become established.

[18]The fact that the paper used rather ill-defined concepts, which was criticized by the philosopher RICHARD TAYLOR [245], did not influence the impact of the paper.

It is noticeable that in the neurophysiological literature we considered, explicit references to the cybernetic vocabulary (like 'feedback', 'control system' etc.) were rare. One reason for this is certainly that we put more emphasize on the neural coding problem, where this terminology was less necessary in order to explain the phenomena of interest. There was probably also some scepticism concerning the analogy between the living and the non-living, as a comment of J. PRINGLE and V. WILSON in 1952 indicates: "While there is not yet agreement among biologists about the value of some of the analogies which may be drawn between the functioning of the living and the non-living, there can be no doubt that much may be learned from the methods used by control-system engineers in the analysis and synthesis of self-regulating machinery. The neurophysiologist, who is still at the stage of preliminary analysis, can be grateful for any hints that may enable him to plan his experiments economically, and to acquire his results in a form that will make easier the ultimate task of synthesis of an understanding of the working of the whole of the nervous system and, ideally, of the whole of the living body"[185]:221. The integrative study of brains and machines was, as ARBIB pointed out in a retrospect [367]:vii in 1987, abandoned by the mid 1960s. Scientists focussed rather on the development of artificial intelligence – a whole new field of interest for historical studies, which we do not discuss in this thesis.[19]

2.3 Preconditions for the Information Processing Brain

2.3.1 The Scheme for Analysis

In brain research of the 19th and early 20th century, the processes performed by nerves and brains are usually described using terms like 'nerve energy', 'spread of activity', or 'action current' – but not by using terms like 'information processing', 'neural coding' or 'neural computation'. As soon as the expression 'information processing brain' is intended to denote more than a crude analogy and becomes an explicit topic of research, several questions have to be answered by scientists: What is 'information'? How is it represented? What is 'processing' of information? Where does 'processing' take place? What are the proper methods and experimental techniques to analyze neuronal 'information processing'? We order these and related questions in a scheme that subdivides the historical development into six strands. The scheme is two-dimensionally: The first dimension distinguishes between spatial scales: a *micro*-scale (concerning the functional units) and a *macro*-scale (the arrangement of functional units into networks). The second dimension distinguishes between qualitative aspects: *structural* aspects (which entity is doing the processing? where in the brain happens this processing?), *dynamic* aspects (what is processed? how is it processed? is the dynamics driven internally or externally?) and aspects concerning the *methods* used (how should the structure and its dynamics be investigated?). This scheme serves as a map that will guide us through the historical analysis. In this way, we obtain a clearer picture about the transitions in each strand that finally lead to the notion of the 'information processing brain'. The map, however, does not intend to list all activities in brain research at that time. Figure 2.2 shows the map with the strands (topics). However, not all of them are of equal interest for us. In the following, only the micro-level will be discussed in more detail. There, however, some 'boundary topics' that emerge between the strands are also of interest for us:

---

[19] A history of the roots of artificial intelligence was published by HOWARD GARDNER in 1985 [311].

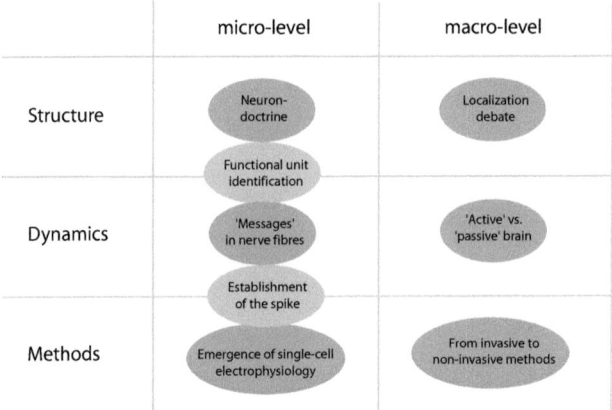

Figure 2.2: A map outlining different strands of the historical development along a spatial scale axis (micro, macro) and a qualitative axis (structure, dynamics, methods). Topics at the borders of each strand are indicated by light-grey ellipses.

The question of how to link measured neuronal dynamics with single-cell-activity (structure-dynamics boundary) and the establishment of the spike as a basic entity for describing the dynamics based on new measurement instruments (dynamics-methods boundary).

We conclude this section with three remarks concerning the strands on the macro-level: The neuro-anatomical localization of cognitive functions was the subject of widespread discussions long before the 20th century. This discussion is well analyzed in the history of brain research [296, 299, 345] and we only make a brief comment on KARL SPENCER LASHLEY.[20] His contribution is known as the 'mass action' hypothesis [141], which he developed in the 1920s in order to understand learning processes. LASHLEY saw a connection between the mass of functionally effective cerebral tissue and the rate of learning. His main point was that, in the case of complex behavior, there is no specialization within an area of the cortex. Rather, the dynamic formation of cell groups for different, especially 'higher' functional tasks, becomes the relevant process. Later, he related this concept to an essentially random connectivity of (cortical) neurons and he was thus one of the promoters of a 'statistical perspective' in respect to the brain (i.e. neural connectivity, see section 3.2.4).

The second remark concerns the differentiation between an 'active' and a 'passive' brain: In the 1950s and 1960s, the first half of the 20th century's history of brain research was seen as a time when the brain was considered to be a passive entity that is only *reacting* to external stimuli [184]:1. The 'switch' from a passive to an active brain has been interpreted

---

[20]See [299]: chapter 10 and the references therein, and [290] for a contemporary (1961) review

as a remarkable event in favor of a new, 'cybernetic' view of the brain, which has to be 'active' in order to perform processes like goal-oriented behavior (see for example the *Note by the Editors* of the *Macy conference* proceedings of 1952, [261]:xvi). The main exponent, to whom the 'passive' view was attributed, was CHARLES SCOTT SHERRINGTON, who presented in 1906 in his monograph *The Integrative Action of the Nervous System* the view of the brain as a 'reflex-machine' [229]. However, such a 'switch' from a 'passive' to an 'active' brain is not so easy to pin down in the history of brain research. The neurophysiologist MARY BRAZIER noted in a review in 1957 that there were indeed several early indicators of 'intrinsic activity' in the brain [44]:224. Electrical measurements and later the introduction of the electroencephalogram (EEG) by HANS BERGER in the 1920s [297] further indicated that there is intrinsic activity in the brain, which led the German neurologist KURT GOLDSTEIN in 1934 to formulate a 'first principle' of a theoretical analysis of the nervous system: "Das System befindet sich nie in Ruhe, sondern in einer dauernden Erregung" [103]:69. Consequently, the transition we are interested in, did not concern the 'passive-active' separation *per se*, but the explanation of the latter on a micro-scale. For example, early theoretical speculations by LAWRENCE KUBIE in 1930 about the nature of spontaneous activity introduced the concept of excitation moving in closed circuits [135]:167. Later, the concept of 'spontaneous activity' has been interpreted as an aspect that makes neurons 'unreliable' in a technical sense.

The third remark concerns the methods used on the macro level. The main experimental techniques used in neurophysiology until around the 1920s can be summarized as 'stimulation' and 'ablation' – the mechanical, electrical or chemical stimulation of brain tissue or the removal of parts of the brain. Although one might think that these two techniques are entirely independent, their application before 1800 usually had the same effect [272]:435. This arose from the fact that the nervous tissue was commonly stimulated by sticking a needle or scalpel into it, pinching or compressing it, or by applying a necrotizing solution to it. These rather primitive experimental techniques were dramatically improved by the development of electrical stimulation and the beginning of electrophysiology – a historically well documented improvement in brain research.[21] Later, a shift to non-invasive methods can be observed – an important example is the EEG, introduced in the 1920s by HANS BERGER [297]. For our purposes, however, a different kind of 'noninvasive methodology' will become important, the modelling approach (see section 3.2.6).

2.3.2 The Neuron Doctrine

The emergence and the establishment of the neuron doctrine is a historically well-analyzed process, that extends over many decades.[22] There was a dispute over whether the brain should be considered as a fully connected network (a *syncytium*) or as a network of discrete units (the neurons) connected by synapses.[23] This dispute ranged over several decades, until GEORGE PALADE and SANFORD PALAY showed in 1954 using electron microscopy, that no protoplasmic continuity is present at the synaptic junction of neurons. For our purpose, however, we don't have to outline this discussion. The main point is, that in the 1930s, the

---

[21]For a short introduction see [331]:477-479. Later, also 'involuntary ablations' as a result of battlefield injuries in the First World War became a major source of knowledge [103].

[22]An analysis on the formation of the neuron doctrine has been provided by G.M. SHEPHERD in 1991 [340].

[23]The term 'synapse' was introduced in 1897 by CHARLES SCOTT SHERRINGTON, see [111], or [331]:479-480.

neuron doctrine was accepted by the majority of the neurophysiologists[24] and probably by all researchers that tried to develop models of neural systems (e.g. NICOLAS RASHEVSKY in the 1930s). Although the detailed mechanics of signal generation and propagation in neurons, and signal transmission via synapses between neurons, were not known, the 'neuron doctrine' provided the structural basis for applying the information vocabulary, as there must be a clearly identifiable entity that performs 'information processing' or 'coding' or that can act as a 'channel'. Such an entity can then also be related to technical units that may serve as 'artificial neurons' in artefacts that model the brain. Also later, in the process of integrating the information vocabulary in neuroscience, the assumption that the neuron is the basic unit of information processing was seldom challenged.

The acceptance of the neuron doctrine led to a new problem for the functional analysis of neurons performed by electrophysiologists: how can the measured activity be attributed to single neurons? This attribution has to include a theoretical hypothesis about the 'normal firing behavior' of single neurons, as different measurements (i.e. irregular versus regular firing) were obtained. When EDGAR ADRIAN and YNGVE ZOTTERMAN investigated this matter in the 1920s, they suggested that *regularity* of firing indicates that a single neuron has been measured. They derived this criterion from empirical observations as well as theoretical considerations. Some of their recordings of peripheral fibres originating from end-organs (sensory nerve cells such as stretch receptors) in muscles displayed a regular[25] (periodic) firing pattern. They explained this observation as follows: "(...) the responses occur so regularly that it is quite impossible that they should be produced by two or more end-organs acting independently" [10]:156. And furthermore: "(...) it is not surprising that an end-organ should produce a regular series of discharges under a steady stimulation: it would have been much more so had the discharge been irregular" [10]:157. Certainly, the argument that regular firing indicates the measurement of a single nerve fibre (and thus a single end organ) is problematic: first, one cannot exclude the possibility that some single fibres fire irregularly – a problem that ADRIAN and ZOTTERMAN were aware of [11]:478, but that did not lead to a revision of the criterion. Second, ADRIAN and ZOTTERMAN also described the phenomenon of adaptation – the decrease in firing frequency when a constant stimulus is applied to the end organ. For attributing a certain measurement to a single neuron, adaptation could become a problem, although they did not discuss it explicitly. In fact, regular firing was also considered in later publications in the 1930s as the criterion to identify single fibres in peripheral nerves [7]:599, [117]:284,286.

For ADRIAN, the regular firing of an end-organ was not only a criterion for single fibre identification, but also an indicator for the reliability of the functioning of the sensory system – thus it was also supported by theoretical considerations. He argued against those who took the "machine-like regularity of behavior" of the frog's muscle spindle as an argument that the working of the nervous system cannot be described by mechanical descriptions: "But those who dislike mechanism might well retort that the sense organs would be of little use to the body if they could not be trusted to give the same message for the same stimulus. The receptor and effector apparatus must respond with rigid precision if the central nervous system is to be in full control, and it is only in the central nervous system that our models

---

[24]In the 1930s, there was still some opposition to the neuron doctrine, especially in continental Europe, as JOHN SZENTÁGOTHAI describes in an autobiographical retrospection [342].

[25]A sequence of interspike intervals (ISI) is denoted as regular by ADRIAN and ZOTTERMAN, when the ISIs varied less than 10% [11]:479.

may be of questionable value" [6]:27. In summary, the neuron as the basic functional unit that performs what was later called 'information processing' was established. However, its identification in an experiment relied upon theoretical considerations about the 'machine-like regularity' of the neuron. These considerations were justified by the argument, that neurons act as reliable transmitters of 'messages'. This constrained the criterion for single neuron identification to neurons of the (mostly sensory) periphery. In other words, not only the practicability (i.e. the larger size and better accessibility of the neurons), but also theoretical reasons restricted the first single unit experiments to peripheral neurons.

2.3.3 Spikes and Messages

The second requirement for applying the information vocabulary to neural systems was the identification of entities that indicate the dynamics of the system. It is not surprising that these entities were found in the electrical domain. The development of instruments (see next section) that were able to measure fast events in neural systems on a millisecond time scale allowed the identification of an entity that may serve the purpose of communication: the 'nerve impulse', 'action potential' or 'spike' – a phenomenon that was previously called an 'action current'.[26] These entities allowed the abstraction of the 'spike train' as a set of discrete events in time. Furthermore, they were the basis for attributing a 'digital character' to the nervous system. The characteristic properties of such impulses – electrical excitability, uniform conduction rate of impulses under uniform conditions, all-or-none response, and absolute refractoriness during response [33] – were discovered step by step in the decades up to the 1930s. According to EDGAR ADRIAN, it was the work of FRANCIS GOTCH, a physiologist in Oxford, and KEITH LUKAS [146], carried out at the beginning of the 20th century, that "gave us for the first time a clear idea of what may be called the functional value of the nervous impulse" [8]:13-14. They reported two main features of nerve impulses: first, they showed that there must be a finite time interval between the impulses (this was later called 'refractory period'). Second, they speculated that the only valuable information of the impulse is its existence or absence, whereas the amplitude of the impulse is not carrying any information – the first time neural activity obtained a 'digital character'. This speculation was later extended to the so-called 'all-or-nothing law', whose introduction in the 1920s is attributed to ADRIAN.

This functional importance of nerve impulses drastically simplified the analysis of processes in neural systems, as ADRIAN noted in a retrospection in 1947: "(...) if all nerve impulses are alike and all messages are made up of them, then it is at least probable that all the different qualities of sensation which we experience must be evoked by a simple type of material change" [4]:14. Thus, the processes in the neural system became simple and attractive to be investigated, as sequences of spikes are considered as carriers of the rich universe of human experiences and thought. The nerve impulse was established as the "basis and only basis of all nervous communication" [4]:12. The consolidation of the 'nerve impulse' introduced a new framework for analyzing the neuronal effect that result from sensory events. These effects became time series showing the occurrence of impulses, and changes in these

---

[26]Today, the term 'spike' is generally used, whereas in the period of the establishment of the spike the term 'nerve impulse' was common and gradually replaced 'action current'. The term 'action potential' was used when the mechanism of spike generation was the object of analysis.

series can be related to changes in the stimulus condition. In this way, the idea of a 'message', that is embedded in these series, is born. ADRIAN is considered by JUSTIN GARSON to be the first who introduced terms like 'message' and 'information' in order to explain neuronal processes [312] (see also section 3.2.2, paragraph 'neural information'). In his book *The basis of sensation* that appeared in 1928, he formulated this idea as follows:

> "Sensation is aroused by the messages which are transmitted through the nerves from the sense organs to the brain, and this [his book] is a description of the nature of the sensory message and the way in which it can be recorded and analyzed" [8]:5.

In this framework, the activity of neurons becomes a tractable subject of investigation, as its "(...) sensory messages are scarcely more complex than a succession of dots in the Morse Code" [6]:12. ADRIAN started to investigate these 'messages' in more detail in the 1920s, when he found two characteristics: the frequency of the impulses increases with the stimulus-intensity (the load on the muscle) and it decreases in time if the stimulus-intensity was hold fixed. The first observation was later called 'frequency coding'[27], according to which the frequency reflects the intensity of a stimulus. The second observation is denoted by ADRIAN as 'adaptation'. This is considered as a means by which the nervous system can maintain efficiency, as it would be inefficient when a sensor signals a high but constant stimulus intensity with a high firing frequency. Not the signalling of the intensity *per se* but the signalling of *changes* in intensity are considered to be relevant for the organism.

In his monograph of 1932, ADRIAN started to speculate further about the nature of the messages by expanding the linguistic connotation of the term 'message': "(...) the message may be like a succession of numbers, or words, sentences" [6]:17 and by asking, whether the messages are related to the "character" of the impulses [6]:56. Moreover, ADRIAN's experimental setup allowed him to ask new questions: the first question related to the classification of neuronal discharge according to several firing patters. In 1930, ADRIAN proposed three classes of firing: continuous and regular firing, irregular firing at lower frequencies, and grouped discharge [7]:598. A second question was whether impulses of different neurons might be synchronized [7]:603-604. A third question was to investigate neuronal firing under different states of activity, e.g. awake vs. anesthetized [5]. Finally, systems other than somatosensory receptors also became objects of investigation. KEFFER HARTLINE and C. GRAHAM, for example, applied Adrian's experimental setup in 1932 to the visual senses, the photo-receptors of the retina of *limulus* [117].

To what extent the problematic character of the 'message-analogy' was expounded in the neurophysiology-community up to the 1930s was not investigated by us in detail. We actually found only one critical remark, but this has been expressed by a notability. SHERRINGTON advised in his monograph *Man on his Nature* against an uncritical use of the term and put the neural activity back in his general framework of the nervous system as a complex reflex-apparatus:

---

[27]From today's perspective, ADRIAN is considered as the originator of the concept of 'rate coding' [651]:7. This is, however, not completely correct, as today's concept of rate coding (see section 5.1.2) not only states the existence of an integration time scale T_{rate} much larger than the time scale T_{spike} of a single spike, but also that the arrangements of spikes in time intervals smaller than T_{rate} is irrelevant for coding. In ADRIAN's concept, the latter point was not fulfilled, as regularity of firing was also the criterion to assign spikes to single fibres.

> "We sometimes call these electrical potentials 'messages', but we have to bear
> in mind that they are not messages in the sense of meaningful symbols. To call
> them 'signals' presupposes an interpreter, but there is nothing to read 'signals'
> any more than 'messages'. The signals travel by simply disturbing electrically
> the next piece of their route. All is purely mechanical" [228]:168.

ADRIAN was aware of this critique [4]:48, but still proposed to use the term 'message', albeit with some prudence. The main problem for him was in any case the interpretation of the 'message' by the central nervous system: "They [neurons] have a fairly simple mechanism when we treat them as individuals; their behaviour in the mass may be quite another story, but this is for future work to decide" [6]:94. This is the problem that has to be solved in order to attack ADRIAN's final goal, "(...) to find out how the activity of the brain is related to the activity of the mind" [4]:2. An indication that ADRIAN still had a rather vague concept of 'information' is given by comparing his use of this term with the use of the term 'message'. In 1928, he seemed to have distinguished these two terms: the 'message' was attached to the activity of a single neuron, whereas the 'information' of the whole sensation is represented in the activity of many neurons. For the latter, one has to "think in terms of areas containing many receptors and not in terms of the single receptor when we are trying to estimate what sort of information reaches the central nervous system" [8]:98. This differentiation between 'message' and 'information' is, however, lost in his later monograph of 1947, where he used these two terms interchangeably [4]:28. This loss in differentiation indicates that Adrian still used these terms in a rather unprecise way.

Although the nerve impulse became in the 1930s the basic entity that reflects the relevant processes within nervous systems, a new complication also arose in this period, focussed on the conditions that lead to nerve impulses. Such experiments were performed in particular by the American physiologists E.A. BLAIR and JOSEPH ERLANGER [35, 36, 37]), who found a remarkable variability of neuronal response. When they exposed several neurons to the same stimulus condition, they found that "fibers composing a peripheral nerve range widely in their reactivities" [36]:524. This variability concerned the threshold of irritability as well as the latency of the response. Furthermore, the same neuron may also respond variably to the same stimulation. Similar findings had been published one year earlier, in 1932, by A.-M. MONNIER and H.H. JASPER working at the *Sorbonne* in Paris [162]. The variability shows up in the form of the action potential, the latency, or the sequence of action potentials produced by the stimulation. The Belgian physiologist CHARLES PECHER was interested in these phenomena and took the findings of BLAIR and ERLANGER as an inspiration to *statistically* analyze the fluctuations between different nerve fibers. He stimulated two close fibres of the same nerve in the frog close to the threshold of irritability. In this way, he could be sure that the fluctuations measured were not a result of fluctuations in the stimulation device. He showed that "les fluctuations d'excitabilité des fibres d'un nerf se produisent indépendamment dans chacune d'elles. Ces fluctuations sont désordnonnées" [173]:842. In 1939, he published a more detailed analysis, where he also considered fluctuations of irritability in single nerve cells as well as fluctuations in latency [172]. Now it seems, that – although one has found a clearly identifiable marker of the dynamics of the neuron, the 'spike' – the neuron expresses *intrinsic* variability.

PECHER as well as BLAIR and ERLANGER speculated about possible causes of this variability. Whereas the latter only offered speculations about "spontaneous variations" in the neuronal threshold of excitation [35]:316, PECHER provided a detailed analysis of the phenomenon. He first considered several possibilities of how the experimental setup (change in electrode position, temperature etc. could have induced the variability. Each possibility was discussed and rejected. Then he considered the hypothesis that only a small number of molecules is involved in the mechanism of excitation, whose fluctuations cause the variability. Although he was not in the position to attack this problem experimentally, he suggested that the mechanism of excitation has intrinsically stochastic elements [172]:149-150.

In summary, the 'nervous impulse' was established in the 1930s as the legitimate entity that describes the dynamics of neural systems on a small scale. This entity has furthermore gained the role of forming 'messages' that are transmitted by nerve fibres. The generation of nervous impulses by neurons, however, had an element of variability in the sense that their presence or absence at a certain time cannot be guaranteed when comparing similar neurons stimulated with equal stimuli or even comparing the responses of the same neuron at different times. In this way, a basis for applying concepts such as 'code' (the 'message' encodes a stimulus) and 'noise' (the cause for variability in firing) was established .

2.3.4 Measuring Single Fibres

"The history of electrophysiology has been decided by the history of electric recording instruments" [6]:2. This remark of ADRIAN, made in 1932, accentuates the decisive importance of the instruments available for electrophysiologists – an aspect that he emphasized in all of his three monographs of 1928, 1932 and 1947. In *The Basis of Sensation* [8], ADRIAN discussed in a whole chapter the history of recording instruments and also in the landmark publication series in 1926, the first [9] of the three papers dedicated to the messages of sensory fibres was entirely devoted to the build-up of his measurement instrument. The challenge that has to be solved such that nervous impulses can be measured reliably and interpreted as 'messages' consists of three steps. First, the instrument must be able to follow the fast transients of an impulse. Second, the instrument must measure weak potential changes. Third, the instrument must store the measurement in a useful format. The first challenge had been addressed in the beginning of the 20th century, when the introduction of LIPPMANN's capillary electro-meter allowed the first direct measurement of an action potential in a muscle by JOHN BURDON-SANDERSON in 1903. The string-galvanometer, introduced by the Dutch physiologist WILHELM EINTHOVEN, allowed a further improvement of measurement techniques [331]:478. ALEXANDER FORBES and ALAN GREGG were among the first in 1915 to publish measurements of action currents in nerves using the string galvanometer [87]. In their measurement (of a group of nerve fibres), however, the action current is still far from a discrete event (see Fig. 2.3, left) and expanded over almost 20 ms. The instruments were not able to measure the (much weaker) electrical activity of single units as well as the fast transients involved when nerve impulses occur – the latter due to the considerable inertia of the instruments.

A solution of this problem was developed not in neurophysiology, but in communication engineering [8]:39. The three-electrode valve, which was developed on a large scale in the First World War, became the basis for amplifiers that fulfilled the requirement for measuring electrical events in nerves. The first proposals to use valve amplifiers date back to the First World War. According to FORBES and THACHER, the American physiologist H.B. WILLIAMS proposed in the spring of 1916 the use of electron tubes to amplify action

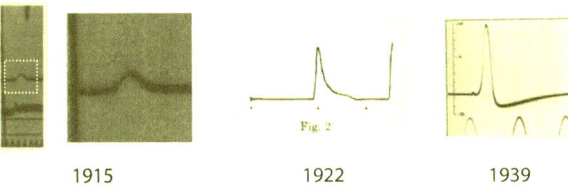

Figure 2.3: The emergence of the spike: Left: The single spike in 1915 using the string galvanometer and no amplification (Top line: excursion of the string; second line: time of stimulation; bottom line: oscillation of a tuning fork for time measurement, one period = 10 ms, measured by FORBES/GREGG, [87]). Middle: The single spike in 1922 using amplification and the oscilloscope. The curve is a drawing from the screen, the dots mark a time interval of 1 ms (measured by GASSER/ERLANGER [88]). Right: The first intracellularly measured action potential of 1939, the time marker indicate 2 ms intervals (giant squid axon, measured by HODGKIN/HUXLEY [123]).

currents in the nervous system [86]:409. Soon after the war was over, valve-amplifiers were applied to physiological research by FORBES in the United States, by DALY in England and by HÖBER in Germany [8]:42. The lack of a suitable instrument to display the recordings, however, made these first attempts impractical for investigating neuronal activity in detail. This problem was attacked by HERBERT SPENCER GASSER and JOSEPH ERLANGER at the beginning of the 1920s. They combined the valve amplifier with a cathode ray oscillograph, whose picture "may be drawn or photographed" [88]:523. The impulse, however, still emerged from a whole nerve (see Fig. 2.3, middle). It was ADRIAN, who obtained in the mid 1920s the first measurements of single nerve fibres. The first measurement of an action potential, where the electrode pas put *inside* a giant squid axon, was performed in 1939 by ALAN L. HODGKIN and ANDREW F. HUXLEY [123] (see Fig. 2.3, right).

Thus, at the beginning of the 1920s, the electrical activity could be measured and displayed – but storage of the data was still rather difficult, as the picture displayed by the oscillograph was usually too faint to be photographed properly. As ADRIAN remarked, this experimental setup was not able to investigate the 'stimulus-response' problem on the neuronal level, as "(...) the cathode ray oscillograph can only be used in experiments where the same sequence of action currents can be repeated over and over again and it is not suitable for recording an irregular series of action currents such as are produced by the activity of the central nervous system" [9]:49-50. One had to find a way to record and store *time series* of neural measurements, which are the basis of any discussion relating sensory events to neural events. A spike train does not immediately follow from the establishment of the single spike. ADRIAN solved this task. He combined the capillary electrometer with the valve-amplifier. The capillary electrometer basically consists of mercury in a glass capillary, in which the surface tension depends on the potential. Potential changes lead to a up and down movement of the mercury in the capillary. By moving a photographic plate or film behind the capillary, the desired time series is obtained. As is obvious when looking at Fig. 2.4 (top), the spike trains obtained by Adrian's measurement instrument in 1926 were still rather difficult to

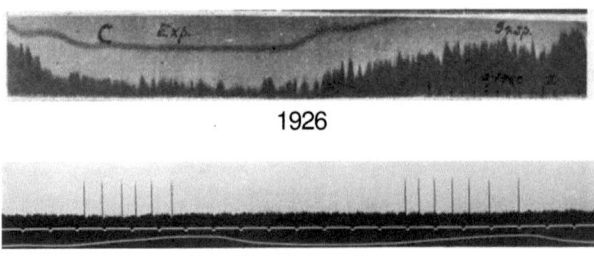

Figure 2.4: The emergence of the spike train: Above: Spike train measured by ADRIAN in 1926 [9]. Below: 'modern' appearance of a spike train, measured by BRONK/FERGUSON in 1934 [46].

interpret. It seems, however, that the electrophysiologists improved the measurement in a rather short time. A publication of 1934 by D.W. BRONK and L.K. FERGUSON not only contained pictures of spike trains where the impulses are clearly identifiable, but they also did not discuss in any detail the way the measurement was performed. This indicates, that the recording of spike trains was at that time already based on an established methodology (Fig. 2.4 bottom).

The measurement made by ADRIAN and other neurophysiologists were still extracellular measurements. Several important improvements of this technology fall in the period after 1940, such as the development of first sharp glass electrodes for intracellular single cell measurement in 1949 by G. LING and RALPH GERARD [145],[28] the development of the voltage-clamp technique in the same year by KENNETH COLE (a technique to stabilize the membrane potential and to measure the ion fluxes involved in action potential generation [535]:152) and the first intracellular measurements of spinal motor-neurons by JOHN ECCLES in 1953 and of cortical neurons by Charles Phillip in 1956 [299]:271-272. These improvements also allowed a more precise electrical *stimulation* of neurons – also an important aspect in order to be able to investigate the input-output-behavior of neurons more precisely. Despite these improvements that were still to come, in the 1930s, a technique was available to investigate neuronal processes on a micro-scale, which was the main *technical* precondition for the 'information processing' and 'neural coding' debate that followed later.

---

[28] The measurement of HODGKIN/HUXLEY of 1939 [123] were performed in the giant axon of a squid, which are much larger than neurons in the central nervous system.

Chapter 3

The Birth of the Information Processing Brain

This chapter outlines the major scientific developments that led to the notion of an 'information processing brain'. Within our scheme, the focus will be on the 'micro-dynamics', which have been increasingly described using the information vocabulary, and the 'macro-dynamics', where the brain-computer analogy emerged. This development was accompanied by a critical attitude of many neurophysiologists towards the usefulness of information theoretic concepts in neuroscience. We complement this analysis with comments on the discussion of the role of the neuron, its integration in (model) networks, and on the emerging influence of statistical and modelling methods in order to understand neurons and neuronal networks.

3.1 1940-1970: Overview

What happened between 1940 and 1970 in order that the parlance of the 'information processing brain' became widespread? Using our general scheme introduced in section 2.3.1, we identify the following topics that will be investigated in detail in this chapter:

- *Micro-structure:* After the neuron became the accepted basic structural unit of the nervous system, the question of what role neurons should play emerged. We focus on the question of whether it has been considered as a 'reliable' or an 'unreliable' element.

- *Micro-dynamics:* The parlance of 'messages' led to the question of whether the communication scheme provided by information theory can be adapted to the nervous system. We investigate how the information vocabulary was used in a neuroscience context.

- *Micro-methodology:* The improved measurement technology used for measuring single units led to the question of how to deal with the data. We demonstrate how statistical methods have increased in importance.

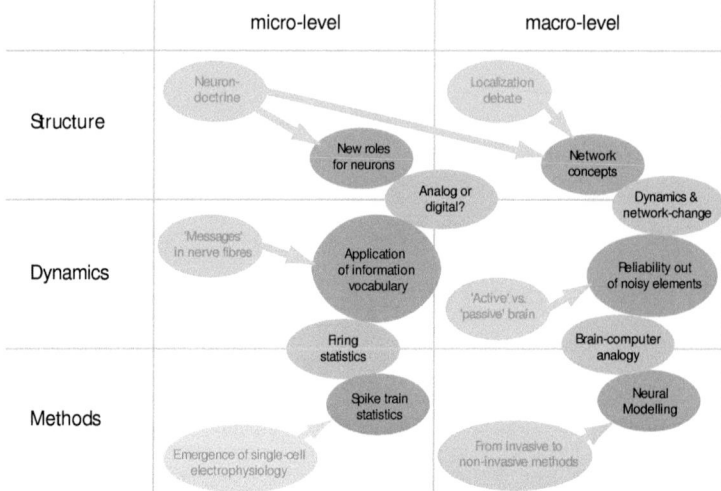

Figure 3.1: A map outlining the main transitions in the different strands of the historical development along a spatial scale axis (micro-macro) and a qualitative dimension (structure, dynamics, methods). Topics at the borders of each strand are indicated by light-grey spots.

- *Macro-structure:* The establishment of the neuron doctrine led to the question of how these neurons are arranged. We focus on network-topologies proposed to resolve the problem.

- *Macro-dynamics:* The variability of neural firing as well as the concept of spontaneous activity led to the question of whether reliable system behavior can be obtained using 'unreliable' components. This discussion led to the brain-computer analogy.

- *Macro-methodology:* In the period under investigation, a large increase in the number of models of neurons and neuronal networks can be observed. This reflects an extension of the 'non-invasive' investigation of the brain using simulation techniques.

Again, at the boundaries of the different 'areas' of our map, several questions emerge. At the structure-dynamics-boundary, one may localize the question of whether the brain should be considered mainly an 'analogue' or a 'digital' device – or a mixture of both. This topic, however, has only been discussed for a rather short period as a problem *per se* and soon was integrated in other types of discussion, e.g. the coding problem or the brain-computer analogy. We therefore forgo discussing this aspect.[1] Furthermore, on the macro-scale, the question of the influence of the dynamics on the network found an influential answer – HEBB's rule [119]. This aspect is too copious and will not be discussed by us. Finally, on the dynamics-methodology-boundary, one finds the question, which statistical model should be chosen to describe the firing of neurons, and the famous problem, to what extent the analogy between brain and computer is useful for understanding the functioning (not the structure) of the brain. These aspects will be included in the discussion on spike train statistics and the topic 'reliability out of noisy elements'.

3.2 1940-1970: Detailed Analysis

3.2.1 A new Role for the Neuron?

Also after 1940, measuring single nerve fibres was still technically demanding. The results of research concerning neuronal variability in the 1930s (section 2.3.3), however, induced a transition in the sense that 'unstable' measurements are not the result of imperfect instruments, but express an intrinsic variability of neuronal firing. This change in viewpoint had several consequences:

- The question emerged as to which processes cause the 'noise'. This could only be studied parallel to the analysis of the processes that produce action potentials and that transmit them across the synapse. This aspect will be discussed in section 3.2.2.

- The standpoint evolved that the responses of the nervous systems to stimuli could only be investigated from a statistical point of view (see this section and 3.2.3).

- Modelers were confronted with the question of how to express this statistical component – in the element itself or in the connectivity of the system (see section 3.2.4).

- Finally, the emerging 'brain-computer analogy' was confronted with the question of how a system composed of unreliable elements could show stable behavior. This aspect will be discussed in section 3.2.5.

An integrative view of the 'noisy neuron' and the consequences that follow for the methodology applied to them, was published in 1968 by DELISLE BURNS in his monograph *The Uncertain Nervous System* [57]. His main point was that "a meaningful statement can only be made about the relation between stimulus and response in terms of the *probability* that the unit will respond to the *average* test stimulus." The neural signals have to be treated as "signals in a noisy communication system" [57]:18. This viewpoint also shifted the burden of being the elementary *functional* unit away from the single neuron to groups of neurons, that could provide the necessary averaging function. This opinion was, however,

---

[1] The digital notion is usually referred to the 'all-or-none law', according to which, in the words of MARY BRAZIER (1964), "a neuron either was or was not active – a yes-no proposition of great attraction for all who think in the yes-no format of relays" [42]:88. In that sense, the digital notion was assumed by the model of MCCULLOCH-PITTS of 1943 (see section 3.2.6). The 'digital' notion was a way to simplify modelling, whereas real neurons have quickly been characterized as entities, where analogue and digital aspects were important. The significance of this distinction was also questioned as early as in 1950 at the 7th *Macy* conference [92].

not undisputed in the neurophysiology community. Two exponents, HORACE BARLOW and
THEODORE BULLOCK, who were leading figures in the early neural coding discussion, presented arguments and theoretical models that object to the view of BURNS and would later
(after 1970) become important concepts in neuroscience.

BULLOCK argued in 1969 at the *Jacques Loeb Memorial Lecture* explicitly against BURNS:
"The postulate, that neurons are basically so unreliable that only a probabilistic statement
about their message content is meaningful, is inherently nearly impossible to establish and
has probably not been established yet in any single case" [49]:566-567. He presented several
theoretical reasons as well as empirical evidence for this statement: First, variations in
response may indicate that the state of the system has changed in a biologically significant
way. Second, one may measure an irrelevant output variable or assume the wrong input: For
example, if latency precision is relevant, then variability in the interspike interval sequence
is not relevant. In the visual system, cells are known that follow other parameters than light
intensity, and keeping intensity constant does not mean than one has to expect the same
output. Third, the noise may be of significant physiological value. Thus "irregularity should
not be called noise without a showing or argument that in the context of the function
of the system it is irrelevant and undesirable" [49]:569. He then lists several examples
of reliability in the nervous system and concludes that "the proposition that the nervous
system operates mainly probabilistically with unreliable components is an unprovable article
of belief" [49]:581. Thus, for BULLOCK, the neuron may be a complex entity – according to
him the "complexity-within-unity" was the price for the consolidation of the neuron doctrine,
as he already stated in 1959 [55]:1002 – but certainly not an unreliable one.

BARLOW had a role for the neuron in mind that also could not cope with unreliability. He
can be considered as the father of the 'feature detector' idea, whereby the activity of single
neurons codes for perceptually significant events (see also [598]). This idea traces back to his
1953 publication on *Summation and inhibition in the frog's retina*, where he discusses the
response to moving stimuli of so-called 'on-off' and 'off' units (ganglion cells) in the retina
of a frog. In this paper he noted that "it is difficult to avoid the conclusion that the 'on-off'
units are matched to this stimulus and act as 'fly detectors'" [29]:86. This conception was
consistent with his considerations on redundancy reduction that is supposed to happen along
the sensory pathway and that he expressed at several conferences (e.g. [26, 28]). The idea
of redundancy reduction basically states that the same sensory information is represented
in the activity of fewer and fewer neurons and was developed by BARLOW in the late 1950s
and 1960s. In 1969, he already found a rather concise formulation of what later would
become the 'grandmother cell hypothesis': "Thus, what seems to happen is that the number
of neurons available to represent the information increases at the higher level, but these
neurons have such highly specific trigger features that any given neuron is only activated
infrequently: the information comes to be represented by fewer and fewer impulses among
more and more fibers. (..) In fact, the type of coding we are talking about is taking us
in the direction of representing the whole or, at least, a large part of the sensory input by
activity in one or a small number of neurons selected from among a very large population"
[24]:224. He opposed not only the concept of the 'unreliable neuron', but also the concept in
which a neuron only plays the role of a single component performing logical functions that
are provided by the 'neural circuitry' in the MCCULLOCH sense [24]:223 (see section 3.2.6).
The formulation of BARLOW's 'single neuron doctrine' in 1972 was then just a logical step,
where the single neuron has become an important entity to be analyzed in detail: "Thus we

gain support from this neuropsychical comparison for the concept of a neuron as a reliable element capable of performing a responsible role in our mental life" [23]:223.[2]

3.2.2 Applying the Information Vocabulary

In the decades after the Second World War, the usage of the terms 'information', 'noise' and 'coding' in papers dealing with the nervous system increased steadily (see Fig. 4.1.b). It is obvious to relate this development with the emergence of information theory and cybernetics.[3] Of particular interest was the question whether the concepts introduced by SHANNON's general communication scheme can be adapted to questions relevant in neuroscience. How can 'biological' concepts like 'stimulus' and 'response' be related to an engineering terminology speaking about 'information' and 'message', 'channel' and 'noise', 'coding' and 'reliable communication'? In neurophysiology, these questions were usually discussed on the micro-level. The main reason is presumably that the entities on the macro-level like the behavioral response that could be related to the concepts of SHANNON's scheme, are much harder to quantify than the entities on the micro-level (i.e. the spikes). In this section, we discuss in detail, how the dynamics on the micro-level – the processes performed by (usually) single neurons – was related to the terms 'channel', 'code', 'noise' and 'information'. The main focus is put on the concept of 'code' by presenting several examples of codes.

The Neural Channel: The first practical problem of relevance for neurophysiologists that emerged out of SHANNON's approach was to calculate the channel capacity (also called 'information-capacity') of sensory systems and single neurons. As early as in 1950, HOMER JACOBSON published in *Nature* an estimation of the channel capacity of the human ear (8000 bits/second for a single ear) [127], followed one year later by an estimate for the human eye (4.3×10^6 bits/second for a single eye) [126]. Although he also tried to calculate an estimate for single sensory neurons, he was only able to calculate the information capacity indirectly based on the number of neurons (= number of nerve fibres in the optic or acoustic nerve) involved in the systems. He obtained very low numbers of only a few bits/second per neuron.

As the practical problems in estimating the neuronal channel capacity were considerable, and as it was unclear what the information really *is* that the neuron receives with its dendrites and passes along its axon, the discussion about neuron channel capacity became predominantly a theoretical discussion. This discussion was closely related to the question of what 'code' the neuron might use. WARREN MCCULLOCH was one of the protagonists in these early attempts to estimate the channel capacity. His intention, however, encompassed much more than just obtaining a number. He saw the neuron as an entity that evolved in order to *maximize* information capacity. At the third conference on information theory

---

[2]The 'single neuron doctrine' was in accordance to the concept of the 'command neuron'. H. ATWOOD and C. WIERSMA described this concept in 1967 as follows: "At present, the labelling of an interneuron as a 'command fibre' signifies only that its activity calls into play a readily recognizable behavioral act. The term is of convenience for descriptive purposes and presupposes very little concerning the details of the mechanisms by which the motor axons are excited" [21]:259. Later, the notion of the 'command neuron' became an influential concept in invertebrate neuroscience, implying that the neuron has some critical function in the generation of normally occurring behavior [549].

[3]Contemporary commentators already drew this conclusion. A.A. VERVEEN and H.E. DERKSEN, for example, noted in 1965: "The expressions 'coding' and 'noise' have been introduced in neurophysiology since the advent of information theory" [255]:153.

in London (see chapter 4) in 1955 he replied to a critical statement questioning the use of calculating the neuronal channel capacity:

> "I presume, the real objection here is to my supposition of a purpose in maximizing the flow of information – I, personally, know no objection to teleology provided the mechanism is specified, but my faith in evolution antecedes information theory. All living things have to find energetic *Lebensraum* between highly organized energy of light from the sun and ultimate heat-death – *Warmetod*. (...) Since the nervous system has contact with the totality of the world (...) by means of signals, I thoroughly expect that when we understand its ineluctable limitation, we will find that it maximizes the transmission of information by proper coding" [273]:343.

For MCCULLOCH, calculating the channel capacity was a way to show that the 'information perspective' allows to understand what neurons do. His first publication on this subject in 1952, together with DONALD MACKAY, was thus not discussing the channel capacity *per se*, but the authors tried to find an argument for the encoding that a neuron performs. That the spike is the relevant entity to look at, was clear to them:

> "Now a neuronal impulse carries information, in the sense that it indicates something of the state of affairs at its point of origin. It is able, moreover, on arriving at an axonal termination, to affect the probability of occurrence of an impulse in the axon of the cell with which its 'synapses'. Thus, whatever else does or does not 'cross a synapse', it is safe to say that information does" [150]:128.

MCCULLOCH and MACKAY did not calculate the channel capacity of the neuron in the sense that they made an input-output comparison, but they focussed on the information content of a spike train by comparing two encoding schemes: One, called 'binary modulation', is the scheme that derived from the famous MCCULLOCH-PITTS-paper *A logical calculus of the ideas immanent in nervous activity* [159] of 1943. In modern terminology, the spike train is considered as a bitstring with bin-width $\Delta \tau$ and its information capacity is just given by $1/\Delta \tau$. The second coding scheme, called 'pulse interval modulation', estimates the information capacity based on the assumption that the varying *intervals* between spikes carry the information. Without outlining the detailed argument,[4] MACKAY and MCCULLOCH concluded that a neuron operating on a pulse-interval modulation code could transmit several times more information per second than a binary modulation system and estimated the maximal information capacity as 1000-3000 bits/second. This result was not taken as a 'proof' for the latter coding scheme: "Perhaps the most realistic conclusion is a negative one. The thesis that the central nervous system 'ought' to work on a binary basis rather than on a time-modulation basis receives no support from considerations of efficiency as far as synaptic circuits of moderate complexity are concerned" [150]:135. Their result, however, served as an argument to theoretically investigate the question of channel capacity based on the argument of efficiency.

---

[4]There was indeed an error in the calculation of MACKAY and MCCULLOCH, which was corrected by them in a later publication in 1956 [273]:336. Some further assumptions, like a uniform probability distribution for the interspike intervals, were questioned later as well.

An interesting theoretical objection to the approach of MCCULLOCH and MACKAY has been provided in 1960 by ANATOL RAPOPORT and WILLIAM HORVATH. Both had close connection to the RASHEVSKY school (see section 3.2.6) and investigated the problem from a real-number perspective. For MCCULLOCH, the spike train was considered in both coding schemes as a 'digital' string, coarse-grained by the absolute refractory period of firing. On the other hand, for RAPOPORT, the spike train should be described by real numbers and the impossibility of determining these numbers exactly is represented by a perturbing function. In this way, they obtained even higher channel capacities of up to 4000 bits/second [193]. The focus of RAPOPORT and HORVATH was not to argue for a certain coding scheme, but for a certain mathematical approach.

It was clear that a purely theoretical investigation of the channel capacity problem was not satisfying. In 1955, at the third conference on information theory in London, MCCULLOCH, together with PATRICK WALL, JEROME LETTVIN and WALTER PITTS, re-evaluated the problem of neuronal channel capacity by including experimental data [273]. They were interested in whether other physiological effects beside the refractory period may limit the information capacity – and indeed found some. The maximum sustained frequency of firing was overestimated, and bursts may appear, that set a constraint on the possible ordering of the spikes in time. Numerical estimations of the information capacity that emerged out of the experiments were presented at the conference, but not included in the proceedings. The main argument in the paper was, however, not to provide a correct estimation. Rather, they discussed the importance of information theoretic concepts for neurophysiological problems by replying to doubts that were obviously around in the neurophysiological community, expressed in the introduction of the paper: "It was assumed that the nervous system had not evolved in such a way as to treat information with the highest efficiency within its physical limitations, also it was concluded that no maximum principle of predictive value could be found. If that was so, information theory could not have the power as a predictive weapon in neurophysiology, which the principle of entropy has in thermodynamics" [273]:329. Although MCCULLOCH made clear in the discussion of the paper (see quotation above) that he did not agree with this view, other neurophysiologists again cast doubts on the whole approach. I.C. WHITFIELD, who worked in the auditory system of cats, noted: "The random nature of the sequences [of pulses] would seem to indicate that no information is being carried in terms of individual pulse position. Calculations of channel capacity based on such concepts would appear to be invalid for the auditory system at any rate" [273]:343.

In the 1960s, several experimentalists adopted the question. In the journal *Kybernetik*, a debate between OTTO-JOACHIM GRÜSSER and HORACE BARLOW emerged concerning a formula to calculate the information capacity, where the parameters refer to experimentally measurable entities. GRÜSSER and collaborators attacked the problem by first estimating the coding properties of the system. They measured the relation between the strength of a stimulation and the emerging spike train in the visual system. Due to the increase of statistical fluctuations in spike trains in neurons along the visual pathway, they rejected the 'optimal code' (pulse interval modulation):

> "Wir haben nachgewiesen, dass eine erhebliche statistische Fluktuation neuronaler Entladungsmuster unter völlig konstanten Lichtreizbedingungen besteht, (...) und dass die *statistische Ungewißheit der Signalübertragung zunimmt, je mehr Neuronen zwischen dem Eingang und dem Ausgang liegen.* (...) Der 'Code'

für die übermittelte Information besteht [in ihm] lediglich aus einer bestimmten Anzahl von Entladungen, wobei die *Einzel*intervalle der Signale nicht wichtig sind" [109]:189.

Based on these conclusions, they calculated a much lesser neural channel capacity of 20-120 bits per second. In this coding scheme ('rate coding' in today's terminology), a new time scale relevant for the functioning of the nervous system emerged: the integration time: "Aus sinnesphysiologischen Daten muß man für den zentralen Summationsprozeß eine Zeitkonstante zwischen 15 und 80 msec fordern" [110]:210. Shortly after the publication of GRÜSSERS result, BARLOW pointed out a mathematical error in the formula used. But this was not the major point of his critique. He pointed out the problems that emerge when a new time constant – the integration time – enters the game, and he was uneasy with the conclusion concerning *coding* GRÜSSER made:

> "I agree that the mean frequency of impulses appears to be the main information-bearing feature of nervous messages, especially in the peripheral neurones of an afferent chain. But there is enormous potential information capacity in the detailed timing of impulses, and the possibility that it is utilized should not be neglected, especially in view of the facts of sound localization" [25].

Again, the idea of optimality is connected to the question of coding.[5] It became clear, that the channel capacity is difficult to estimate and cannot be separated from the question of what 'code' the neuron actually uses. In 1967, RICHARD STEIN provided a detailed analysis of the problem based on the assumption of a frequency code, but by taking into account the different variants of statistical dependencies between succeeding ISIs [238]. The numbers obtained for neural channel capacity in this way were much lower than the original numbers of MCCULLOCH or RAPOPORT. STEIN also set a counter point against MCCULLOCHS original idea of optimality, but kept information theory as a possible tool for neurophysiology: "(...) the aim of the biological organism is maximum survival, not information transmission. However, once one understands how the nervous system codes the patterns which are necessary for its survival, information theory should provide a very useful way to measure efficiency and to compare different sensory modalities and different levels of the nervous system" [238]:825.

As a culmination of the channel capacity discussion, we mention the publication of GEORG FÄRBER, who provided in 1968 an extensive overview of several empirical studies on the channel capacity of neurons in a variety of sensory systems. Almost all of them listed channel capacities of the order of 10-100 bits/second [77]:21. Furthermore, a remark of FÄRBER indicated that neurophysiologists often questioned the relevance of such investigations [77]:28. This sceptical attitude presumably expresses the general critique within the neuroscience community towards the usefulness of concepts of information theory within neuroscience. This explains, why in the 1970s the question of neuronal channel capacity was only occasionally discussed, as far as we can conclude based on our literature study. It is, however, interesting to note that one of today's exponents of the neural coding debate,

---

[5]It is interesting to note that optimality in terms of energy was probably not considered at that time, at least based on our literature review. The earliest considerations of the efficiency in terms of energy per spike we found dates back to 1974 [180].

MOSHE ABELES, was at the beginning of his scientific career also interested in the neuronal channel capacity [3].[6] The channel capacity discussion obviously has kept its role as a motivator for the neural coding problem.

The Neural Code: The notion of a '(neural) code' did not appear in the neurophysiological literature we investigated before the onset of information theory. The term was also not used in ADRIAN's main publications of the 1920s and 1930s on sensory messages, although A.E. FESSARD expressed in 1962 at the Leiden symposium *Information processing in the nervous system*, that the notion of a code in the nervous system grew out of the experiments of ADRIAN [91]:416. After the onset of information theory and cybernetics at the end of the 1940s, this changed. The first written entry we found on neural coding traces back to the *Macy conference* of 1950, where RALPH GERARD presented his talk on *Some problems concerning digital notions in the central nervous system* [92]. In the discussion, JOHN VON NEUMANN used the term 'coded' in context to neural messages at first – but in a sense that questioned its value for describing neural processes:

> "It seems to me that we do not know at this moment to what extent coded messages are used in the nervous system. It certainly appears that other types of messages are used, too; hormonal messages, which have a 'continuum' and not a 'coded' character, play an important role and go to all parts of the body. Apart from individual messages, certain sequences of messages might also have a coded character. (...) The last question that arises in this context is whether any of the coded ways in which messages are sent operate in any manner similar to our digital system" [259]:20-21.

Here, VON NEUMANN refers to the 'code' in the sense defined by information theory. The following discussion on coding among the *Macy conference* participants, however, did not lead to a satisfactory answer on the meaning of 'coding' in the context of the nervous system. Rather, GREGORY BATESON mentioned the "obscure" character of the concept of a code in the nervous system [259]:26 and J.H. BIGELOW finally criticized that the term 'code' is not defined in the discussion [259]:35.

Although the critique of the unclear meaning of the term 'code' in relation to neuronal processes was a constant companion of the following years, the terminology of information theory became more and more popular in neurophysiology in the 1950s. HORACE BARLOW may serve as an example. In 1953 in his seminal paper on *Summation and inhibition in the retina* BARLOW considered the optic nerve as a "communication channel" and the retina as a "filter rejecting unwanted information and passing useful information" [29]:70/87. He assimilated the information vocabulary in his later publications (see e.g. [26]:222-223,225). BARLOW emphasized that the sensory pathway should not only be understood as a transmitter of information, but also as an information *processor*. The sensory systems filter out those stimuli that serve as 'passwords' for specific responses of the organism. These 'passwords' are the relevant information, and he formulated in 1961 the hypothesis, "that sensory

---

[6] ABELES showed, using the pulse interval modulation assumption, that the POISSON distribution rather than the uniform distribution of intervals is optimal for information transmission – a result that had already been published in 1963 by the German engineer H. MARKO, see [77]:18. Furthermore, as early as 1957, MC-CULLOCH mentioned the work of ALEXANDER ANDREW, who showed that changing the uniform distribution of intervals into one where shorter ones have higher probability increases channel capacity [156]:191.

relays recode sensory messages so that their redundancy is reduced but comparatively little information is lost" [26]:225. In this publication, his concept of coding is compatible with the information theoretic concept, as codewords of frequent events should be shorter than codewords for rare events: "The principle of recoding is to find what messages are expected on the basis of past experience and then to allot outputs with few impulses to these expected inputs, reserving the outputs with many impulses for the unusual or unexpected inputs" [26]:230.

In the late 1950s and 1960s, 'neural coding' became a relevant issue in the developing neuroscience. This is indicated by the growth in number of 'candidate codes'. In the 1950s, basically four codes were considered: the classical 'labelled line' code, the 'frequency code' based on ADRIAN's findings and the two proposals discussed in the channel capacity problem, the 'binary modulation' and the 'pulse interval modulation' codes.[7] Experimentally, the first and the second candidates were actually supported best; also because the notion of *what* is encoded – the modality or the intensity of a stimulus – is clear. Later, however, timing aspects were increasingly considered to be important. In the discussion about neuronal channel capacity, the postulate of 'maximizing information transmission' led to the idea, that the information could be coded in intervals and patterns of intervals. RAPOPORT, one of the exponents in the channel capacity discussion, expressed the attractiveness of this concept at the symposium on *Information processing in the nervous system* 1962 in Leiden as follows:

> "This idea [a pattern code] is very attractive to those who would think of the operation of the nervous system in the language of digital computers, because a fixed temporal pattern, although in principle subject to a continuous deformation, has a strong resemblance to a digital code. It is in fact a generalization of the Morse code. It is also like a template of a key. The discovery of such patterns would immediately pose a challenging decoding problem" [192]:21-22.

The timing of nerve impulses had already been investigated before this theoretical argument was used[8], but it was not discussed within a coding framework. In 1950, however, CORNELIUS WIERSMA and R. ADAMS published evidence that solely the timing of impulses by unchanged frequency of firing may have a functional role in the nervous system. They investigated to what extent the slow and the fast response of the claw muscles are determined by the sequence of the time intervals between the impulses. They found, that an

---

[7]Interestingly JOHN VON NEUMANN had already proposed in 1950 at the 7th *Macy conference*, that the *correlation* between spike trains could serve as a potential information carrier: "For neural messages transmitted by the sequences of impulses, as far as we can localize the state of the transmitted information at all, it is encoded in the time rate of these impulses. If this is all there is to it, then it is a very imperfect digital system. As far as I know, however, nobody has so far investigated the next plausible vehicle of information: the correlations and time relationships that may exist between trains of impulses that pass through several neural channels concurrently" [259]:21. In 1958, he recapitulated this proposal in his monograph *The Computer and the Brain* [266]:80. However, VON NEUMANN's proposal was rarely investigated at that time, because such correlations were difficult to measure and because the statistical difficulties involved in analyzing higher-order correlations are considerable.

[8]For example H. BLASCHKO and colleagues investigated the phenomenon of two types of responses of the claw motor system of *crustacea* (fast and very strong, or slow, sustained and of moderate strength) and showed that a single additional stimulation impulse may change the response from the first to the second type [38].

arrangement of impulses in 'double-shocks' can give a much more pronounced reaction than the same number of impulses arranged regularly. Although the effect was not stable (in some fibres, the effect only occurred rarely and also the strength of the fast response was unequal), the result was immediately put in a 'neural coding' context:

> "The conditions which prevail during stimulations with double shocks are almost certainly far removed from any occurring under normal circumstances. These experiments serve, however, to show that synaptic structures may be profoundly influenced by impulse-spacing. This is of general interest, because it seems possible that impulse patterns do play a role in transmission in the central nervous system" [281]:31.

Later, in 1953, WIERSMA (together with S. RIPLEY) investigated the phenomenon further by systematically varying the spacing between the impulses. They showed that the spacing effect is a "real junctional effect occurring at the postulated facilitatory loci between the nerve impulse and the muscular contraction" [200]:12. They classified the nerve fibres associated with the motor control of the claw as either 'pattern-sensitive' or 'pattern-insensitive'. They also discuss the possible gain of pattern sensitive systems, especially in maintaining maximal economy of the neuromuscular junction, because only the spacing of the impulses could determine which motor response is generated. Furthermore, they speculated on the general benefit of spike patterns: "Pattern sensitive and pattern insensitive synapses might conceivably be responsible for the 'unscrambling' of complex information arriving in one sensory axon" [200]:15.[9]

The results of WIERSMA were often considered as an important example indicating a possible role for patterns, although the results were obtained at the motor periphery of an invertebrate. The term 'spatio-temporal pattern' became an attractive concept (see for example WEDDELL's review of 1955 [276]:132) and people started to look for other systems, where one might find a relevant influence of the timing of spikes. Most of these studies were not successful in finding a functional role for patterns.[10] This, however, did not stop

---

[9]In 1965, D. WILSON and W. DAVIS demonstrated that the *natural* stimulation of the *crustacea* motor system indeed showed the pattern characteristics that had been found using artificial stimulation [283]. In 1969, H. GILLARY and D. KENNEDY showed, that the discharge of *crustacea* motor-neurons in a pattern of repetitive bursts can be more effective in causing contractions than an impulse train of the same average frequency [100]. The discussion on this matter has been continued by DONALD WILSON (and others) under the heading of the 'catch property' of muscles in several animals [282, 56]. Also WILSON was interested in this topic in the 1960s, using insects as model animals [285, 283].

[10]For example WILLIAM UTTAL stimulated the *ulnar* nerve of human subjects extracorporeally with a triplet of three pulses in different temporal arrangements but with inter-pulse intervals such that the stimuli were fused into a single sensation. The experiment showed, that the intensity of the perception was only related to the amplitude of the individual pulses, but not to the temporal microstructure [253, 252]. W. DOUGLAS and colleagues investigated in 1956, to what extent the effector system (the aortic nerve) of the vertebrate baroreceptor system (a sensory system that controls the blood pressure and the frequency of heart beating) is influenced by the arrangements of the spikes in time. Earlier investigations claimed that the arrangement in bursts, which are phase-locked to the heart beat (the natural stimulus of the baro-receptors), is the most appropriate stimulus for the effector [73]. When systematically changing the arrangements of the spikes, but keeping the mean frequency the same, DOUGLAS did not find this effect [72]:242. In 1960, at the conference on information theory in London, LAWRENCE VIERNSTEIN and ROBERT GROSSMAN presented results from neurons of the cochlear nucleus of cats indicating that the patterns of discharge were highly irregular despite an unchanging stimulus (continuous tone) [256]. They concluded, that the results exclude any 'pattern code', referring specifically to the proposal of a pulse interval modulation code by MACKAY

the discussion about pattern codes, as the 'theoretical advantage' of patterns for solving a neural coding task was too tempting. In 1962, R. MELZACK and P.D. WALL elaborated the pattern concept in the context of somesthesis by eight propositions. Their central idea was, that the receptors are not transmitting modality-specific information (in other words, they opposed a labelled line code for the somesthetic system) but produce multi train patterns of nerve impulses. These patterns can be 'read' by higher order neurons using a whole set of variations in presynaptic axonal arborization, in synaptic strength, temporal summation, coincidence detection and adaption. Thus, each discriminable somesthetic perception is represented by a unique pattern of nerve impulses [160]. Also in 1962, at the conference on *Neural theory and modelling*, RICHARD REISS emphasized the promising possibility of spike pattern codes, as they would enable the nervous system to take "full advantage of the information capacity of pulse interval coding" [199]:134. The main problem, he thought, is that "one of the most striking features of the neurophysiological literature, from the theoretical standpoint, is the absence of an adequate terminology for describing (and therefore thinking about) pulse patterns" [199]:135-136. At the same conference, LEON D. HARMON (an engineer) also discussed the possible important role of temporal patterns for neural coding, but he also made clear that the neurophysiology community is divided on this matter: "A bit of controversy among physiologists is apparent here, some believing strongly that discrete-interval information is important, others feeling that in most cases average frequency information alone is utilized" [115]:11. This sceptical attitude is well mirrored by JOHN ECCLES note, stated at the 1962 symposium *Information processing in the nervous system* in Leiden: "There is a growing belief that significant information is carried in coded form by this temporal pattern [of impulse discharges]. However, it must be recognized that this temporal pattern is likely to be lost, or at least smeared, when several lines are converging on a neurone, each with its own temporal pattern of impulses, and I would think that several serially arranged smears give complete erasure of a pattern" [91]:142. Also BULLOCK discussed at the same symposium the possibility of pattern codes – but noted that the empirical basis for analyzing this aspect is still rather poor [53].

The next years were dominated by extending the empirical basis on this matter.[11] MARY BRAZIER mentioned at the 1962 conference on *Information storage and neural control* that "a great deal of work in many laboratories is currently being devoted to pulse-interval analysis of the message set up by stimulation of receptors" [43]:235. One important study in this respect was provided by J. SEGUNDO, G. MOORE, L. STENSAAS and T. BULLOCK in 1963. They studied the sensitivity of neurons in *Aplysia* to spike patterns using the same approach of RIPLEY and WIERSMA. They tested a whole set of patterns (the firing frequency was kept constant) and investigated the probability of firing of the postsynaptic neuron dependending on the pattern type. The focus was, however, the detailed mechanism of spike generation in the postsynaptic neuron based on the induced excitatory postsynaptic potentials (EPSPs). They showed that changes in the timing of the input produce definite changes in the magnitude of the output, both in terms of depolarization induced and of spikes evoked. The following properties contribute to timing-dependence: temporal summation of

---

and McCULLOCH (see above).

[11]The number of neural modelling studies on this matter was much smaller. Beside the 'resonant network' of RICHARD REISS [199], we only found one model study performed in 1967 by LESTER FEHMI and THEODORE BULLOCK. Their neural network model was able to discriminate between different sequences of impulses with the same average frequency [80].

successive EPSPs, interaction, and post-spike excitability. This result has been put in the context of information processing. They called spike patterns that are effective on the post synaptic neuron, 'words'; and the "optimal production of spikes will depend on adequate timing or 'words' " [224]. They furthermore claimed that such a sensitivity to timing would be biologically advantageous, especially in areas of sensory convergence, for it provides an additional coding parameter. In a later paper in 1965, the analysis was refined using a more sophisticated statistical apparatus [223]. In this contribution, however, they did not discuss the coding aspect. Many other studies investigated specific timing aspects in a variety of systems: the electro-sensory systems in electric fish [112], the *cerebellum* [40], the auditory system [203, 121, 202] – just to mention a few.

This growing number of studies had a profound effect on the neural coding discussion. By integrating the possibility that the precise timing of a spike may bear information, the number of 'codes' increased. In BULLOCK's and ADRIAN HORRIDGE's monumental work on *Structure and Function in the Nervous Systems of Invertebrates* of 1965, five basic types of 'code' were listed [52]:273. In 1967 at the first *Intensive Study Program* of the NRP, VERNON MOUNTCASTLE already presented seven possible 'codes' (he called the listing not comprehensive) [165]: Four variants of a frequency code (in single axons, in populations, time derivatives of the frequency profile, and frequency modulation), a code based on the internal structure of the spike train (pattern code), coincidence-gating and labelled lines. Note, that now also variant of population codes (frequency code in populations, coincidence-gating) were mentioned. MOUNTCASTLE also expressed the warning that code relations discovered by the experimenter may be irrelevant for the brain. In 1968, Bullock presented a list of 10 'codes' [50]: Four variants of a frequency code (the most recent interval, the weighted average over some time interval, the frequency increment expressed as additional impulses per unit time, and frequency modulation), three codes where exact timing matters (the instance of occurrence, temporal microstructure, and shift in latency or phase) two 'statistical codes' (interval variation (changes in ISI distributions), and probability of missing, e.g. in periodic firing), and burst code (spike number in a burst or duration of a burst). MOUNTCASTLE's as well as BULLOCK's listings indicate the tendency, to relate any possible information-bearing aspect of a spike train to the coding problem. This tendency reached its cumulation in the *Neurosciences Research Program* work session on *Neural Coding*, held on January 21-23 1968. The organizer was BULLOCK and the session gathered the major figures in the neural coding debate at that time.[12] The initial question of the symposium – "Is the code of the brain about to be broken?" [177]:225 – was only mentioned as a *pro forma* general aims, as the participants did "(...) not believe that there is a single 'neural code' (...) that can be cracked" [177]:231. Rather, the work session started with a whole set of questions, that reflect the disparate character of what is called the 'coding problem':

- What is meant by a code in the context of the nervous system?
- What modes of representation are theoretically plausible?
- How is a candidate code established?
- How can signal and noise be distinguished?

---

[12] Among the participants were: HORACE BARLOW, GEORGE GERSTEIN, LEON HARMON, DONALD MACKAY, VERNON MOUNTCASTLE, DONALD PERKEL, JOSÉ SEGUNDO and RICHARD STEIN

- Are there rules or principles about the kinds of codes?

- What do 'resolution', 'noise', 'ambiguity', 'redundancy', and 'reliability' mean?

The work session report starts with a discussion of the notion of 'coding'. Several possibilities are considered: A cryptographic code as a mapping between symbol sets (the genetic code served as an example), the disparate meanings of 'code' in the field of digital computers ('code' as a repertoire of elementary instructions, 'code' as program text, or 'code' as a set of rules by which data are represented electrically within the workings of the computer). These codes display a "rigid, mechanical aspect" and can be 'cracked' – but the 'neural code' seems not to be of this kind. The term is rather considered to reflect all signal-processing activities of neurons – or in the words of BULLOCK: "To ask how the sequences [of impulses] are generated and what features of the sequences are read by the decoding cell, i.e. every postsynaptic cell, is equivalent to asking what might be the code or codes; what is the language of the nerves" [51]351. This relativization is not only the result of the several theoretical difficulties that emerged in a system where timing aspects (delays, coincidence etc.) are important, the basis set of symbols is rather unclear, transmission and 'computation' seems to be involved and where a change in medium (as in the genetic code) is not clearly seen except in primary receptors and effectors. It is furthermore the variety of the experimental physiological findings of the last decade that have entered the coding debate:

> "The representation and transformation of information in nervous systems goes on continually, and any division of the 'message' into discrete 'symbols' is arbitrary, doing a greater or lesser degree of violence to the physiological reality. Moreover, many different kinds of representations and transformations are employed by different parts of the nervous system, by different species, and perhaps to some extent by different individuals within a species, or by a given individual at different times (labile coding) It also seems clear that various 'encoding' and 'decoding' schemes are adaptive; they are especially suited to their individual roles in the functioning of the organism. It follows, then, that we cannot investigate 'coding' or representation of information in nervous systems in general but must begin by studying a multitude of specific examples with sufficient thoroughness and compass to provide a strong foundation for subsequent generalization about modes and properties of 'neural coding' [177]:231.

In other words, the situation has become very complicated, which is reflected by the 43 'candidate codes' that are listed in the appendix of the session report. Some of the candidates are highly speculative and lack an empirical basis.[13] Especially 'pattern codes' were considered rather sceptically, as findings for pattern codes were "rare (...) because the

---

[13] An interesting proposal was made by BERNARD STREHLER, a biologist who later became a key figure in aging research and the recent pattern coding debate (see section 6.5). He proposed a detailed model of axonal delay-lines, where spike pattern codes could make sense for neuronal information processing: "In the present model, the concept is developed that the various patterns impinging from the environment through the sense organs are coded (either by the sensors themselves or at a more central station) in the form of unique time sequences of pulses; and, further, that the 'selection' of appropriate responses is achieved through decoder systems that operate analogously to tRNA function and consist of single cells which are the individual decoder elements.(...) The striking feature of the model is its essential simplicity and its analogy to well-established mechanisms which are known to operate at the molecular-genetic level" [243]:587.

demands are stringent for suitable preparations" [177]:257, and because spike patterns "make high demands on precision of timing and are therefore susceptible to deterioration by noise" [177]:259. In the report, more than 200 publications are listed that fall into the field of 'neural coding' (the list was explicitly called non-exhaustive). Despite this obvious 'complexification' of the problem, a general scheme of 'neural communication' was proposed at the work session, involving a *referent* (some aspect of a physical signal), the *transformation* of the signal, the *transmission* of the transformed signal, and the *interpretation* of the transformed signal. Thus, the concept of 'neural coding' involves the clarification and embedding of all information-related processes for a neuronal (mostly sensory) system in a general scheme.

The problem of 'neural coding' was addressed explicitly within the NRP only at the end of the 1960s. Beside the work session of 1968, *The Second Study Program* of 1969 included several papers on neural coding, where the problem was further subdivided: SEGUNDO defined a code as a mapping between two classes of physical events [222]. PERKEL emphasized, that one has to differ between a 'neuron-centered' and the 'observer-centered' study of neural coding [176] (the former was later called the 'organism point-of-view', see section 5.1.1.).[14] C.A. TERZUOLO and STEIN finally emphasized the use of proper statistics and proposed to use the terminology of signal-processing (i.e. considering the frequency space) for attacking the problem of neural coding [246, 237]. As a last witness for this complexification of the neural coding debate, we call WILLIAM UTTAL. He, together with MADELON KRISTOFF, stated, that "the important point is that at each stage in the chain of nervous transmission the same information may be presented by almost unrecognizable different patterns of physical and electrochemical energies. Thus, when we talk about 'the' nervous code, it is absolutely critical to define the transmission link or level about which we are speaking. (...) The code must be broken for each of the separate recording stages. (...) The specific level of code under investigation in the neurophysiological laboratory is simply a function of where the electrode is placed" [251]:263. Investigating neural coding must therefore consider many different possibilities, as "sensory coding, in all cases, must be considered to be a multidimensional process with overlapping and redundant codes found at all levels" [250]:364.

Neural Noise: The electrophysiological studies in the 1930s indicated that neurons show a certain variability in their responses. This phenomenon motivated two different kinds of questions: The neurophysiologists started to become interested in the causes of this variability – on the one hand, because such studies help to investigate the mechanism of action potential generation; on the other hand, because the explanation of the variability may also shed light on the causes for spontaneous activity. The modelers, however, were confronted with the question of how the variability of neuronal response can be linked with the stability of the behavior of the whole system. If the building blocks of a technological system (like a computer) showed such a variability, the system would not work. Within this context 'variability' is interpreted as 'unreliability'. The question of building 'reliable' systems out of 'unreliable' elements became thus a central question in the emerging automata

---

[14]The distinction between 'sign' and 'code', proposed by W.R. UTTAL in 1973, may be mentioned here. If an observed regularity in nature only has a meaning for an external observer, then the regularity is called a 'sign' – e.g. when diagnosing a disease or when doing a weather forecast. If the regularity has a meaning for the system in the sense that it is involved in some identifiable functional role, such a regularity is called a 'code' [249]. All codes employ signs but not all signs are necessarily involved in codes – they are epiphenomena.

science, as we will discuss in more detail in section 3.2.5.

In this section, we discuss, how neurophysiologists attacked the problem and how the concept of 'biological noise' developed.[15] The concept of 'noise' appeared in the 1950s where sensory systems (especially vision) were studied from an information theoretic point-of-view. The term 'noise' was used in the SHANNON sense as a disturber of a communication process, which is for example exemplified by RICHARD FITZHUG's description of the problem: "In the process of coding, the statistical fluctuations in the durations of the impulse intervals represent noise in engineering terminology" [85]:933. To some extent, the brain becomes a communication engineer, as "the problem of the analysis of a nerve fiber message by the brain is similar to the engineering problem of detection of a signal in a noisy communication channel" [85]:939.[16] The term 'noise', however, also appeared in the 1950s in another context: in studies that focused on synaptic processes. PAUL FATT and BERNHARD KATZ used the term 'biological noise' explicitly in a *Nature* publication of 1950 [79], where they presented an observation, that later became fundamental for explaining synaptic transmission. They reported, that the membrane potential of a muscle fibre near an endplate (the synapse between motor neuron and muscle cell, with acetylcholine as neurotransmitter) showed constant fluctuations. The amplitude of the fluctuations indicate "a local mechanism by which acetylcholine is released at random moments, in fairly large quantities; and the most plausible explanation is the occurrence of excitation at individual nerve terminals, evoked by spontaneous fluctuations of their membrane potential" [79]:598.[17] Thus, the 'noise' in neural systems is not only an effect that emerges as a result of an artificial stimulation experiment. It can also be observed in a (more or less) undisturbed system *and* it has a physiological effect even when the system is not stimulated. 'Biological noise' could thus provide a mechanism for spontaneous activity. FATT and KATZ, however, did not intend to interpret this result in the sense that the neural system may act in some 'random' way due to noise. Rather, they proposed in 1952 to use the term 'random' with caution: "This random sequence of miniature potentials should, however, be interpreted with some caution. It evidently means that in this particular set of observations there was no noticeable interaction between the various contributing units, but it does not prove that the constituent units themselves discharge in a completely random manner" [78]:123.

The spontaneous release of synaptic transmitters as well as failures in synaptic transmission provided a first basis for investigating the cause of neural noise. In the 1960s, however, there was no general agreement on the influence of noise on neural systems, and even on the meaning of the term itself. This is shown by the discussions in several conferences in the early 1960s. At the Leiden symposium *Information processing in the nervous system* in 1962, the problem of noise has been discussed repeatedly. BERNSTEIN defined 'noise' just as "anything that is not significant to the system [91]:42. BIGELOW remarked, that the difference between noise and signal is relative to the frame of reference [91]:43 SCHEFFLER, finally, even found a functional role for noise as "background noise is a fine thing to regulate the sensitivity of a system" [91]:274. This short overview indicates the nonuniform use of the

---

[15]For today's concept of neuronal noise we refer to section 6.3.

[16]In 1957, KUFFLER described the task of the brain in almost the same terms: "The analysis of the nerve fiber message by the brain is similar to the engineering problem of detection of a signal in a noisy communication channel" [136]:699.

[17]The term 'random' means in this context, that the distribution of time intervals between the discharges measured in the muscle fibre have POISSON characteristics.

concept of noise. Also MARY BRAZIER observed at the 1962 conference on *Information storage and neural control* this vagueness in the use of the term 'noise': "(...) neurophysiologists use this term in the vernacular rather than in its critically defined sense. This is because we do not usually apply the criteria for randomness when speaking of biological noise. As a matter of fact, many use the term 'noise' in quite the opposite sense from that defined by mathematical theory. In the neurophysiological journals, we frequently find 'noise' used to describe disorderly, unpredictable activity in which no regularity can be detected" [43]:232. And at the same conference, SALZBERG pointed out, that "there are many things which people refer to as noise that are quite different from one another" [210]:22. This confusion indicated, that the phenomenon itself should be described more precisely. Some technical-oriented researchers became interested in the properties of neural noise in spectral terms. A.A. VERVEEN and H.E. DERKSEN published in 1965 precise measurements of the membrane voltage fluctuations and found a $1/f$ relationship (f stands for the frequency in the spectrum) in the noise power spectrum – a type of power spectrum which "is well known in physics and technology. It is called by different names such as current noise, flicker noise or excess noise. It is found in carbon resistors, carbon microphones, (...)., vacuum tubes, (...) [and] semiconductors" [255]:155. As they could put the noise problem back in the technical context, where noise was well studied, they hoped to find an explanation for noise in terms of ion conduction on a molecular scale. One year later, they indeed were able to show that the $1/f$ noise is related to the flux of potassium ions [71].

Although this finding may explain what in the 1930s had been hypothesized as 'threshold fluctuation', the neurophysiologists were more interested in the question of which cause of noise should be considered as the most relevant one in terms of firing variability. WILLIAM CALVIN and CHARLES STEVENS presented an answer in 1967 and 1968 and concluded, that synaptic noise, and not threshold fluctuations, is the main source for firing variability [65, 64]. This was also supported by modelers like RICHARD STEIN in the mid 1960s [240, 239]. Thus, at the end of the 1960s, a clearer understanding of the main causes of noise (synaptic effects) and its main effects (variability in firing) was reached. CALVIN and STEVENS further noted, that noise should not be considered as an "imperfection in the machinery of the nervous system which could be eliminated or at least minimized by proper 'design' of spike-generating membrane; rather, it is inherent in the integration process" [64]:585.

The communication *ansatz* for explaining the role of noise is thus considered as a wrong perspective on the problem. Rather, the question should be asked, whether the neural system could derive any *functional gain* from noise. Neurophysiologists had already considered this possibility in the 1950s. L.R. GREGORY presented these different perspectives in 1961: "It is well known that all communication systems are ultimately limited by random noise, which tends to cause confusion between signals. It seems impossible that the nervous system can be an exception, and so it is hardly a discovery that there is 'noise' in nerve fibres, and in the brain. (...) It is interesting in this connection that GRANIT (1955) has recently summarized the evidence for random firing of the optic nerve but has not interpreted this as a background 'noise' level against which visual signals must be discriminated, but rather regards is as necessary for keeping the higher centers active. (...) Given the engineering viewpoint, we should ask how the system is designed to minimize the effect of the background noise, and this is quite a different sort of question, leading to quite different experiments" [105]:328-329. In 1966, LAWRENCE PINNEO expressed this possible new role for noise very clearly, attacking those scientists (explicitly M. TREISMAN and DONALD HEBB) who considered noise as a disturbance for important functional aspects for the nervous system like perception and learning:

> "These two papers [of TREISMAN and HEBB] illustrate widely held misconception of brain function, namely, that the spontaneous random or background discharge of neurons has little or no functional value; that is, this activity has no functional value for the organism and therefore is noise in the communications sense of the word" [182]:242.

He listed several findings that indicate a functional role for 'noise'. For example, the spontaneous activity in the visual system is highest during dark adaption, when visual sensitivity is greatest and lowest under conditions of high illumination when sensitivity is poorest. Noise must thus be seen as something that helps the system and not as something that disturbs it. The concept of 'noise' in the SHANNON sense was thus opposed in neuroscience and led to the opinion that noise may have indeed an important role for neurons. Or as J.S. GRIFFITH noted in 1971: "We must therefore be prepared to contemplate the possibility that, in many brain neurones, the potential variability of output which is presumably there as a consequence of the mechanism of quantization of synaptic transmission (or possibly for other reasons) might be 'deliberately' exaggerated to almost any extent under normal operating conditions. If so, then it would not be reasonable that such variability should necessarily be called 'unreliability' " [106]:27.

Neural Information: According to an analysis by JUSTIN GARSON [312], ADRIAN used the term 'information' at first in a 'modern' sense, i.e. relating the term to the content of sequences of nerve impulses. In this way, combined with the establishment of the spike train and of its mathematical abstraction (a series of event-times), the basis for a quantitative notion of 'information' in the sense of SHANNON's information measure has been set. In the 1950s, the hopes for a successful application of SHANNON's approach were still intact. For example in 1958 FITZHUG expressed this hope as follows: "This paper may stimulate productive application of communication theory to the largely unknown processes of detection and integration in the nervous system" [84]:691. This hope was not fulfilled. Rather, explicit definitions of 'information' in a neuronal context are very rare in the literature of the period we investigated. If they are provided, they remain rather vague (e.g. R.W. GERARD defined the term in his opening address at the 1962 Leiden conference as "the organization or pattern of input and output" [91]:4) or explicitly undefined. The NRP work session report on *Neural Coding* remarked on this point at the beginning, that the term 'information' is not defined but is used "in its more colloquial non-SHANNON sense; the information we discuss is not necessarily measurable in bits" [177]:227 (footnote). What was the reason for this failure to determine a precise, quantitative definition of 'neural information'?

This failure to find a generally accepted and precise definition of 'neural information' went along with an increasing scepticism towards the usefulness of information theory for neuroscience, as well as for biology in general.[18] Compared to the development in molecular

---

[18] A prominent example ist the failure of HENRY QUASTLER to establish a theory of biological information applicable to molecular biology in the 1950s. QUASTLER, an Austrian scientist who emigrated at the onset of the Second World War to the United States, was committed to the general usefulness of information theory [191]. However, as LILY KAY showed, he failed to create an experimental setup in order to investigate his

biology, however, the disillusioning statements were found some years later, although "neurophysiologists have been prominent among those who wished to explore the potentialities [of information theory] for their field" [43]:230, as MARY BRAZIER pointed out in 1962. This critical attitude was expressed in various conferences at the end of the 1950s and the beginning of the 1960's. At the 1959 conference on *Self-organizing systems*, MCCULLOCH expressed his critique against using the information theoretic concepts of redundancies of code and channel in order to understand the reliability of biological systems [155]:265. At the 1962 conference on *Information storage and neural control*, the usefulness of information theory was discussed several times. BERNARD SALZBERG, for example, expressed this critical attitude as follows:

> "With reference to the theme of this symposium, one might say that information theory provides insight for analyzing and improving storage and communication processes, but does not unravel the bewildering complexities associated with significance, meaning, or value judgement. From my personal experience with the problems of physiological signal analysis, this fact lies at the core of the difficulties which the life sciences face in applying information theory to their problem" [210]:16.

The difficulties when applying an information theoretic terminology to processes in neuronal systems were increasingly considered in the 1960s, also by prominent advocates of information theory. ANATOL RAPOPORT noted at the 1962 Leiden conference that "information is much more elusive than either matter or energy; its measurement is much more dependent on certain frames of reference, definitions and conventions" [192]:16. DONALD MACKAY for his part presented in 1965 a whole set of difficulties when applying information theory to neurobiological problems: "The first [consideration] concerns the difficulty, in a living organism, of *identifying the categories* in terms of which to analyze a stimulus for purposes of information measurement.(..) The second is that even when we have decided on our categories, the probabilities p_i are often impossible to determine with any confidence. (...) The third consideration is that in many neural processes, the engineer's ability to record for optimal efficiency has no parallel, so that the relevance of Shannon's measure is correspondingly reduced" [147]:637. MARY BRAZIER commented on this situation in 1962, saying that "it is so difficult to define information measures for ensembles in biology that most biologists who use information theory usually do not attempt to do so in a quantitative

---

claims and, in combination with his premature death, this meant that his quantitative studies were later forgotten. The second of the two conferences on *Information Theory in Biology* organized by QUASTLER in 1956, reflected this disenchantment towards the use of information theory for problems in molecular biology. The recordings of the final round table discussion started with the words: "Information theory is very strong on the negative side, i.e. in demonstrating what cannot be done; on the positive side its application to the study of living things has not produced many results so far; it has not yet led to the discovery of new facts, nor has its application to known facts been tested in critical experiments. To date, a definite and valid judgment on the value of information theory in biology is not possible" [189]:399. Also MCCULLOCH expressed doubts on the usefulness of information theory for problems in biological communication, when answering the invitation from YOCKEY to the second Symposium on Information Theory in Biology in 1956 (quoted after Kay [323]:175-176). A comment made ten years later by DONALD MACKAY serves as a last example: "Those who remember the high hopes generated among biologists and others, that *information theory* would provide a new *interlingua* for science, may be somewhat surprised that so few applications of SHANNON's mathematics have been made in Biology, and that so few even of these have proved worthwhile" [147]:637.

way. Generally, they do not actually measure the information; and hence, they fail to exploit the full potentialities of the theory" [43]:240. At the end of the 1960s the role of information theory for understanding neuronal systems is apparently considered as basically useless by many neuroscientists. K.N. LEIBOVIC expressed this rather clearly in the final session at the 1969 symposium on *Information Processing in The Nervous System*:

> "I submit that 'information theory' in the sense of SHANNON or WIENER is too restrictive for our purposes. (..) SHANNON's theory is, of course, quite relevant to problems of signal transmission, including channel capacity and coding. But, when one is dealing with biological information in the broad sense, it is well not to identify 'information' with a narrow definition from engineering science" [142]:335-336.

Similar statements are available from several conferences. For example a short meeting report in *Science* about a Japanese-US meeting on spike train analysis in 1967 stressed that "the consensus of opinions appeared to be that although information theory has played an important role in the development of modern communication techniques, it presently has very limited applications to biological systems" [274]:1025. GEORGE MOORE, an important promotor of statistical methods in neuroscience, wrote in a review in 1966: "The promise held out to students of the nervous system by information theory (statistical communication theory) since the publication nearly two decades ago of the classical monograph of SHANNON & WEAVER has been realized to a disappointingly small extent. (...) At best, in fact, the concepts and constructs of information theory are metaphors when applied to the nervous system, and the questionable aptness of these metaphors lies at the heart of the difficulty. It is thus important to exercise great caution in the identification of the formal entities defined by information theory (e.g. the source, the source alphabet, the channel, etc.) with the realities encountered in neurophysiological investigations" [163]:507.

What was the reason for this change of mind? Certainly, the difficult methodological problem that arose when applying the theory to biological systems was one cause. Furthermore, the neurophysiologists were not interested in the kind of answers information theory may provide. As DONALD MACKAY had said at the 1962 Leiden conference: "The impression I get from my friends who are neurophysiologists, however, is that the bulk of things that really baffle them at this stage are not waiting for a numerical answer; and that in nine cases out of ten what is needed is more an aid to intuition, based perhaps on the qualitative notions of information theory (...) rather than the development of mathematical theories leading to numerical predictions" [42]:92. Also MARY BRAZIER answered the question (the title of her talk in 1962) *How can models from information theory be used in neurophysiology?* by referring to the future:

> "In closing, let me say that the application of quantitative information theory to neurophysiology lies largely in the future. Possibly a partial answer to the question in the title of this paper is that if information theory has not led to the uncovering of many new facts in neurophysiology, it may have led to many new ideas" [43]:240.

The failure to find a precise notion of 'neural information' therefore reflects the loss in credibility of information theory within neuroscience. The neuroscientists of the 1950s and

1960s were not able to relate SHANNON's concept of information with their measurements. Later, several attempts to restore the credibility of information theory followed, for example by REINHARD ECKHORN and BERTRAM PÖPEL [74].[19] Although we did not investigate the more recent attempts to use concepts of information theory in neuroscience systematically, it is noticeable that the theory has regained its reputation within neuroscience [398, 651]. We suppose, that this accompanies the 'second boom' of theoretical neuroscience (see section 4.1), in which detailed and large-scale computational modelling provided a new tool for investigating neuronal systems. Within such a framework, the signal as well as the physiological boundary conditions are unter full control of the experimenter and allow a precise application of the theoretical apparatus provided by information theory.

3.2.3 The Emergence of a Toolbox for Spike Train Analysis

The development of single cell recording, starting in the 1920s, substantially increased the amount of data available to investigate the neuronal system on a micro-scale. A statistical analysis of the data, however, was still rather difficult, because data generation was very time consuming. Photographic film was the medium used to record and store data obtained from neurophysiological measurements [63] and numerical data were usually obtained by manually reviewing the film, e.g. to obtain sequences of interspike interval (ISI) times. About eight man hours of work were required to analyze 2-3 minutes of recordings in this way [58]:243. Not until the late 1950s, when methods for automated generation of numerical data as well as computers for performing the statistical analysis were available for neurophysiology laboratories [96], was a statistical analysis of neurophysiological data feasible. It is thus not surprising that only very few papers used histograms before 1960. Probably the first histogram was published by BRINK and colleagues in 1946 [45], followed by some contribution by the Japanese neurophysiologist HAGIWARA in 1949 and 1950 (see [163], the diagrams were not named 'histograms').

As soon as the technology allowed a statistical treatment of the data obtained by single cell electrophysiology, the need to look at the data from a statistical perspective was emphasized (see also section 3.2.1). GEORGE GERSTEIN, who pioneered the application of statistical methods for spike train analysis, promoted in his *Science* paper of 1960 the use of statistics to detect stimulus-response relations "buried in a background of spontaneous activity" and argued against the "tedious 'hand and eye' measurements of strip-film records" [95]:1811. In the same year, he published, together with NELSON KIANG, the first 'method paper' on spike train analysis, introducing the ISI and the post-stimulus time histogram in a formal way [96].[20] Two years later, R.W. RODIECK, together with KIANG and GERSTEIN, extended the set of available statistical methods by emphasizing that "it is clear, that casual visual inspection will not provide adequate descriptions of the spike trains" [201]:355. The interest in a statistical approach to neuronal data increased significantly in the following years. The first review paper, provided by GEORGE MOORE, DONALD PERKEL and JOSÉ SEGUNDO in 1966, listed more than 130 references on this matter – the majority dated from 1960 to 1965.

---

[19]Their contribution started with the sentence: "Many neurophysiologists believe that the application of information theory to neuronal data is not meaningful" [74]:191.

[20]For a classification of histograms we refer to section 7.2.1.

This period of rapid development of statistical methods for neuronal data led to some confusion. A stronger theoretical underpinning was required. PERKEL, GERSTEIN and MOORE noted in 1967 several problems: "(a) inconsistency of nomenclature and notation in this field, (b) difficulties, not always well enough appreciated, in assigning measures of statistical significance to experimental findings, (c) presentations of experimental data in several forms that are in fact mathematically derivable from each other, and (d) the risk of attributing particular physiological significance to results that illustrate purely mathematical theorems of that are more plausibly attributable to change effects" [178]:393. They grounded the statistical methods for spike train analysis in the theory of stochastic point processes, which serves now as the general theoretical framework [178, 179].[21] The computer power available at that time did, however, not allow the analysis of large scale data sets of simultaneously recorded neurons (proposed e.g. by WALTER ROSENBLITH [206]:538 in 1959 and introduced by GERSTEIN and W.A. CLARK [98] in 1964) and higher order correlations. This is supposedly a reason that explains the focus on single cell electrophysiology in the 1970s, expressed by the 'single neuron doctrine' of BARLOW.

The emerging statistical viewpoint within neuroscience not only provided the tool set for practically applying the theoretical apparatus of information theory (e.g. for estimating probability distributions), it also set new constraints on the execution of neurophysiological experiments. An important requirement is *stationarity*, i.e. the probability distribution describing the firing of the neuron should not change in time. This is a strong requirement for biological systems, and it is usually granted that the system under investigation can only fulfill it to some extent. Furthermore, the statistical approach determined new entities that are considered as the 'true' response of the neuron (the 'mean response') – an aspect we already discussed in section 3.2.1. Finally, the question emerged, as to whether one can find a *stochastic model* that describes neuronal firing. As we will discuss this problem further in the scientific part of this thesis (see section 6.2.2), we add some more remarks on this aspect. The first studies in this respect can be traced back to the early 1950s and were performed by Japanese researchers: in 1950 by YASUJI KATSUKI and colleagues [130] and in 1954 by SUSUMU HAGIWARA [113]. Both studies considered the interspike interval distributions as emerging from a random (POISSON) process, although the amount of data was rather poor. We did not find further studies in this respect – probably due to the high investment in work such an analysis needed. But in the 1960s, several authors discussed this point – and a rather conflicting picture emerged: Some studies confirmed the early Japanese findings: In 1961, R.G. GROSSMAN and L.J. VIERNSTEIN reported the results of 31 neurons measured in the cochlear nucleus of the cat (spontaneous activity) – 30 of them showed POISSON characteristics in their discharges [108]. In 1963, T.J. BISCOE and A. TAYLOR investigated the firing of chemoreceptor afferent fibres from the cat carotid body. By analyzing the ISI distribution and calculating the coefficient of variation, they concluded that the spike trains satisfy two criteria for a random series – a more prudent statement

---

[21]In this respect it is interesting to note that this framework emerged out of an industrial problem: the failure and replacement of technical components, such as electrical light bulbs. This problem lead to the development of renewal theory, the study of 'self-renewing aggregates'. As it became clear that this theory is applicable for serval classes of problems, especially in combination with increasing interest in operations research, the theory was popularized in the early 1960s by the British statistician D.R. COX in his monograph *Renewal Theory* [70]. COX later provided – together with P.A.W. LEWIS – a second monograph on the *Statistical Analysis of Series of Events* [69]. Both monographs became standard theoretical references for spike train analysis.

[31]:343. In 1965, D.R. SMITH and G.K. SMITH analyzed the spontaneous firing of cortical neurons in the unanesthetized cat forebrain, which was isolated from the remaining nervous system by a cut across the midbrain at the *tentorium cerebelli* [234]. Two groups of ISI distributions emerged: the larger group of cells (25 neurons) show an exponential decay, and the smaller group (15 neurons) a "nonlinear" distribution (in the log-plot). The latter has been explained as the overlap of two POISSON processes with different rates.

The POISSON process for explaining the properties of spontaneous (and evoked) firing was attractive, as the statistical model is simple and mathematically easy to handle. However, the majority of investigations concerning this issue in the 1960s showed a much more complex situation. In 1962, RODIEK and colleagues used this problem to present their statistical toolkit for spike train analysis [201]. They displayed the results obtained from four units (cat cochlear nucleus, spontaneous activity), which they considered as representatives of four classes of firing, each class consisted of about $\frac{1}{4}$ of all neurons studied (the total number was not provided). One showed POISSON characteristics in the ISI histogram. The second unit showed pacemaker activity (firing of periodicity 1 with quasi GAUSSian time jitter). The third neuron showed burst-firing characteristics (two length scales). And the last neuron was characterized as "unimodal, asymmetric and non-exponential" (from today's perspective it probably has long-tail characteristics). Their results differ "markedly" from the earlier results of GROSSMANN and VIERNSTEIN (see above), as they noted. This classification of neuronal firing into several types of distributions was found in several other studies (1964 by P.O. BISHOP and colleagues [32], in the same year by A.V.M. HERZ and colleagues [120], in 1965 by V. BRAITENBERG and colleagues [41] and in 1966 by L.M. AITKIN and colleagues [12]. Furthermore, several other studies failed to find POISSON firing characteristics in their data (in 1964 by EDWARD EVARTS [76], in the same year by GIAN POGGIO and LAWRENCE VIERNSTEIN [183], in 1964 by V.E. AMASSIAN and colleagues [13] and in 1966 by H. NAKAHAMA and colleagues [168]. This brief overview indicates that the POISSON model fails to describe the majority of the data. Thus, the question emerges, how and why have these early investigations been overruled by today's opinion, whereby the POISSON model is considered as the adequate statistical model for neuronal firing. We discuss this matter in more detail in section 6.2.2.

3.2.4 The (Random) Network

The use of the term 'network' in the neurophysiological literature indicates the intention to understand the functioning of the nervous system by analyzing how the single units are connected [299]:349. We did not investigate when this perspective began to become a relevant one within the neurophysiological community.[22] Our literature review indicates that this belief was already widespread before 1940 and the main problem was rather *what* the topology of the connections looked like. The early modelling studies of the RASHEVSKY school in the 1930s as well as the MCCULLOCH-PITTS approach (see section 3.2.6) were based on a precisely defined connectivity between neurons (see Fig. 3.2). Within the modeler community, this view was challenged in the 1940s. The brain was increasingly considered as a 'statistical entity' in the sense that a certain randomness in the connectivity has to be

---

[22] A rather modern concept of 'neuronal networks' was already known in the 19th century. The Austrian physiologist SIGMUND EXNER presented in 1894 a detailed concept of neuronal nets, which anticipates many aspects of HEBB's concept (see [299]: section 1.3).

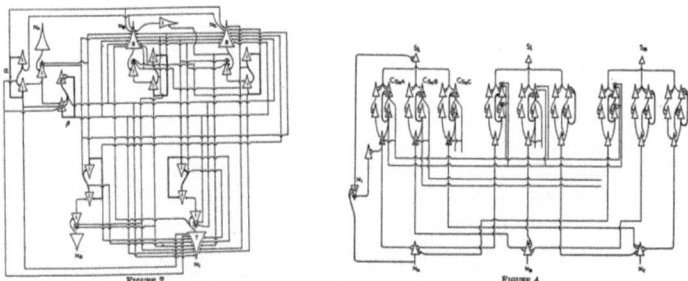

Figure 3.2: Examples of network-structures from early neural modelling studies. RASHEVSKY's neural topology for logical thinking in 1946 (derived from [197]).

assumed. This led to the question of the implications of this view for the functioning of the system.

Probably the earliest approach to a statistical perspective in this respect derives from the work of WALTER PITTS, as NORBERT WIENER remarked at a 1946 conference: "The anatomical picture of the cortex suggests that we may usefully employ statistical methods in the study of its function. This work has been taken up brilliantly by Walter Pitts" [279]:217. This work of PITTS is not available,[23] but there were other researchers from the RASHEVSKY school – ANATOL RAPOPORT and ALFONSO SHIMBEL – who developed a tool to investigate this issue: the random net. In 1948, they published two papers on this issue, where a neural net is taken to consist of a semi-infinite chain of neurons with connections distributed according to a certain probability distribution of the length of the axons [194, 195]. They discussed the question, as to whether steady states are possible in such random nets and provided conditions for the existence of such steady states. This work was clearly intended to promote the importance of randomness in the connections between neurons because "most of the neural nets thus far considered are so highly specific in their structure that the failure of even one neuron to function renders the entire mechanism useless" [232]:41. In 1950, SHIMBEL wrote that the "most serious shortcoming of the method of MCCULLOCH-PITTS lies in the fact that it assumes an extremely detailed knowledge of the neural pattern under consideration" [231]:242. As an alternative approach, "the notion of probability in studying the structure and function of the nervous system" [231]:243 must be considered. Also at the famous *Hixon* symposium in 1948, this point was discussed. LASHELY said: "The anatomical variability [of the nervous system] is so great as to preclude, I believe, any theory which assumes regularity and precision of anatomical arrangements. Such facts lead me to believe that theories of neuron interaction must be couched, not in terms of the activity of individual cells but in terms of mass relations among the cells" [128]:70. The zoologist PAUL

---

[23]Probably because PITTS destroyed a large part of his work as a result of a deep psychological crisis in the 1950s, see [325].

WEISS expressed a similar opinion: "The study of the developed nervous system, with which the anatomist, physiologist, and psychologist are usually working, suggests a high degree of precision in the arrangement of the constituent elements, but it must be realized that this impression is illusory. The organizational stability of a performance of the nervous system is much greater than the precision of the underlying structural apparatus" [128]:72-73.

The requirement for randomness concerning neuronal connectivity was soon adopted by the modeler community, although they had now to solve the problem of how the requirement for a random assembly of elements can be linked with stable dynamics: ROSS ASHBY analyzed this problem in 1950 using a purely mathematical model and came to the conclusion that "randomly assembled dynamical systems have only a small chance of being stable. (...) So it is clear, for instance, that in a discussion of cerebral dynamics one may not postulate a randomly assembled system unless one is prepared to accept the implied instability" [20]:481. ASHBY, however, changed his opinion on this matter, as two years later he declared in his monograph *Design for a Brain* randomness in connectivity to be a *necessary* property of neuronal systems: "Any system which shows adaption must (...) be assembled largely at random, so that its details are determined not individually but statistically" [19]:v.

In the 1950s, random connections in neural nets forming the cerebral cortex became an assumption for several important models: KARL PRIBRAM mentioned in a review in 1960 in this respect the models of HEBB, LICKLIDER, MACKAY, MILNER, and UTTLEY [184]:26-27. A.M. UTTLEY emphasized the need for a statistical approach very clearly: "There is reason to believe that many of the complexities of the cerebral cortex are due to an organization that is fundamentally of a statistical nature" [233]. He used this approach to build a card-sorter machine, where the connections between the input elements and the indicators are randomly assigned. His machine was able to classify punch-cards, which he took as an indication that similar processes may be at work in the nervous system: "Assuming that connections are formed as a result of this chance proximity, it can be shown that a classification system will arise to some extent" [254]:482.

An influential contribution was provided by R.L. BEURLE in 1956 [30]. He discussed the problem of how activity in a randomly connected net of model neurons spreads, and what happens if some property of the individual cell – like size, extent of the axon, dendritic structure or threshold – changes with repeated use. BEURLE's work is based on the assumption, that the "apparently very large random factor in the distribution of neurons and the way in which their dendrites and axon fibres spread amongst each other" is a marked feature of certain parts of the cortex [30]:56. He also sets his work in opposition to the automata approach (see next section): "The comparisons that have been made in the past between the properties of computing machines and living organisms are interesting when made in relation to abstract fundamental concepts, but less productive when details are considered. This is because, in the detailed design of most large computing machines, components are connected together exactly according to some specification, and have a precise function in relation to all other components. This specific organization in a computer contrasts with the apparently very large random factor in the interconnexions between neurons in many parts of the cortex" [30]:83. At that time, there was not much statistical information available about the structure of the cortex, so, as BEURLE notes, one might still have the opinion that the random structure of the mass of cells he considered does not bear a very close relationship to the actual physiological structure. The main point, however, is that simple and ordered forms of activity-spread can be observed in such a network: plane, spherical and circular waves and vortex effects. There was, however, also the problem that his network of simple model neuron often tended either to become completely quiescent or completely active. This was interpreted that the mass has a sensitive threshold for stimulation and may act as an on/off switch comparable to the shifting of attention in an organism. By implementing modifications of the elements, trial and error learning, conditioned responses, and the ability to regenerate internally a sequence of past events can be observed. Thus, a connection between ordered structures of behavior resulting from a probabilistic microstructure was established – or in the words of BEURLE: "(...) some of the basic forms of behavior of living organisms can be simulated by a mass of simple units without the necessity of postulating a large degree of specific organization of these units" [30]:83. This is a major insight that inspired later probabilistic approaches to the nervous system.

However, the problem of how a highly connected, but randomly connected aggregate of neurons is able to maintain a state of intermediate activity was not satisfyingly solved. In 1962, ASHBY, VON FOERSTER and WAKER provided a more realistic analysis of this matter: again the network evolved in either a completely quiet or a completely active state [18]. One year later, J.S. GRIFFITH offered a solution in his study using the same model of ASHBY and colleagues. He showed that the desired stability could be achieved by allowing either *some* order in connectivity on a large scale or by introducing inhibitory connections [107]. The assumption of random connections later also served as a guiding principle for neural networks, as fo example FRANK ROSENBLATT pointed out [204]. This aspect, however, will not be discussed by us.

3.2.5 The Active and Reliable Brain

Many different empirical and theoretical investigations in the period from 1940 to 1970 focussed on the activity of the whole brain or large parts of it, for example the studies on learning by DONALD HEBB, or the use of the EEG. We, however, will focus only one aspect: how the notion of a 'spontaneous active' brain was consolidated in that period and to what extent this led to the problem of how 'reliable' system behavior could result from 'unreliable' components. The latter was discussed in the emerging theory of automata and computer science and became an important aspect of the brain-computer analogy. In the 1950s and 1960s, a line has been drawn between a period before the 1940s, when the brain was considered as 'passive reflex machine', and the time after the 1940s, when the brain had become an 'active device'. In DELISLE BURNS's monograph on *The uncertain nervous system* that appeared 1968, this point of view is well-documented: "Until some twenty years ago, attempts to explain function in the intact nervous system were necessarily dependent upon the results of experiments employing relatively large stimulating and recording electrodes. (...) As a result, the physiologist unwittingly left his readers with an impression that central neurones only discharged when instructed to do so by a neurophysiologist – the system appeared to be an immense assembly of wholly predictable relationships. The widespread, recent use of micro-electrodes has done much to alter this picture. One is now forced to think of most, if not all, central neurones as continually active" [57]:162. We showed (see section 2.3.1) that such a sharp transition line cannot be drawn, as there were experimental findings and theoretical models in which 'spontaneous activity' was playing a role, well before the 1940s. However, the 'active brain' became the dominant idea after 1940. An important promotor of this point of view is certainly KARL SPENCER LASHLEY. He expressed this opinion very clearly at his talk on *The Problem of Serial Order in Behavior* at the *Hixon* symposium in 1948:

> "Neurological theory has been dominated by the belief that the neurons of the central nervous system are in an inactive or resting state for the greater part of the time (...). Such a view is incompatible both with the widespread effects of stimulation which can be demonstrated by changes in tonus and also with recent evidence from electrical recording of nervous activity. (...) It is probably not far from the truth to say that every nerve cell of the cerebral cortex is involved in thousands of different reactions. The cortex must be regarded as a great network of reverberatory circuits, constantly active" [139]:131.

LASHLEY contrasted his concept with the more static model of MCCULLOCH and PITTS (see next section), where neurons are embedded in networks performing logical functions. He rather proposed that "the analysis must be extended to the properties of such nets, the way in which they are broken up into reactive patterns in the spread of excitation, to give, for example, directional propagation or its equivalent. I strongly suspect that many phenomena of generalization, both sensory and conceptual, are products, not of simple switching, but of interaction of complex patterns of organization within such systems" [139]:134. LASHLEY emphasized this point in the discussion following his talk: "I agree thoroughly with Dr. MCCULLOCH that the transmission of excitation by the individual neuron is the basic principle of nervous organization. However, the nervous activity underlying any bit of behavior must involve so many neurons that the action of any one cell can have little influence upon the whole. I have come to feel that we must conceive the nervous activity in terms of the interplay of impulses in a network of millions of active cells" [139]:145

In the 1950s, the 'spontaneous activity' of several neural (sensory) system was investigated, where the term 'spontaneous' is usually understood as denoting the "activity observed in the experimental situation in the absence of deliberate stimulation" [151]:377 (see also [136]:697). DONALD HEBB made clear that the term 'spontaneous' would not involve any non-mechanistic, metaphysical forces acting within the system: "Spontaneity of firing by central neural cells is not philosophic indeterminacy, as some writers have thought; the 'spontaneity' means only that synaptic stimulation is not the sole cause of firing" [119]. These studies in sensory systems have usually been put in the context of the 'noise' problem. In that sense, spontaneous activity was one aspect of noise that disturbs information transmission (see section 3.2.2). S.W. KUFFLER, however, presented an alternative view in the sense that the representation of information is just the modulation of the spontaneous activity [136]:699. This line of argumentation leads to the discussion on the functional relevance of noise (see section 3.2.2). In the review of PRIBRAM of 1960 [184], the studies made in BURNS' laboratory in the late 1950s [59] were considered as the ones that consolidated the concept of 'spontaneous activity'. In these studies, brain tissue was completely isolated neurally from other nervous tissue. In that case, the neurons are indeed quiescent, but also easily aroused to prolonged activity: "Hence, at 'rest', they [the neurons] may be conceived to be just below the excitatory level for continuous self-excitation" [184]:2. In 1962, BURNS, together with G.K. SMITH published another study demonstrating that in the unstimulated (cat) brain, almost all neurons exhibit spontaneous activity [58]. Thus, spontaneous activity was an accepted phenomenon in the 1960s.

For the theoretically oriented neuroscientists, these findings posed a problem: How can the neural system function when the neurons themselves are unreliable in the sense that they show variable response and spontaneous activity. For logical nets of the MCCULLOCH-PITTS-type, this was indeed a problem, although there were some attempts in the 1960s to assess the stability of these nets in noisy conditions [39]. The interesting point is, however, that spontaneous activity, as well as the problem of random connectivity discussed earlier, was soon put in a *technological* context. GRIFFITH for example, who provided a solution of the stability-problem of random nets (see prior section), declared in 1963 that "such masses [of randomly connected cells] are potentially useful for the construction of artificial brain-like systems" [107]:308. The fascinating question for the emerging 'bionics' community in the early 1960s was, as VON FOERSTER mentioned: "What are the structures which make complex systems immune against errors of all kinds?" [257]:x. A main figure in this discussion was JOHN VON NEUMANN, who attacked the problem of the *Synthesis of reliable organisms from unreliable components* (the title of his 1956-paper [267]) in a sophisticated way. As a major problem concerning the construction of such systems he identified "not so much that incorrect information will be obtained, but rather that irrelevant results will be produced" [267]:62. Not the single error is relevant, but the fact that errors accumulate such that the long-run behavior of the system is unreliable. To solve this problem, he introduced the 'multiple line trick' or 'multiplexing' – which later led to the theory of redundancy in computer design [181] –, a 'majority rule', and a 'restoring organ'.[24] In this way, the system operates along several parallel pathways and is able to compare after each processing step the outcomes of the parallelly processed information and, using the result of this comparison, to maintain the correct outcome. The system is organized in such a manner, "that a malfunction of the whole automaton cannot be caused by the malfunction of a single component, or of a small number of components, but only by the malfunctioning of a large number of them" [267]:70. VON NEUMANN's solution, however, could only deal with unreliability expressed in the components, but not in their connectivity. In 1963, S. WINOGRAD and J.D. COWAN included the latter aspect and came to a similar conclusion to GRIFFITH, whereas randomness in connectivity within certain modules should be combined with a specified connectivity pattern between the modules [286]:84-85. In this way, the question of how component-variability ('unreliability' of neurons), spontaneous activity and a certain degree of randomness in connectivity can still allow the functioning of biological systems was taken as an inspiration for improving technological systems, which is indeed a new development in the history of brain research.

This inspiration found its reverberation in the brain-computer analogy – the dominating technological metaphor for the brain in the 20th century. The brain-computer analogy was a *functional* analogy[25] and included the epistemic principle of cybernetics that one may learn something about the brain when one constructs artefacts that perform like brains. The latter point is illustrated by a statement of the German cyberneticist Karl Steinbuch in his

---

[24]In 1957, also HENRY QUASTLER noted the importance of redundancy: "It is possible that there is a message for engineers in the organization of biological computers. Traditionally, the engineer's ideal is to eliminate redundancy and to perfect elementary reliability, and this is achieved during a more or less extended period of collecting case histories and correcting specific shortcomings. As systems become very complicated, this method becomes less and less feasible" [190]:194. The solution is thus to combine redundancy with low elementary reliability.

[25]One finds also references that claim a structural analogy, see [167]:6. This, however, does not reflect a general opinion.

monograph *Automat und Mensch*: "Unsere Einsicht in die Funktion unseres Denksystems ist gering. Wenn nun plötzlich durch Automaten vergleichbare Eigenschaften erzeugt werden können ("künstliche Intelligenz"), erschliesst sich dem forschenden Geist ein neuer Weg zum Verständnis des Menschen: Nämlich über das Verständnis der Automaten" [241]:v. In the following, we only discuss those aspects of the brain-computer analogy that were closely connected to neurophysiological considerations and we neglect the (much broader) discussion about the brain-computer analogy in artificial intelligence and the emerging cognitive sciences.

NORBERT WIENER was probably the first, around 1940 to explicitly compare features of the electronic computer and the brain, as ASPRAY stated [295]:124-125. This analogy soon found friends in other fields. The Harvard psychologist and historian of psychology, EDWIN BORING, considered WIENER's suggestion that all functions of the brain might be duplicated by electrical systems in a letter in 1944 as "very attractive" [310]:247. The brain-computer analogy propagated fast in the cybernetic community. Especially MCCULLOCH was fascinated. His motivation was not only to promote the idea of the brain as a logical machine, which was a consequence of his modelling approach, but, in 1949, he was also referring to the brain as an inspiration for *building* computing machines: "Computer machine designers would be glad to exchange their best relays for nerve cells" [158]:493. At the *Hixon* symposium in 1948, MCCULLOCH not only stressed the technological aspect: "Neurons are cheap and plentiful. (...) They operate with comparatively little energy. (...) VON NEUMANN would be happy to have their like for the same cost in his robot, so he could match a human brain with 10 million tubes; but it would take Niagara Falls to supply the current and the Niagara River to carry away the heat" [157]:54-55.[26] He also emphasized the point that the brain-computer analogy contains a powerful vision for society: "The former revolution replaced muscles by engines and was limited by the law of the conservation of energy, or of mass-energy. The new revolution threatens us, the thinkers, with technological unemployment, for it will replace brains with machines limited by the law that entropy never decreases" [157]:42. DONALD MACKAY, a companion of MCCULLOCH in this respect, saw also a link in direction of psychiatry. He hoped, that the knowledge gained by using this analogy may help to establish a "working link between the concepts of psychiatry and those of physiology and anatomy.(...) The considerable effort going into this theoretical model-making is justified chiefly by the hope that out of it may come a way of describing the thinking process, sufficiently close to psychiatric realities to be useful in diagnosis, yet sufficiently operational and objective to allow the physiologist to make his maximum contribution to the study and treatment of mental illness" [149]:266-267.

In the small community [295], in which the brain-computer analogy was developed, its limitations had been realized. JOHN VON NEUMANN was probably the most careful in this respect. At the *Hixon* symposium, he stated: "It is very obvious that the brain differs from all artificial automata that we know; for instance, in the ability to reconstruct itself

---

[26] VON NEUMANN, however, would probably not have supported this analogy. In 1950, at the 7th *Macy conference*, he commented: "There has been a strong temptation to see the neuron as an elementary unit, in the sense in which computing elements, such as electromechanical relays of vacuum tubes, are being used within a computing machine. The entire behavior of a neuron can then be described by a few simple rules regulating the relationship between a moderate number of input and output stimuli. The available evidence, however, is not in favor of this. The individual neuron is probably already a rather complicated subunit, and a complete characterization of its response to stimuli, or, more precisely, to systems of stimuli, is a quite involved affair" [259]:22.

(as in the case of mechanical damage). It is always characterized by a very great flexibility in the sense that animals with look reasonably alike and do the same thing, may do it by rather different cerebral mechanisms. Furthermore, though all humans belong to the same category and do the same things, outwardly, in some cases they are using different cerebral mechanisms for the same things, so there seems to be a flexibility of pathways" [128]:109. In his talk at the symposium, JOHN VON NEUMANN also took the analogy as an inspiration for building a new type of automata. These systems should not be designed down to every detail, but only "on some general principles which concern it, plus a machine which can put these into effect, and will construct the ultimate automaton and do it in such a way that you yourself don't know any more what the automaton will be" [128]:110. A (first) cumulation of this program was the publication of the anthology *Automata Studies* in 1956, whose list of contributors included the major protagonists in the field [225].[27]

In the following years, the *differences* between brains and computers became the focus of research. Beside the points already mentioned above made by VON NEUMANN, the intrinsic activity of neural systems have also become a point that distinguishes them from computers. THEODORE BULLOCK noted in 1961: "Nervous systems are not like present computers, even complex ones, but have oscillators and built-in stored patterns; they do not give outputs predictable by their inputs or externally controlled 'instructions' " [54]:65. The issue of the brain-computer analogy was also disputed at a work session of the *Neurosciences Research Program* in 1964 about *Mathematical concepts of central nervous system function*. H. GOLDSTINE provided an overview of four main differences between brains and computers: 1) Computers function either analogically or digitally, but nervous systems appear to function by a combination of the two methods. 2) Parts of the nervous system utilize frequency-modulation coding rather than the binary coding of the computer. 3) The level of mathematical precision of the nervous system is considerably less than that of a computer. 4) Computers do not share with the brain its structural redundancy [236]:116 These obvious differences as well as a critical attitude of many neurophysiologists toward modelling studies led the cybernetics community to take this critique on board by using one of two strategies, either weakening the analogy, or stressing that a 'new' concept of computation will be necessary to understand biological computation. In his 1961 monograph with the provocative title *The Brain as a Computer*, F.H. GEORGE acknowledged that "the brain is clearly a vastly complicated system, and there is an obvious naivety – doubtless irritating to a neurophysiologist – in such statements as, 'just a switching device', 'the eyes are like a television scanning system', 'the brain is a complex digital (and analogue) computer', and so one" [89]:310. He concluded that "we still have a very long way to go before we can fit all the empirical facts together, especially at the molecular level, to supply anything like *the* working model we need" [89]:380. ROSS ASHBY was in favor of the second strategy. In his 1966 review on the use of mathematical models and computer technology in neuroscience, he wrote: "What has been found by computer studies in the last ten years thus suggests that one central problem of the 'higher' functions in the brain is to discover its methods for processing information. The methods we know today tend to be inefficient to 'astronomical' degree: the living brain may well know better" [17]:104. Some attempts to 'copy' the brain in this respect had already been undertaken at that time. One obvious example is the de-

---

[27]The contributors were: W.R. ASHBY, J.T. CULBERSON, M.D. DAVIS, S.C. CLEENE, K. DE LEEUW, D.M. MACKAY, J. MCCARTHY, M.L. MINSKY, E.F. MOORE, C.E. SHANNON, N. SHAPIRO, A.M. UTTLEY and J. VON NEUMANN.

velopment of neural nets – whose history we cannot provide in this thesis. Another example is the founding of the *Biological Computer Laboratory* (BCL) by HEINZ VON FOERSTER at the *University of Illinois* in January 1958.[28] The BCL was in its initial phase an active promoter of concepts like 'self-organization' and 'bionics' (this word was coined in 1958 at a conference organized by the BCL by JOHN E. STEELE) and turned later to the social sciences. A complete history of how brain research inspired computer technology, however, remains to be written.

3.2.6 The Modelling Approach

Models were a permanent companion of brain research. The number of available models, however, increased considerably in the 1950s and 1960s. In 1964 LEON HARMON identified more than 50 models that were created from 1940 to 1964 [115]:20. A literature review on this subject in 1968 listed more than 500 articles devoted to neural modelling [144]:247. We therefore only briefly discuss the contributions of the RASHEVSKY school and the famous proposal made by MCCULLOCH and PITTS in 1943. For a short contemporary review of 1966 we refer to the publication of HARMON and LEWIS [114]:519-530. Three causes are mentioned when the usefulness of models is discussed [114]:516. First, models serve as a kind of 'summary' of the current knowledge about the system under investigation. It is thus a prerequisite for a model that it is accurate both quantitatively and qualitatively with respect to the phenomenon under investigation up to the degree of precision that has been chosen by the investigator. Second, with validity tentatively established, one may attempt to discover new properties of the model, that might, for example, lead to new experiments. Third, the model may serve as a tool to test hypotheses about the system more rapidly and economically than direct physiological measurements permit.

The emergence of the computer as a research tool, however, supplemented this list. First, it changed the perspective towards the processes in the nervous system by putting an emphasis on the information processing aspects of the nervous system. HARMON and LEWIS wrote in their 1966 review: "The advent of digital- and analog-computer technology, well established by the mid-1950s, added new dimensions to the foundations on which neurophysiological research is based. Nervous systems began to be considered more and more explicitly as processors of information, literally as biological computers" [114]:530. Second, the computer allowed a new, 'synthetic' approach towards the understanding of neural systems. FRANK ROSENBLATT, the inventor of the *Perceptron*[29], expressed this view explicitly in his monograph *Principles of Neurodynamics*. He called this approach the 'genotypic approach', which "begins with a set of rules for generating a class of physical systems, and then attempts to analyze their performance under characteristic experimental conditions to determine their common functional properties. The results of such experiments are then compared with similar observations on biological systems, in the hopes of finding a behavioral correspondence" [204]:11.

The use of the computer as a modelling tool was not undisputed in the modeler community. At the 1964 NRP working session on *Mathematical concepts of central nervous system*

---

[28] ALBERT MÜLLER has provided a historical overview of the BCL [329].

[29] The *Perceptron* is probably the most famous neural net. Although the *Perceptron* is today used mainly as a tool for pattern recognition, this was not the objective of ROSENBLATT: "A perceptron is first and foremost a brain model, not an invention for pattern recognition" [204]:vi.

function, there was a debate about the role of the computer in neural modelling during which several objections to the computer as a modelling tool were raised. It was noted that a certain amount of elegance is lost in the process of computer programming, and that – in contrast with the use of classical mathematical methods – the use of the computer tends to restrict the freedom of modelers [236]:145. At this session, MICHAEL ARBIB did not agree and even predicted an influence of the computer on mathematics: "We may imagine that in the mathematics of the future, computer programs will form an integral part of mathematical proofs" [236]:155. The classical mathematical modelling approach has been promoted by the RASHEVSKY school (consisting, beside others, of ALSTON HOUSEHOLDER, HERBERT LANDAHL and ROBERT ROSEN – also PITTS was for a short time member of RASHEVSKY's group.). NICOLAS RASHEVSKY – an Ukrainian who emigrated in 1924 to the United States – was an important figure in the development of mathematical biology, although his scientific program was not successful and was even considered as mere theoretical speculation without grounding in empirical facts [309]:83. RASHEVSKY had indeed an ambitious plan: his aim was the "building-up of a systematic mathematical biology, similar in its structure and aims to mathematical physics" [196]:vii. He considered many different fields of biology in his work. One aspect was a theory about excitation in neurons and its propagation in neural nets. For doing this, he based his approach on the methodological ideal of physics:

> "We start with a study of highly idealized systems, which, at first, may even not have any counterpart in real nature. This point must be particularly emphasized. The objection may be raised against such an approach, because such systems have no connection with reality and therefore any conclusions drawn about such idealized systems cannot be applied to real ones. Yet this is exactly what has been, and always is, done in physics" [196]:1.

As EVELYN FOX KELLER showed in her historical analysis of the theory of developmental biology [309], this conception of theory had little appreciation in biology at that time. This was the main reason that RASHEVSKY's attempt to build up a 'theoretical biology' grounded on mathematics and physics failed and discredited the idea of 'theoretical biology' for quite some time. RASHEVSKY's work on neural systems, however, was more appreciated – but mostly by early workers in artificial intelligence (for example HERBERT SIMON and MARVIN MINSKY [309]:88) rather than by neurophysiologists. The reason is again that he was not able to relate his theoretical findings to experiments that could be used to test these findings. RASHEVSKY's approach is, however, interesting for several reasons. First, he created a mathematical model of excitation in nerves using two coupled first-order differential equations ([196]: chapter XXIV). He and his followers thus worked in the world of *continuous* mathematics, attempting to extend their modelling efforts to include large systems of nerves and phenomena such as perception and discrimination [144]:253. This approach has to be distinguished from the model proposed by MCCULLOCH and PITTS (see below), which is founded in discrete mathematics. RASHEVSKY realized this difference and also considered the discrete approach as problematic: "The method, though powerful, has one serious limitation: It requires that all neurons fire at intervals which are integral multiples of the synapse delay. In our notation it requires that they all fire at times which are represented by natural numbers" [196]:540. Second, RASHEVSKY usually based his models on quite rigid network structures (see also section 3.2.4). He was aware of this problem, and considered the random net approach of his collaborators RAPOPORT and SHIMBEL in the foreword to his monograph *Mathematical Biophysics* in 1948 as "very important work" [196]:xx. Third, the RASHEVSKY school was committed to a view according to which 'principles' should be found in order to explain the brain [124]:114-115. The idea, that 'new principles' or a 'new mathematics' is necessary in order to understand the brain is stated quite regularly at that time – e.g. by the Italian theoretician E.R. CAIANIELLO, who published in 1961 a very abstract 'whole brain model' based on so-called 'neuronic equations' explaining sleep, creativity (and much more) [62]. He stressed the role of mathematics for further developments in neural modelling: "It is my conviction that the most urgent task at hand is the creation of concepts, equations, and problems, which offer a *natural* way of dealing with the structure, function, and control of a neuronal medium, just as tensor calculus did for general relativity theory" [61]:101.[30]

The most influential model that was emerged since 1940, also has a link to the RASHEVSKY school, as WALTER PITTS was working for some time with RASHEVSKY. PITTS, however, published *A logical calculus of the ideas immanent in nervous activity* together with WARREN McCULLOCH [159] . In the following, we only briefly describe this model, as TARA ABRAHAM and GUALTIERO PICCININI have recently provided a detailed analysis of this matter [292, 335] and the importance of this paper for the research program of McCULLOCH has been discussed by LILY KAY [322]. The paper describes the nervous system as a system that performs logical functions and claims (without proving it)[31] that the formal system described is able to compute all numbers which are also computable by a TURING machine. The paper uses a modern concept of computation in the sense of TURING, which led McCULLOCH to the conclusion (stated in 1949) that "the brain is a logical machine" [158]:492. Furthermore, the simplifying assumptions (the all-or-none 'law' and the discretization of time) in the paper supported an information perspective towards the neural system in the sense that spike trains can be considered as binary strings. The paper was accompanied by a short extension, where HERBERT LANDAHL, together with McCULLOCH and PITTS, introduced a statistical viewpoint in the sense that the logical relations among the actions of neurons in a net have been converted into statistical relations among the frequencies of their impulses ([137] see also [124]:Part 3). The neurophysiology community largely ignored the paper, and even logicians had their doubts about the quality of the contribution.[32] On the other hand, the paper had a tremendous theoretical influence among neural modelers as well as computer scientists (see section 4.3).[33] The automata approach towards modelling

---

[30]CLAUDE SHANNON remarked in 1964 concerning this point that "perhaps the call for a 'new mathematics' reflects only that the problem is new to mathematicians" [236]:142. As a 'new' approach, however, that has been introduced into neural modelling at the end of the 1960s, one may mention statistical mechanics. Some early attempts in this direction were made by WIENER and PITTS, as J.D. Cowan remarked in 1967. He found it remarkable, "that in the 20 years since, with the exception of WIENER's own rather abstract work on Hamiltonians and neuromechanics, there is no trace of Hamiltonians and Statistical Mechanics in the field" [68]:186.

[31]The assumption that the paper had proven that the model is equivalent to a universal TURING machine is considered as a "widespread misinterpretation" [153]:xviii in the foreword by SEYMOUR PAPERT to McCULLOCH's monograph *Embodiments of Mind*, although McCULLOCH himself propagated this misconception [154]:368 (page numbering according to [153], see also [335]).

[32]FREDERIC FITCH wrote in his review on the paper: "Symbolic logic is used freely, and indeed rather carelessly. The notation is unnecessary complicated, and the number of misprints is so great that the reviewer was unable to decipher several crucial parts of the paper. (...) In any case there is no rigorous construction of a 'logical calculus' " [83]:49.

[33]Consider also the introduction to the re-issued paper of McCULLOCH-PITTS [294]:15-17.

neuronal systems is considered as a direct result of the work of MCCULLOCH and PITTS [142]: vii, and even JOHN VON NEUMANN utilized the MCCULLOCH-PITTS notation for the logical design of the EDVAC computer (according to GOLDSTINE [236]:117).[34] The model itself, however, was soon considered as an inadequate description of neuronal processes.

---

[34]GOLDSTINE's comment is not quite correct, as VON NEUMANN did not use the notation in the same way as MCCULLOCH and PITTS. In his famous *First Draft of a Report on the EDVAC* of 1945 [269], he made a close neuron analogy in which he presented the neuron as a purely digital device, in contrast to his later statements on the neuron at the *Macy conferences*. His *First Draft* is considered to be a founding paper on the modern digital computer and it is thus of considerable interest that the biological inspiration was outlined rather extensively in this paper. The first EDVAC, however, was not build according to VON NEUMANN's report, but by the MOORE school design group. For a history of the EDVAC consider [314, 344].

Chapter 4

1940 – 1970: The Dynamics of Research

This chapter complements the historical analysis with bibliometric and sciento-metric investigations. We we analyze the increase of the number of publications and journals since the Second World War. We focus on major conferences and investigate, to what extent they can be grouped in thematic clusters, which serve for identifying protagonists that distributed ideas on neural coding and information processing in different scientific communities. Based on a citation analysis, the influence of selected protagonists will be further investigated. We conclude this chapter by a summary of the historical analysis.

4.1 Journals and Publications

Journals serve as an important communication tool within scientific communities. New topics are accompanied by the founding of new journals or by an increase in the number of articles on the new topic. We investigate this matter for the time period since the Second World War (the onsets of the databases used are 1945 and 1965) up to the present time.

Methods: We investigate the number of journals and publications using the *MedLine* and the *ISI Web of Knowledge* databases.[1] Both databases are hosted in the United States and introduce a well-known bias towards mainstream and anglophone journals (latter will become important in the citation analysis). Referencing and possible search strategies are not equivalent in both databases. To estimate the number of neuroscience papers ('neuro-papers') dealing explicitly with neurons or neural systems

---

[1] *MedLine* is a public database hosted by the *National Institutes of Health* of the United States. Access via http://www.ncbi.nlm.nih.gov/entrez/query.fcgi. The database covers papers associated with biomedical research. The *ISI Web of Knowledge* is a integrated Web-based platform containing the *Science Citation Index Expanded*, the *Social Sciences Citation Index* and the *Arts & Humanities Citation Index* and is hosted by *The Thomson Corporation*. Access (subscription necessary) via http://go5.isiknowledge.com/portal.cgi/

in either database, we used the following search strategies: In the *ISI Web of Knowledge* database, the 'general search' interface was used and we looked for papers that contained the truncated search term 'neur*' in their title for periods of fife years, beginning in 1945 (onset of the database). The constraint 'Title only' was necessary, as papers up to 1991 are only accessible via title search, whereas papers after 1991 are accessible by full-text-search (abstracts). In *MedLine* (onset of the database: 1965), the search involves titles and abstracts (if available) and truncation search is not possible. We used the boolean search expression 'neuron OR neural'. The number of 'coding' and 'information processing' papers was only estimated for *MedLine*, as words indicating such papers tend more to show up in abstracts than in titles. We used the following boolean search expressions: '(neuron OR neural) AND (code OR coding)' for neural coding papers, and '(neuron OR neural) AND (information AND processing)' for neural information processing papers. To compare the results for the period 1991-2004, a similar analysis has been performed using the *ISI Web of Knowlede* database. They confirmed the results obtained by the *MedLine* analysis (results not shown). We estimated the number of neuroscientists by the members of the *Society of Neuroscience* (SfN)[2] – the oldest and largest institution in neuroscience, founded in 1968 under the leadership of the psychologist NEIL MILLER, the biochemist (and later behavioral scientist) RALPH GERARD and the neurophysiologist VERNON MOUNTCASTLE [305]:347.

We estimated the growth of the number of 'neuroscience journals' using the *ISI Web of Knowledge* database and by including all journals classified as 'neuroscience' journals that hat an impact factor of at least 2 in the *Journal Citation Report* of 2003.[3] We complemented this list with journals known to be important for neuroscience or theoretical neuroscience from the categories 'behavior', 'biophysics', 'computer systems: artificial intelligence', 'computer systems: cybernetics', 'computer systems: interdisciplinary applications', 'multidisciplinary sciences' and 'physiology'. For 'theoretical neuroscience' journals, the impact-factor criterium has not been applied. The latter journals are (in brackets: impact factor, year of founding, journal country; ordered according to time of funding): *Bulletin of Mathematical Biophysics* (1.468, 1938, USA), since 1972: *Bulletin of Mathematical Biology*; *Biophysical Journal* (4.463, 1960, USA); *Kybernetik* (1.933, 1961, Germany), since 1975: *Biological Cybernetics*; *Journal of Theoretical Biology* (1.550, 1961, USA); *BioSystems* (0.971, 1967, England); *IEEE Transactions on Systems, Man and Cybernetics* (1.029 (B), 1971, USA): separated later in three parts A,B,C; *Neural Networks* (1.774, 1988, England); *Neural Computation* (2.747, 1989, USA); Neurocomputing (0.592, 1989, Netherlands); *Network: Computation on Neu-

---

[2] Access via www.sfn.org
[3] The following journals listed in our journal count (see below) were present or were founded during 1940-1970 (in brackets: year of founding, today's journal country): *Proceedings of the Royal Society of London* (1841, England), *Nature* (1869, England), *Proceedings of the National Academy of Sciences USA* (1877, USA), *Brain* (1878, England), *Journal of Physiology – London* (1878, England), *Behavioral Neuroscience* (1887, USA), *Clinical Neurophysiology* (1897, Ireland), *Journal of Comparative Neurology* (1891, USA), *Science* (1895, USA), *American Journal of Physiology* (1898, USA), *Journal of General Physiology* (1918, USA), *Physiological Review* (1921, USA), *Journal of Experimental Biology* (1927, England), *Journal of Neurophysiology* (1938, USA), *Annual Review of Physiology* (1939, USA), *Journal of Neuropathology and Experimental Neurology* (1942, USA), *Journal of Neurochemistry* (1956, England), *Experimental Neurology* (1959, USA), *Biophysical Journal* (1960, USA), *Acta Neuropathologica* (1961, Germany), *Kybernetik* (1961, Germany), *Journal of Theoretical Biology* (1961, USA), *Neuropharmacology* (1962, England), *Neuropsychologia* (1963, England), *Journal of Neurological Sciences* (1964, Netherlands), *Psychophysiology* (1964, USA), *Biological Psychiatry* (1965, USA), *Neuroendocrinology* (1965, Switzerland), *Cortex* (1965, Italy), *Experimental Brain Research* (1966, Germany), *BioSystems* (1967, England), *Frontiers in Neuroendocrinology* (1969, USA), *Psychopharmacology* (1969, Germany) and *Journal of Neurobiology* (1969, USA).

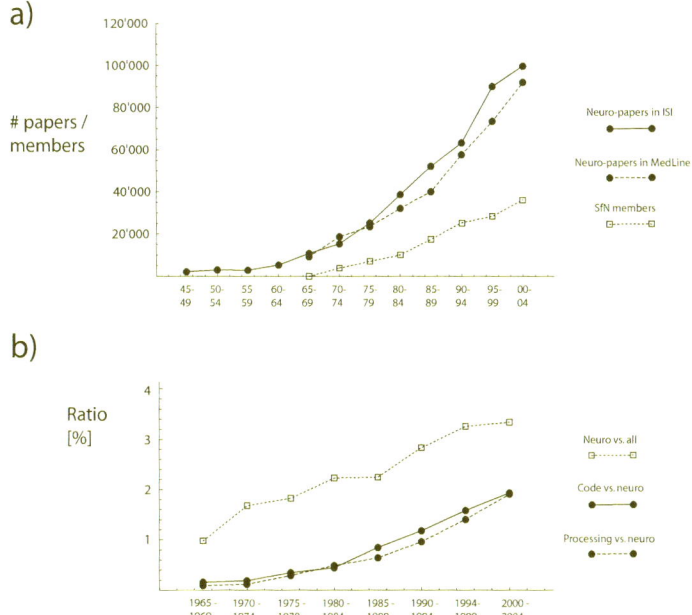

Figure 4.1: The neuroscience boom since the Second World War: a) Increase of the number of 'neuro-papers' contained in the *MedLine* and the *ISI Web of Knowledge* databases. b) Relative increase of the number of 'neuro-papers' and of 'neural coding' and 'neural information processing' papers in *MedLine* (see text).

ral Systems (2.208, 1990, England); *IEEE Transactions on Neural Networks* (1.666, 1990, USA); *Neural Computation Applications* (0.449, 1992, USA); *Journal of Computational Neuroscience* (2.776, 1994, USA); *Neural Processing Letters* (0.631, 1994, Belgium); *Neuroinformatics* (-, 2003, USA). In this way, a total of 130 'important journals' has been obtained. For each journal, its funding year (= the year of publishing of the first issue) and its present home country (publisher) have been evaluated.

The bibliometric analysis confirms the expected increase of publication activity in neuroscience: both databases show a similar increment in absolute number, which is furthermore comparable to the growth of the number of scientists (Fig. 4.1.a). The ratio of all 'neuro-papers' compared to all papers indexed in *MedLine* demonstrate an increasing interest in

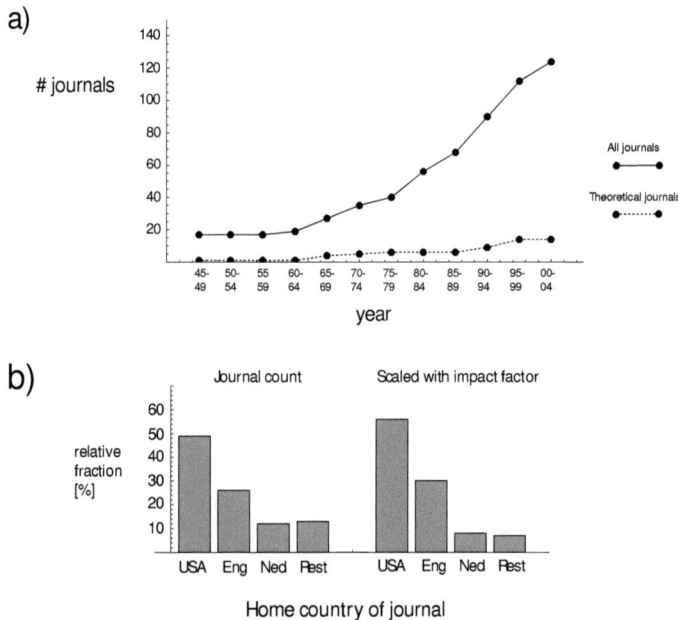

Figure 4.2: Journal analysis: a) Cumulative histogram of the founding years of 'important' journals. b) Geographical distribution of journal home countries.

neuroscience within biological and medical research in general. Finally, the ratio of all 'neural coding papers' and 'neural information processing papers' compared to all 'neuro-papers' point out an increasing interest in explicitly addressing neural coding or neural information processing as a topic of research (Fig. 4.1.b). Remind, that the absolute numbers obtained by the analysis should be taken with a pinch of salt, as the search strategies used are not exhaustive. However, this first approximation on the dynamics of research in neuroscience indicates that analyzing the brain by explicitly focussing on its information processing capacities and using an information vocabulary is performed by a rather small minority within the neuroscience research community. The large majority of research in neuroscience is presumably focussing either biomedical or bio-molecular aspects of the brain or neurons, which is supposedly not in conflict with an 'information processing brain' perspective, but does not use this vocabulary for their analysis.

Conference name, source	Date	Place	# Part.	Abbr.
1. *Macy conference*, [333]:383	08./09. March 1946	New York	21	M1[†]
2. *Macy conference*, ASC	17./18. October 1946	New York	25	M2[†]
3. *Macy conference*, ASC	13./14. March 1947	New York	26	M3[†]
4. *Macy conference*, ASC	23./24. October 1947	New York	23	M4[†]
5. *Macy conference*, ASC	Spring 1948	New York	28	M5[†]
6. *Macy conference*, [258]	24./25. March 1949	New York	23	M6
7. *Macy conference*, [259]	23./24. March 1950	New York	24	M7
8. *Macy conference*, [260]	15./16. March 1951	New York	23	M8
9. *Macy conference*, [262]	20./21. March 1952	New York	27	M9
10. *Macy conference*, [263]	22.-24. April 1953	Princeton, NJ	28	M10

Table 4.1: Conferences of the Macy-cluster. Abbreviation: ASC: American Society for Cybernetics. The exact date of the 5th *Macy conference* could not be evaluated. Abbreviation marked with [†] indicate members of a stable sub-cluster (see Fig. 4.3).

The journal count analysis corroborates our findings (Fig. 4.2.a). The geographical distribution of the journal countries (the home country of the journal publisher) indicates the dominance of the United States – an aspect that is even pronounced, when the number of journals is scaled with their impact factor (Fig. 4.2.b). The remarkably high importance of the Netherlands can mainly be attributed to the publisher *Elsevier*. Of particular interest are the times of founding of 'theoretical neuroscience journals', what occurred in two phases: in the period 1960-1971 and the period 1988-1994. This indicates, that there are two phases where the theoretical interest in neuronal information processing increased considerably. This supports our choice of the period of investigation. The second phase indicates a revival of the topic in the mid 1980s. This revival has several characteristics:[4] An increasing interest in the physical basis of computational processes (promoted e.g. by RICHARD FEYNMAN and CARVER MEAD), the revival of the neural network approach after the introduction of the HOPFIELD-network, and the *Carmel* symposium *Computational Neuroscience* in 1987. In 1988, TERRENCE SEJNOWSKI, CHRISTOPH KOCH and PATRICIA CHURCHLAND reviewed the research program of the new field 'computational neuroscience' [665] in an article in *Science*. In the 1990s, several new institutes in computational neuroscience (or neuroinformatics, the term used in Europe) have been founded. This second phase is not subject of our historical analysis.

4.2 Conferences

Conferences are important scientific marked places, where new ideas and concepts are presented. The analysis of our historical sources led to the identification of several conferences, where neural coding and related issues had been discussed. In this section, we group them according to similarities such that 'clusters' of conferences represent scientific communities. This serves as a basis to identify persons (next section), that are present in several clusters and serve as 'ambassadors' of scientific ideas. We investigated the major conferences in the fields cybernetics (with focus on the *Macy conferences*), information theory in relation to biology and neuroscience in relation to the *Neurosciences Research Program* as well as con-

---

[4] RODNEY DOUGLAS, Institute of Neuroinformatics, Zurich: personal communication.

ferences that focus neural modelling and theoretical aspects. We did, however, not intend to analyze all conferences in the fields mentioned that happened between 1940 and 1970, as this would be far too ambitious. We have chosen the following conferences: For cybernetics, we mainly focussed on the *Macy conferences*[5] (10 conferences). For neuroscience, we focussed on the activity of the *Neurosciences Research Program* in the 1960s and early 1970s, which are published in a series of proceedings and monographs (45 work sessions and 3 *Intensive Study Programs*).[6] Figure 4.3.a provides an overview of the distribution of the conferences over the time of investigation, indicating, that the majority of the conferences (especially the NRP work session) were located in the 1960s. As we were not interested in a detailed analysis of clustering of topics within the NRP sessions, the 45 NRP work sessions have been aggregated into eight groups according to the participant lists of the *Neuroscience Research Symposium Summaries*. Finally, all conferences, that were found during our search for historical sources, were considered (17 conferences). This lead to a total of 38 'conferences' that were analyzed as described in the methods section below. In a contemporary overview in 1961 by the German cybernetics KARL STEINBUCH [241]:136, more potentially interesting conferences were listed: The first (1949) and the second (1952) *London symposium on information theory*[7], the second and third (1958,1961) *Congrès International de Cybérnetique* in Namur (Belgium), and several conferences about the emerging computer science that have taken place since the mid 1950s, for example the World Conferences of the *International Federation for Information Processing* (IFIP, in 1959, 1962, 1965, 1968 for the period of interest), the *National Conventions of the Institutes of Radio Engineers* (together with the *American Institute of Electrical Engineers* the precursor of the *Institute of Electrical and Electronics Engineers*) and the National Conferences of the *Association of Computing Machinery*.[8]

Methods: To obtain the conference clusters, we use the sequential superparamagnetic clustering paradigm (for details, see section 7.5). Two different types of 'distances'

---

[5]Sources: The re-issued *Macy* Proceedings [332, 333] and the website of the *American Society for Cybernetics* (ASC): http://www.asc-cybernetics.org/foundations/history/MacySummary.htm#Part1. The participant lists given by the ASC and listed in the original publication did not always match. We used the original numbers whenever possible provided by [332]. For the years 1946 (October), 1947 and 1948 we referred to the ASC numbers. The representative of the *Joshia Macy Jr. Foundation*, FRANK FREMONT-SMITH, was included in the count. The *Macy* meeting 1942 on "cerebral inhibition" with 20 participants[261]:xix – among them members of the later core group of the *Macy conferences* like GREGORY BATESON, LAWRENCE FRANK, FRANK FREMONT-SMITH, LAWRENCE KUBIE, WARREN MCCULLOCH, MARGARET MEAD and ARTHUR ROSENBLUETH – was not considered in the analysis. The *Joshia Macy Jr. Foundation Conference Program* also included a series of five conferences on the 'Nerve Impulse' from 1950 to 1954 [166]. Based on our (not enclosing) analysis, these conferences focussed on physiological aspects like action potential generation and not on coding issues.

[6]Source: [212, 213, 214, 215, 216, 217, 218, 219, 220, 187, 221]. The events were called "work session', workshop' or 'symposium' – we use 'work session' as general term. The work sessions usually took place at the MIT [338]. For our analysis, we only considered those work sessions that were published in the seven volumes of the *Neuroscience Research Symposium Summaries*. The fourth *Intensive Study Program* of 1977 has been excluded from the analysis.

[7]These conferences focussed on either mathematical and philosophical aspects or telecommunication aspects [66]:v and did not discuss applications on biological questions.

[8]We took a sample (IFIP-proceedings: http://www.informatik.uni-trier.de/ ley/db/conf/ifip/ and [129]) to check whether neuro-related aspects were discussed to a considerable amount at those conferences. The result indicated, that applications on biological systems and problems played probably only a very remote role within these conferences at that time.

Conference name, source	Date	Place	# Part.	Abbr.
NRS summaries 1 (9), [221]	1963-64	Cambridge, MA	108	N1
NRS summaries 2 (6), [220]	1965-66	Cambridge, MA	102	N2
1. ISP, [187]	July 1966	Boulder, CO	64*	N3
NRS summaries 3 (6), [219]	1967-68	Cambridge, MA	97	N4
NRS summaries 4 (5), [218]	1966-1968	Cambridge, MA	108	N5
NRS summaries 5 (5), [216]	1967-1969	Cambridge, MA	114	N6
2. ISP, [217]	21.07. - 08.08. 1969	Boulder, CO	92*	N7
NRS summaries 6 (5), [215]	1968-70	Cambridge, MA	103	N8
NRS summaries 7 (4), [214]	1970-71	Cambridge, MA	85	N9

Table 4.2: Work sessions and ISP of the NRP-cluster. In brackets: number of work sessions. Abbreviations: NRS: *Neurosciences Research Symposium*. ISP: *Intensive Study Program*. Numbers of participants marked with * indicate the number of contributing authors and not of participants.

between conferences are applied: The first distance measure, the *participant distance*, is given as the size of the intersection of the participant sets of two conferences. In that way, two conferences are 'close' if the same people participate in both conferences. To calculate the participant distance, we created a database containing all names of researchers that were either listed as participants or – if a participant list was not available – listed as contributors to the proceedings. The name data base contains 1481 names, whereas we have checked in three independent runs to verify name identities. For the NRP work sessions, we created aggregated lists according to the 8 volumes of the *Neuroscience Research Symposium Summaries*, covering 4 to 9 work sessions. The pairwise comparison of all conferences leads to the sizes of intersection sets. The obtained numbers are normalized by the largest intersection set, and then 1 is subtracted. The absolute value of the result is taken as the distance between two conferences. Thus, a distance 0 indicates that exactly the same participants were present in both conferences, and a distance 1 indicates, that no person was in both conferences. The resulting distance matrix is used as input for the clustering algorithm.

The second distance measure, the *topic distance*, follows the idea, that two conferences are close if the relative number of contributions that have been assigned to certain topics are similar. For each conference, a six-dimensional 'topic vector' is constructed by counting the number of contributions that fall into the following classes:

- *Theoretic*: Contributions of the fields mathematics, physics, statistics and information theory.

- *Molecular Level*: Contributions that deal with aspects of the molecular biology of cells/neurons like genetic studies, protein studies, and membrane studies.

- *Neuron Level*: Contributions that deal with very small neuronal networks, single neurons or parts of neurons, e.g. electrophysiological input-output-studies, spike train analysis, work on synapses, coding properties of single neurons or small neuronal networks, influence of chemicals on single cells, as well as biomorphic modelling on the single neuron level.

- *Neuronal System*: Contributions that deal with brains and neuronal networks on a 'system-level' and is based on, or is in close connection to, experimental work in biological systems , e.g. sensory and motor systems, developmental aspects, EEG studies, neuroanatomy, as well as biomorphic modelling on a systems level.

- *Behavioral*: Contributions dealing with social systems, language, learning and behavior without linking this analysis with experimental work in neuronal systems.

Conference name, source	Date	Place	# Part.	Abbr.
The Hixon Symposium, [128]	20.-25. Sept. 1948	Pasadena, CA	19	T1
Information Theory in Biology, [191]	Summer 1952	Urbana, IL	15*	T2
3. Symposium on Inf. Theory, [66]	12.-16. Sept. 1955	London, UK	50*	T3+
1er congr. int. de cybernétique, [186]	26.-29. June 1956	Namur, Belgium	79*	T4+
Information Theory in Biology, [287]	29.-31. Oct. 1956	Gatlinburg, TN	32*	T5
Mechan. of Thought-Processes, [169]	24.-27. Nov. 1958	Teddington, UK	211	T6+
Self-Organizing Systems, [289]	05.-06. May 1959	Chicago, IL	19*	T7+
4th Symposium on Inf. Theory, [67]	29.08. - 02.09. 1960	London, UK	53*	T8+
Principles of Self-Organization, [257]	08./09. June 1961	Urbana, IL	39	T9+
Inf. Storage and Neural Control, [81]	1962	Houston, TX	16	T10
Cybernetic Problems in Bionics, [170]	03.-05. May 1966	Dayton, OH	69*	T11+
School on Neural Networks, [60]	June 1967	Ravello, Italy	22*	T12+
Inf. Proc. in the Nervous Sys., [142]	21.-24. Oct. 1968	Buffalo, NY	67	T13+

Table 4.3: Theory-Cluster. Numbers of participants marked with * indicate the number of contributing authors and not of participants. Abbreviation marked with + indicate members of a stable sub-cluster (see Fig. 4.3).

- *Cybernetic*: Contributions of the fields automata theory, 'pure' (non-biomorphic) modelling, and theories of control and self-organization

    Each 'topic vector' is normalized by the total number of the Contributions. The distance between two conferences is obtained by calculating the dot product between two vectors, subtracting 1 from the result and taking the absolute value. In this way, the distance 1 indicates that the topic vectors are orthogonal (i.e. the conferences dealt with completely different topics) and a distance 0 indicates, that in both conferences basically the same topics were discussed to the same degree. Again, the resulting distance matrix is used as input for the clustering algorithm. As the topic distance contains more ambiguity (in assigning contributions to classes) than the participant distance, latter will be used for identifying scientific communities (clusters) and former is used for validating the results.

The cluster analysis led to the following result: Using the participant distance, four clusters are identified: The most stable cluster is – as expected – the *Macy* cluster (Table 4.1). The conferences of this cluster are close by construction, as the *Macy conferences* were organized around a large core-group of participants. The *Macy* cluster is subdivided into two sub-clusters, indicating, that the number of visitors in the first five conferences were smaller and that the members of the core group were more disciplined in attending the conferences.

Conference name, source	Date	Place	# Part.	Abbr.
Principles of Sensory Comm., [205]	19.07. 01.08. 1959	Cambridge MA	42	B1
Inf. Proc. in the Nervous Sys., [91]	10.-17. Sept. 1962	Leiden, Netherlands	66	B2
Neural Theory and Mod., [198]	04.-06. Dec. 1962	Ojai, CA	36	B3
1. Int. Symp. on Skin Senses [132]	March 1966	Tallahassee, FL	45^g	B4
3. ISP, [213]	24.07. - 11.08. 1972	Boulder, CO	124*	B5
NRS summaries 8 (5), [212]	1970-72	Cambridge, MA	113	B6

Table 4.4: Remaining conferences of clustering process, called 'biology-cluster'. Numbers of participants marked with * indicate the number of contributing authors and not of participants.

The second stable cluster is the one formed by most of the NRP work sessions and *Intensive Study Programs* (Table 4.2). The late NRP work sessions and study programs are not part of the cluster, using *both* distance measures. This indicates an alternation of generations in combination with a change of focus within the NRP events, happening at the beginning of the 1970s. This result supports our focus on the period 1940-1970. The third emerging cluster is called the 'theory cluster', as it contains the conferences dealing with information theory, cybernetics and neural modelling (Table 4.3). The theory cluster is less stable than the two other clusters, indicating a larger mean distance between the conferences. Furthermore, the theory cluster splits into two moderate stable clusters, one of which contains all cybernetic and information theory conferences, the second containing the two 'information theory in biology' conferences organized by QUASTLER, the *Hixon* symposium and the conference on information storage and neural control. The remaining conferences are contained in a fourth group called 'biology cluster' because they mostly deal with biological topics (Table 4.4). In this group, also the remaining NRP events are contained. Figure 4.3.b gives an overview of the cluster-analysis.

When using the topic distance, the clustering analysis led to a comparable result. In detail, however, some differences emerge (results not displayed). Almost all *Macy conferences* as well as most of the NRP conferences again form clusters. The most stable cluster is, however, a sub-group of the theory cluster, formed by the conferences dealing with self-organization and cybernetics. Again, a 'biology cluster' remains. We also added the two distance matrices and used the result as input for the clustering algorithm. An interesting result of this approach is that the *Hixon* symposium becomes a part of the *Macy* cluster, indicating a closeness in respect to participants and topics. The Hixon symposium is thus in the same line of tradition of the early interdisciplinary conferences of the 1940s and early 1950s.[10] When looking at the distribution of conferences along the time axis, one sees that the number of conferences of the 'theory' or 'biology' cluster, that deal explicitly with neural information processing, was still rather small in the 1950s compared to the 1960s. This development goes in parallel with a change in appreciation of biologists towards the theoretical approach in the 1960s. At the end of the 1950s, the attitude was still rather critical, as the statement of WALTER A. ROSENBLITH, made at the 1959 conference on *Principles of Sensory Communication*, shows: "Responsible workers in the behavioral and life sciences became increasingly squeamish about the one-day symposium in which mathematicians, physicists, and engineers vented frequently the belief that the intelligent application of some rather elementary notions from mathematics and physics should yield spectacular results in the solution of a variety of thorny problems" [205]:v. In the mid 1960s, RICHARD REISS still commented: "Various developments in science and engineering have in recent years combined to produce a surge of activity in neural theory and modelling. As in other interdisciplinary fields, research has been carried forward by a very mixed company of biologists, chemists, engineers, physicists, mathematicians, and computer specialists; these men have brought with them the research philosophies and methodologies peculiar to their own disciplines. The result has been a potpourri of partial theories and models, ideas and data, that is difficult to comprehend" [198]:v. In the end of the 1960s, however, JOHN ECCLES came to a different conclusion: "It is very encouraging for us to find that more

---

[10] The *Hixon* symposium is a historically well-analyzed conference and it is considered as a starting point of cognitive science, see [311].

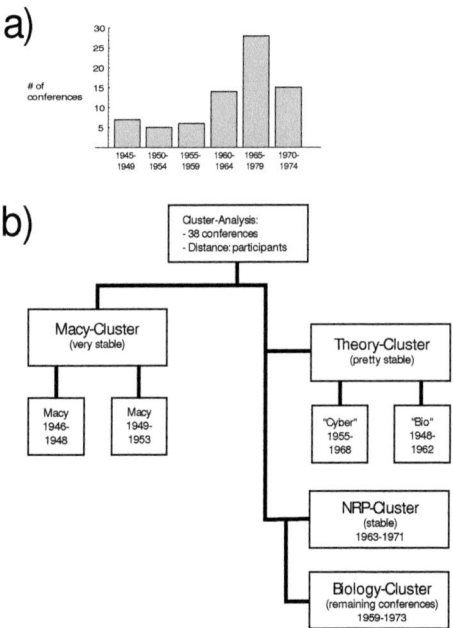

Figure 4.3: Cluster analysis of conferences. a) Histogram of conferences considered in the analysis. b) The result of clustering using the participant distance.

and more they [the model builders] are reading the neurobiological literature with greater understanding" [142]:vi.

The question remains, if there are any distinguished conferences. For this analysis, we considered all 'protagonists' (see next section) and counted their number of appearance in the individual conferences. The result of this analysis is displayed in Figure 4.4. As expected, several well-analyzed conferences – as the *Macy conferences* and the *Hixon* symposium show up – but the 'most important' conference by this respect was the 1962 symposium on *Information Processing in the Nervous System* that took place in Leiden during the XXII international congress of the *International Union of Physiological Sciences*. This symposium, judged as an 'experiment' by R.W. GERARD in his opening address, indicates a raise in credibility of questions of neural information processing, as, "no one has previously shown the temerity, nor has any congress had the courage to invite such temerity, to stage a week-long Symposium, not as an isolated unit but as part of a large Congress" [91]:3.

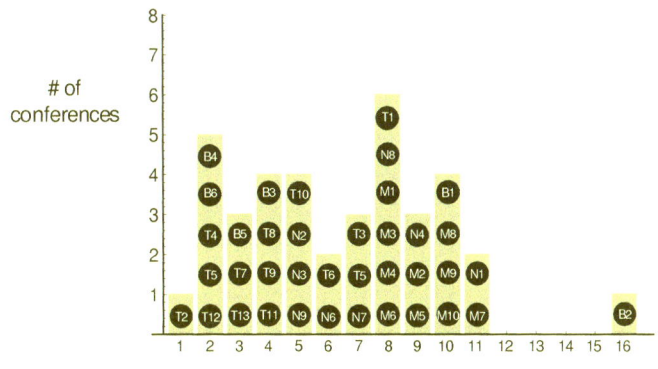

Figure 4.4: Identifying important conferences: Histogram of conferences with 1 to 16 'protagonists' as participants (see text).

The organizers intended to gather three different groups of scientists: neurophysiologists, behavior researchers and 'theorists' (modelling, system theory, computer experts and mathematicians). The entire symposium and its discussions have been published, whereas the transcripts have been minimally altered by the editors. The congress served as an excellent source and has been used extensively in our detailed analysis in the previous chapter.

4.3 Protagonists

The cluster analysis of the conferences serves as a starting point for identifying scientists which are frequently present in the four scientific communities identified. We are interested in the question, whether we can identify protagonists that played a major role in developing and distributing the ideas and concepts related to the 'information processing brain'.

> **Methods:** Using our name database of conferences, we identified all persons that participated in at least 5 conferences to obtain our set of analysis. In this set, we included persons, that were identified in the previous chapter as prominent researchers in our field of interest. These persons are A. RAPOPORT, H.B. BARLOW, M.A. BRAZIER, C. CHERRY, J.P. SEGUNDO and C.E. SHANNON. We counted the number of participations in conferences of each cluster identified using the 'participant distance'. Persons, that were only present in one cluster, are called 'cluster persons'. Persons, present in two clusters, are called 'border crossers'. Persons, present in at least 3 clusters are called 'ambassadors'. We furthermore checked for persons, whose multiple presence in the

NRP- and biology clusters was due to the fact that they participated in the two NRP-conferences that fell in the biology cluster. Those persons represent the continuation of the NRP community into the 1970s and were neglected when identifying 'border crossers' or 'ambassadors'. The members of the set of all 'border crossers' and 'ambassadors' are called 'protagonists'. For all 'protagonists', their year of birth and death, the name of the institution they were affiliated longest for 1940-1960, their origin and their main field of work have been evaluated. For six 'protagonists', whose choice will be justified in the main text, a detailed bibliometric analysis is performed using the *ISI Web of Knowledge* database. First, the number of papers per year present in the database for the period 1945 up to the death of the person as well as the number of citations per year for the period of 1945-2004 have been evaluated. The presence of homonyms has been carefully checked. Furthermore, the publications and citations is analyzed according to the ISI subject categories in order to investigate, in which main fields they published and in which main fields they had the strongest impact. Categories, that contained less than 1% of the citations are excluded from the analysis (REICHARD: 0.5%). The ISI subject categories refer to the journals in which a paper is published. Papers in *Science* and *Nature*, for example, are classified as 'interdisciplinary science'. A publication in such a journal, or a citation within a publication in such a journal, leads to a score. As the analysis led to almost 100 subject categories in which the papers of the protagonists, or the people who cited them, fell, they have been grouped to 8 classes with comparable total number of citations and with a comparable thematic connection:

- **Neuroscience:** Neurosciences.
- **Biology - Systems:** Anatomy & morphology, behavioral sciences, biodiversity, conservation biology, developmental biology, ecology, endocrinology, & metabolism, entomology, environmental sciences, evolutionary biology, marine & freshwater biology, microscopy, oceanography, plant sciences, zoology.
- **Medicine:** Anesthesiology, cardiac & cardiovascular system, clinical neurology, ergonomics, hematology, medicine: general & internal, medicine: research & experimental, nutrition & dietetics, obstetrics & gynecology, oncology, ophthalmology, orthopedics, otorhinolaryngology, parasitology, pathology, peripheral vascular disease, pharmacology & pharmacy, psychiatry, public environmental & occupational health, radiology, nuclear medicine & medical imaging, rehabilitation, sport sciences, surgery, toxicology, tropical medicine.
- **Human Sciences:** Communication, education & educational research, history & philosophy of science, humanities: multidisciplinary, information science & library science, management, operations research & management science, philosophy, psychology, psychology: applied, psychology: biological, psychology: clinical, psychology: experimental, psychology: mathematical, psychology: multidisciplinary, religion, social issues, social sciences: interdisciplinary, social sciences: mathematical methods, sociology.
- **Multidisciplinary Sciences:** Multidisciplinary sciences.
- **Technical Sciences:** Automation & control systems, computer science: artificial intelligence, computer science: cybernetics, computer science: hardware, & architecture, computer science: information systems, computer science: interdisciplinary applications, computer science: software engineering, computer science: theory & methods, engineering: biomedical, engineering: chemical, engineering: electrical & electronic, engineering: industrial, engineering: mechanical,

Macy cluster	NRP cluster	Theory cluster	Biology cluster
10: WH Pitts (T: 2)	8: TH Bullock (B: 4)	4: GA Pask (B: 1)	4: C Pfaffmann (N: 1)
9: H Klüver (T: 1)	8: R Galambos (B: 2)	3: C Cherry (B: 1)	2: MA Brazier (T: 2)
9: G Bateson (T: 1)	8: PA Weiss (T: 1)	3: H Quastler (M: 1)	
7: HL Teuber (N: 4)	5: F Morrell (T: 1)	2: A Rapoport (B: 1)	
6: HW Brosin (B: 1)	5: VB Mountcastle (B: 2)		
6: R Lorente de No (T: 1)	4: DM Wilson (B: 1)		
6: J von Neumann (T: 1)			
3: CE Shannon (N: 1)			

Table 4.5: Border crossers: Persons that participated in conferences of 2 clusters. In brackets: The number of participations in conferences of another cluster. Abbreviations: B: biology cluster, M: *Macy* cluster, N: NRP cluster, T: theory cluster.

engineering: multidisciplinary, instruments & instrumentation, materials science: multidisciplinary, robotics, telecommunications.

- **Basic Sciences:** Chemistry: multidisciplinary, chemistry: physical, electrochemistry, mathematics: applied, mathematics: interdisciplinary applications, mechanics, optics, physics: applied, physics: fluids & plasma, physics: mathematical, physics: multidisciplinary, statistics & probability.
- **Biology – Cells:** Biochemistry & molecular biology, biophysics, cell biology, genetics & heredity, physiology.

The relative numbers of publications and citations, that fall in each of these eight categories, are displayed in a 'spider diagram'.

Our analysis has lead to the following groups of persons. In total, 32 cluster persons (results not shown), 20 border crossers (Table 4.5) and 12 ambassadors (Table 4.6) were identified. As expected, the cluster persons belong to either the *Macy* cluster (the persons of the core group) or the NRP cluster. When considering the *Macy* core group, we see that 24 persons participated in at least 5 *Macy* conferences. Half of them were protagonists, indicating the strong interdisciplinary interest of those scientists. Of special interest are the ambassadors. Although WARREN MCCULLOCH was the 'most busy' conference visitor, the person most evenly present in all four clusters was DONALD MACKAY. However, with

Name	Macy	NRP	Theory	Biology
20: WS McCulloch	10	1	9	0
13: DM MacKay	1	6	4	2
12: RW Gerard	9	0	2	1
10: JH Bigelow	8	1	0	1
9: H von Foerster	5	0	3	1
6: WE Reichardt	0	3	1	2
5: JC Eccles	0	3	1	1
5: LD Harmon	0	2	1	2
5: OG Selfridge	0	1	3	1
5: PD Wall	0	2	1	2
5: CA Wiersma	0	1	1	3
4: HB Barlow	0	1	2	1

Table 4.6: Ambassadors: Persons that participated in conferences of at least 3 clusters.

Name	Birth, death	Place of work	Field
Julian H. Bigelow	1897-1979	I Advanced Study, Princeton	Technical Science
Mary A. Brazier	1908-1988	U California, Los Angeles	Neuroscience
Henry W. Brosin	?	U Pittsburg	Medical Science
Theodore H. Bullock	1915-2005	U California, San Diego	Neuroscience
Robert Galambos	1914*	Walter Reed Army I, Silver Spring MD	Neuroscience
Ralph W. Gerard	1900-1974	U Chicago	Biology: cells
Leon D. Harmon	?	Bell Laboratories, Murray Hill NJ	Technical science
Warren S. McCulloch	1898-1968	U Illinois (later: MIT)	Biology: cells
Frank Morrell	1926-1996	Stanford U	Neuroscience
Vernon B. Mountcastle	1918*	John Hopkins S of Medicine, Baltimore	Neuroscience
Carl Pfaffmann	1913-1994	Brown U, Providence	Human Science
Walter H. Pitts	1923-1969	U Chicago	Basic science
Claude E. Shannon	1916-2001	Bell Laboratories, Murray Hill NJ	Basic science
Cornelius A. Wiersma	1925-2001	CalTech, Pasadena	Neuroscience
Donald M. Wilson	?	U California, Berkeley	Neuroscience

Table 4.7: 'Important people' of American origin. Abbreviations: U: University; I: Institute; S: School. For some persons, no assured source for the date of birth and death have been found.

RALPH GERARD, JULIAN BIGELOW and HEINZ VON FOERSTER, the *Macy* community has accounted (beside MCCULLOCH) three more persons to the group of the ambassadors. Four of the other ambassadors (BARLOW, ECCLES, WALL, WIERSMA) were physiologists by training, whereas the other three (HARMON, REICHARD, SELFRIDGE) descent from a basic science or engineering tradition with some affinity to biological questions (especially WERNER REICHARD).

In total, 32 persons were identified as 'protagonists'. They represent a considerable part of the 'scientific elite' of the 1940s to 1960s – at least in the emerging neuroscience (e.g. ECCLES, GERARD, BARLOW), information and computer science (e.g. SHANNON, SELFRIDGE, VON NEUMANN), and cybernetics (e.g. MCCULLOCH, MACKAY, PASK) as well as people with an outstanding interdisciplinary reference (e.g. BATESON, RAPOPORT, VON FOERSTER). This shows, that scientific questions related to the brain, and neuronal information processing in particular were indeed of great interest of outstanding scientists at that time.

Name	Birth, death	Place of work	Field	Origin
Gregory Bateson	1904-1980	Stanford U	Human sci.	UK (1939)
Heinrich Klüver	1897-1979	U Chicago	Biology: sys.	Germany (1923)
Rafael Lorente de No	1902-1990	Rockefeller I, New York	Biology: cells	Spain (1931?)
Henry Quastler	1908-1963	U Illinois, Urbana	Biology: cells	Austria (1939)
Anatol Rapoport	1911*	U Michigan, Ann Harbor	Multidisc. sci.	Ukraina (1922)
Oliver G. Selfridge	?	MIT, Cambridge	Technical sci.	UK (age of 14)
Hans-Lukas Teuber	1916-1977	New York U	Neurosci.	Germany (1941)
Heinz von Foerster	1912-2002	U Illinois, Urbana	Multidisc. sci.	Austria (1948)
John von Neumann	1903-1957	I Adv. Study, Princeton	Basic sci.	Hungary (1930)
Paul Alfred Weiss	1898-1989	U Chicago	Biology: sys.	Austria (1939?)

Table 4.8: 'Important people', emigrants to the US. Abbreviations: U: University; I: Institute; sci: science. The year in brackets after the country of origin indicates the time of emigration. For some persons, no assured source for the date of birth and death have been found.

Name	Birth, death	Place of work	Field	Origin
Horace B. Barlow	1921*	Cambridge U	Neurosci.	UK
Colin Cherry	1914-1981	Imperial College, London	Basic sci.	UK
John C. Eccles	1903-1997	Australian National U, Canberra	Neurosci.	Australia
Donald M. MacKay	1922-1987	King's College, London	Human sci.	UK
Gordon A. Pask	1928-1996	Systems Research Ltd. London	Multidisc. sci.	UK
Werner E. Reichardt	1924-1992	MPI biol. Kybernetik, Tübingen	Neurosci.	Germany
Patrick D. Wall	1925*	MIT (later: U College, London)	Neurosci.	UK

Table 4.9: 'Important people', non-Americans. Abbreviations: U: University; sci: science.

The classification of the protagonists according to origin leads to three results. First, 25 of 32 persons originate or migrated to the USA (15 persons were born in the United States, Table 4.7). This indicates again, that the geographical origin of the 'information processing brain' is the United States (however, our analysis did not cover any sources in the Soviet Union). Second, the persons that immigrated into the US (10 in total) – mostly due to the war induced transitions – account for a considerable part of the 'American' scientific power in this field (Table 4.8). Third, researchers from the United Kingdom were dominant within the groups of non-Americans (Table 4.9). We choose the following six persons for a citation analysis: WARREN MCCULLOCH, who is the 'ambassador' with the highest number of conference participations, and, as the historical analysis showed, was involved in many important developments that lead to the 'information processing brain' – notably in neural modelling, the neuronal channel capacity discussion and the establishment of the brain-computer analogy. DONALD MACKAY, who was probably the intellectually most comprehensive researcher within the domain, which is reflected in his various publications in many different as well as in the fact, that we has present in conferences of all four clusters. RALPH GERARD, who was descended (as MCCULLOCH) from the *Macy* tradition, was co-organizer of the important 1962 conference in Leiden and was founding member of the *Society of Neuroscience*. As fourth and fifth persons, we chose the two neurophysiologists HORACE BARLOW and THEODORE BULLOCK (latter is not classified as 'ambassador'), who were major figures in the early neural coding debate, as the historical analysis showed. Finally, we have chosen WERNER REICHARDT as sixth person, as he was later an important promotor of cybernetics in Germany and showed up surprisingly frequent in the conferences we investigated. We also performed a citation analysis for HEINZ VON FOERSTER. Due to the comparatively low number of entrances in the *ISI Web of Knowledge* database, we excluded him later from the analysis. The 'ambassadors' BIGELOW, ECCLES, HARMON, SELFRIDGE and WALL, and WIERSMA were not considered as a result of their seldom appearance in the historical analysis, or due to time constraints.

Protagonists – Short Biographies: We provide short biographies of the six persons identified as 'protagonists':[11]

---

[11] Our main sources are: BARLOW: *Laudatio* in honor of the *Australia Prize*, awarded to BARLOW in 1993 (https://sciencegrants.dest.gov.au/SciencePrice/Pages/PreviousPrizeWinners.aspc) and the internet encyclopedia *Wikipedia* (http://www.wikipedia.org). BULLOCK: His autobiography [302]. GERARD: Obituary in *Behavioral Science* [90]. MACKAY: Obituary in *Nature* [326]. MCCULLOCH: dictionary entry of LETTVIN [324] and *Wikipedia*. REICHARDT: Obituary of the *Max-Planck-Society* http://www.kyb.tuebingen.mpg.de/re/obituary.html

- **Horace B. Barlow (1921*):** HORACE BASIL BARLOW was born 1921 as son of ALAN and NORA BARLOW – which were part of the Darwin-Wedgwood family and BARLOW is thus the great-grandson of CHARLES DARWIN. After getting medically qualified at *Cambridge University*, the *Harvard Medical School*, and the *University College Hospital* (1947), he became a research student under ADRIAN (1947-1950) in Cambridge. In the 1960s, he became professor of physiology at the *University of California* in Berkeley. In 1973, he returned to England and became research professor in the physiological laboratory in Cambridge (he retired in 1987). BARLOW's main field of work was visual neurophysiology, where he introduced the concept of a 'feature detector' (see section 3.2.1).
- **Theodore H. Bullock (1915*):** THEODORE HOLMES BULLOCK was born on May 16 1915 in Nanking, China as the son of Presbyterian missionary parents. He made his studies at the *University of California* in Berkeley, where he received his BA in 1936 and his PhD in 1940. His appointments were at the *Yale University School of Medicine* and the *University of Missouri*, until he got a permanent position at the *University of California* in Los Angeles (1946-1966). From 1966 he changed to the *University of California* in San Diego, where he remained professor until 1982. His main field of work was invertebrate neurobiology and his two volume treatise with ADRIAN HORRIDGE, *Structure an Function on the Nervous System of Invertebrates*, became a standard textbook.
- **Ralph W. Gerard (1900-1974):** RALPH WALDO GERARD was born in Harvey, Illinois in 1900. He received his BS in 1919 and his PhD in 1921 at the *University of Chicago* and later, in 1924, the MD degree from the *Rush Medical College*. In 1926 he worked in Europe (England and Germany) and returned in 1927 to the *University of Chicago*, where he remained in the Physiology Department until 1952. He then became Director of Laboratories at the Neuropsychiatric Institute of the Medical School (*University of Illinois*). In 1955, he became professor of behavioral science. In the same year, he helped to found the Medical Health Research Institute at the *Institute of Michigan* in Ann Harbor, where he also became professor of neurophysiology and physiology (1955-63). From 1963 he helped organize the Irvine Campus of the University of California and served as dean of its Graduate Division until his retirement in 1970. His field of interest ranged from neurophysiology to behavioral sciences. He had also a special interest in questions of education and teaching.
- **Donald M. MacKay (1922-1987):** DONALD MACCRIMMON MACKAY was born in 1922. After a training in physics and a three year wartime experience in radar research, he joined in 1946 the *King's College* in London, where he turned his attention to high-speed analogue computers and their possible relevance to the human brain. He was among the first British scientists who appreciated information theory and cybernetics in setting theoretical limits to the performance of the human brain. In 1960, he moved to *Keele University*, where he founded the Department of Communication and Neuroscience. This institution soon became a interdisciplinary research institute of international standing. MACKAY also made several philosophical contributions to the debate about free will and determinism, as well to religious questions, as he was an active churchman. He was also an effective communicator of scientific and philosophical ideas.
- **Warren S. McCulloch (1898-1968):** WARREN STURGIS MCCULLOCH was born in Orange, New Jersey. He first studied medicine in *Yale* and *Columbia University* in New York and received his MD in 1927. Later, he also obtained

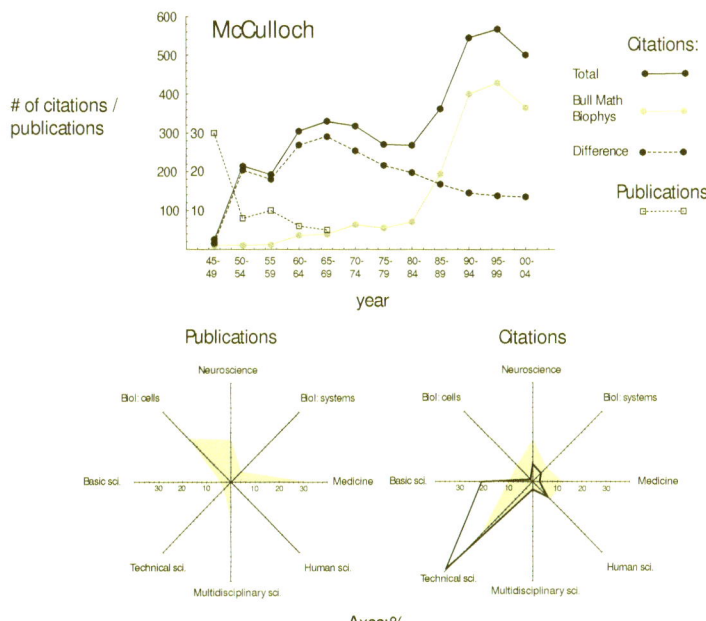

Figure 4.5: Citation analysis for MCCULLOCH. Above, the number of citations (all papers, only [159], difference between both) and the number of publications per 5-year-period since 1945 (note that the number of publications in the graph is scaled with 10 to increase visibility). Below, the classification of the publications of MCCULLOCH and the citations of his work. The distribution of the citations of the MCCULLOCH-PITTS paper is indicated by the grey line (unfilled area).

a training in neurophysiology (1928-1931), studied mathematical physics (1931-1932) and worked as a clinician from 1932 to 1934. Then, he joined the *Yale Laboratory of Neurophysiology*, where he became an assistant professor in 1941. His mentor was DUSSER DE BARENNE. In 1941, MCCULLOCH joined the *Illinois Neuropsychiatric Institute* as associate professor of psychiatry at the *University of Illinois*, where, in 1942, he took WALTER PITTS as student. In the 1940s MCCULLOCH joined the 'teleological society' of WIENER and later became the chairman of the *Macy conferences*. In 1951, he moved his group to the MIT (PITTS was already at the MIT), where he stayed until his death. MCCULLOCH had many fields of interest, besides neurophysiology especially in cybernetics. He

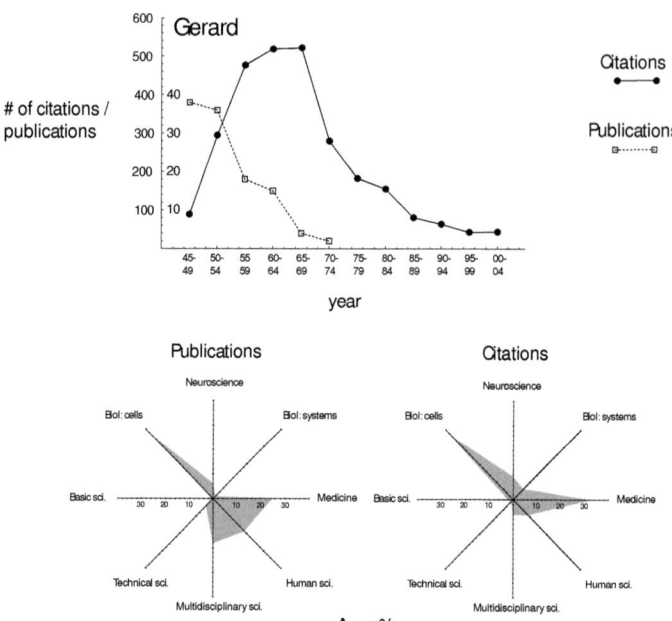

Figure 4.6: Citation analysis for GERARD: Above the citations and publications per 5-year-period. Below, the classification of the publications and citations of GERARD.

was founding member and first president (1967-1968) of the *American Society of Cybernetics*.

- **Werner E. Reichardt (1924-1992):** WERNER REICHARDT was born 1924 in Berlin. Right after his high school graduation in 1941, he was called to the *Luftwaffe* and served in a technical unit developing long distance radio communication for weather forecasts. In 1943, REICHARDT joined a resistance group trying to establish a radio contact with the Western Allies. These activities were discovered at the end of 1944, REICHARDT was arrested and expected his execution. In the last days of the Nazi regime he managed a narrow escape during a rebellion of prisoners and could hide himself in Berlin until the end of the war. After the war, REICHARDT studied physics at the *Technical University* in Berlin and finished his studies with a doctoral thesis in solid state physics in 1952. In the 1950s, he started to work with the visual system of insects. In 1954, he was in-

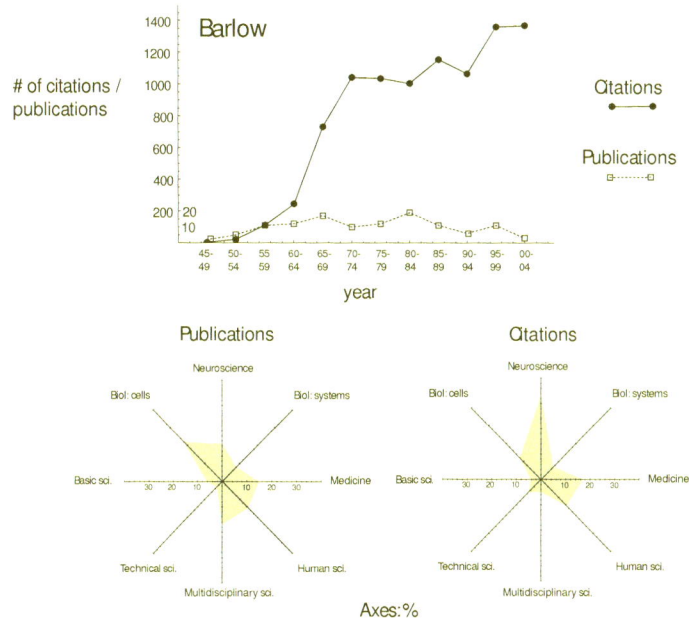

Figure 4.7: Citation analysis for BARLOW. Above the number of citations and publications per 5-year-period. Below the classification of publications and citations according to different fields.

vited by MAX DELBRÜCK to the *California Institute of Technology* in Pasadena. Back to Germany, REICHARDT became in 1955 Research Assistant at the *MPI für physikalische Chemie* in Göttingen. In 1958, a research group for cybernetics was established at the *MPI für Biologie* in Tübingen. In 1960, he became professor in Tübingen. A separate building for his department was opened in 1965 and was transformed into the *Max-Planck-Institute für biologische Kybernetik* in 1968.

The bibliometric analysis for the six protagonists shows a more detailed picture (note, that in the graphs the number of publications are scaled with 10 to increase visibility). It is clearly visible, that the results of the citation analysis of the '*Macy*-persons', MCCULLOCH (Fig. 4.5) and GERARD (Fig. 4.6) are very different, although they worked in quite similar fields and were members of the same generation of scientists. The particularity of MCCULLOCH's citations can be ascribed to the MCCULLOCH-PITTS paper of 1943 [159]. If

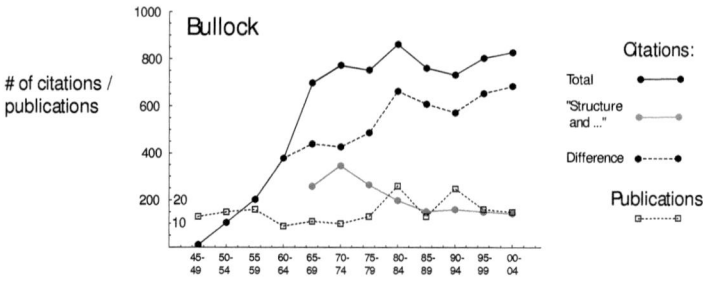
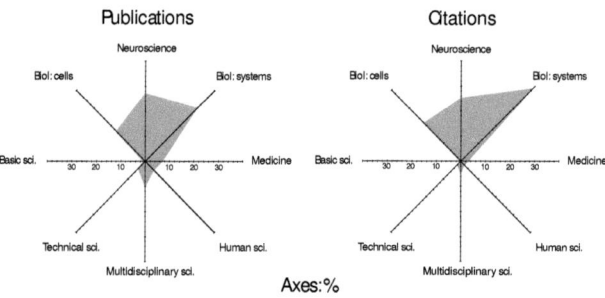

Figure 4.8: Citation analysis for BULLOCK. Above the number of citations and publications per 5-year-period. Below the classification of publications and citations according to different fields.

this paper is excluded from the citations, the number of citations of MCCULLOCH per time (fife-year-period) show a rather classical behavior [313]: a peak (in the mid 1960s, and a first peak in the early 1950s) and a steady decay. Remind, that the location of the peak is also a result of the onset of the database (1945, i.e. earlier citations are not in the database) and must be interpreted carefully – especially for GERARD, MCCULLOCH, and BULLOCK.

If the MCCULLOCH-PITTS-Paper – which accounts for more than 40% of all citations of MCCULLOCH's work – is included, the picture changes dramatically: Since the mid 1980s, the number of citations increases to an amount, which is very uncommon. This shows that the MCCULLOCH-PITTS-paper is considered as a 'founding-paper' for a growing community of scientists – the neural network and (to a lesser extend) the computational neuroscience community. This is shown when analyzing the fields of influence of MCCULLOCH's work, as a strong bias towards the technical sciences (mostly computer science) is visible, largely caused by the citation of the MCCULLOCH-PITTS paper. When furthermore comparing the field of publication activity with the field of citation, one sees that the publications of MCCULLOCH in Medicine (the field with the largest fraction of MCCULLOCH's publications) is not strongly acknowledged. The peak in the field 'multidisciplinary' in the publications when compared with the citations is a general phenomena, that is also visible when looking at other protagonists. This results from publications in interdisciplinary journals (e.g. *Nature* or *Science*), which usually are not again cited in a publication of a multidisciplinary journal, but mostly in special journals. Furthermore, publications in interdisciplinary journals are used to promote own work, such that citations in such contributions often refer to work published in specialized journals.[12] Because of that effect, the number of publications in the field 'multidisciplinary science' is usually larger than the number of citations in that field. The graph of the citation analysis of GERARD shows a classical course: a peak in the mid 1960s, followed by a decay in citations. Also when the fields of publication are compared with the fields of citations, a much more regular picture emerges, as GERARD was recognized basically in the same fields where he published. An exception is his work in psychology (human sciences), which is obviously less well recognized than his work in cellular biology and medicine.

The two biologists of the protagonists investigated – BARLOW and BULLOCK – show the highest number of citations in general. This indicates a different citation culture in biology when comparing with other fields. There are, however, some interesting differences: Almost a quarter of BULLOCK's citations emerge from his well-known standard monograph *Structure and function in the nervous systems of invertebrates* [52], which appeared in 1965 and was very soon acknowledged in the community (Fig. 4.8). BARLOW, on the other hand, has various, well-cited publications, but no single publication that accounts for a significant amount of the total number of citations (Fig. 4.8). BARLOW furthermore publishes not many paper per year – but when he publishes, it is usually well-acknowledged in the community. The citations increase considerably in the mid 1960s, indicating that BARLOW became a prominent scientist in that period. The fact that BARLOW's citations usually fall in the category neuroscience (which is a modern category), together with the still not decreasing curve of citations, indicate, that he is still well acknowledged in neuroscience today. BULLOCK's increasing importance in the mid 1960s is basically ascribed to his monograph, although a first increase in importance is already visible at the beginning of the 1960s. He furthermore shows the most stable publication-citation comparison concerning the distribution of citations within categories. Whereas BARLOW's work has some impact in the technical sciences, human sciences and medicine as well, the work of BULLOCK is restricted to the biological sciences.

The two multidisciplinary European scientists – MACKAY and REICHARDT – display also interesting differences in the citation analysis. For both, the absolute numbers of citations are considerably smaller than those of BARLOW and BULLOCK. At least for REICHARDT, this difference may reflect the bias of the ISI-database towards American and English journals, as a considerable part of his publications are not contained in the ISI database, which distorts the citation analysis.[13] The citation analysis of MACKAY shows a peak in the mid 1970s,

---

[12] URS SCHOEPFLIN, *MPI for the History of Science, Berlin*: personal communication.

[13] The official website of the *Max-Planck-Society* (http://www.kyb.mpg.de/~wreichardt) lists 70 journal articles, 27 conference papers, 24 book chapters and 5 popular scientific publications. The *ISI Web of Knowledge* database contained 59 entries. Conference proceedings, popular science publications are, how-

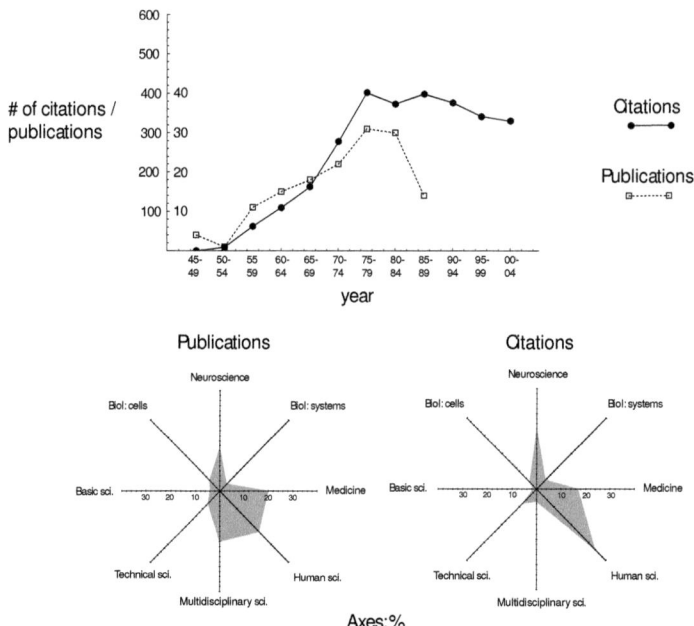

Figure 4.9: Citation analysis for MACKAY: Above the number of citations and publications per 5-year-period. Below the classification of publications and citations according to different fields.

the same time he also published the most (Fig. 4.9). His citations decay slowly, indicating, that he is still rather influential. MACKAY shows publication activity in all fields, which demonstrates his broad interdisciplinary interest. He is most cited in the category human sciences (especially in psychology). REICHARD's citations, finally, show the slowest increase of all six protagonists (Fig. 4.10). His citations reach a peak in the early 1990s (after his death). The analysis of REICHARDT's publications according to category shows a 'double-peak' characteristic for interdisciplinary work, as he published in neuroscience as well as in technical science journals. His major impact, however, lies in the biological sciences.

---

ever, usually not contained in the database. Concerning the analysis of REICHARDT's work, the number of homologes was quite large and needed a considerable amount of work to check individual entries.

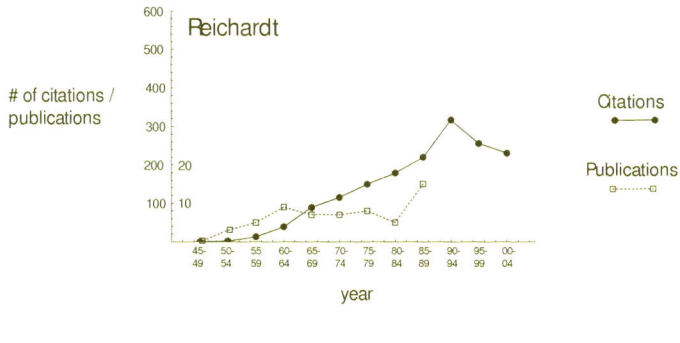

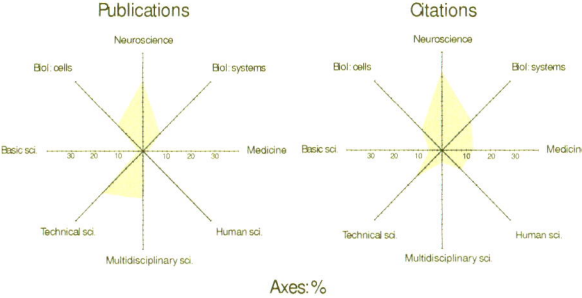

Figure 4.10: Citation analysis for REICHARDT. Above the number of citations and publications per 5-year-period. Below the classification of publications and citations according to different fields.

4.4 Synopsis: Qualitative and Quantitative Analysis

We conclude the historical analysis by a synopsis, where we integrate the results of the historical (qualitative) and scientometric (quantitative) analysis. We shortly review the preconditions of the 'information processing brain' that there established until the 1930s, summarize the main developments in the 1940s to 1960s and list the open achievements and questions, that were set at the beginning of the 1970s. Figure 4.11 provides an overview.

4.4.1 Up to 1940: Preconditions

The conceptualization of the 'information processing brain' needed a clear picture of the activity on the micro scale, in order that an entity is able to perform the 'processing' and another entity is able to represent the 'information'. The first aspect – the neuron as basic processor – was based on the neuron doctrine, which at that time was sufficiently firm (although not proven) to serve as justification. The second aspect needed the emergence of the single, 'digital' spike and the spike train, that can be measured reliably. The spike train and his mathematical abstraction (as series of event-times or 'interspike intervals') then can serve as a bearer of information that is accessible for further (theoretical) analysis. The creation of the spike train required instruments that were able to measure and store the spikes. The American physiologists FORBES, GASSER and ERLANGER and the British physiologists LUCAS and ADRIAN were prominent among those scientists, that enabled this development. ADRIAN had furthermore a distinguished role, as he described the neural communication problem comprehensively and by using the concept of a 'message' that nerve fibres transmit. The establishment of the spike train was, however, not unproblematic. Improvements in measuring spikes, e.g. by PECHER, also showed an intrinsic variability of neuronal firing, that can no more be attributed to unreliable instruments. Furthermore, the mass action hypothesis of LASHLEY advanced the perspective of a 'active brain', whose elements (the neurons) are capable of spontaneous activity. From a technical point of view, these aspects make the neuron 'unreliable'. This had two important consequences for the later development: First, the 'unreliable neuron' hampered the application of the information vocabulary as it was unclear, to what extend one can understand the variability as resulting from noise. Second, the 'unreliable neuron' also served as an inspiration for technology development, which was recognized very early – notably by MCCULLOCH and VON NEUMANN – and the question emerged, how one could build reliable machines out of unreliable components.

In the 1930s, neurons and the brain also became object of a rigid mathematical approach in order to model their behavior. A protagonist in this respect was RASHEVSKY, whose program of mathematical biophysics, especially in the field of cell and molecular biology, later was discredited as being too theoretical and lacking empirical grounding. In neural modelling, however, he (and later his students) insisted on the use of a continuous mathematics instead of the discrete approach of MCCULLOCH/PITTS. In this sense he was more prudent in respect of the mechanization of neural processes. Furthermore, he later supported the statistical approach in respect of neural connectivity. We therefore suppose, that RASHEVSKY can be considered as one precursor of today's dynamical systems and statistical physics approach towards the modelling of neural systems, although the MCCULLOCH/PITTS model had a much larger appreciation within the theoretical community at that time (and now).

4.4.2   1940-1970: Main Developments

A first important step towards the information processing brain was the conceptualization of information, which included the emergence of information theory and cybernetics. However, not all of the persons identified as main protagonists in this respect – WARREN MCCULLOCH, WALTER PITTS, CLAUDE SHANNON, ALAN TURING, JOHN VON NEUMANN and NORBERT WIENER – were equally important in the following historical developments. MCCULLOCH was a decisive figure in the whole development due to several reasons: First, as shown by the bibliometric analysis, he introduced together with PITTS a very influential neural model based on the 'digital character' of the spike and the neuron doctrine, especially for the 'second phase' of interest in theoretical questions of neural information processing (since the 1980s). Second, he was chairman of the *Macy conferences*. Although questions of neural information processing were not in the focus of this conference series, the *Macy* discussions anticipate several aspects, like neural coding and the technical implications of neuronal information processing, that later became important. Third, as our cluster analysis has shown, MCCULLOCH was a busy ambassador of his ideas, especially in the theoretical neuroscience community. Also VON NEUMANN was an important figure, as a shift of his focus towards problems of neural information processing emerged since the 1940s. In particular, he took the inspirational value of neuronal information processing for technological systems very serious (much more than MCCULLOCH) and developed a precise framework to study these questions within the emerging automata theory. His contributions at the *Macy* discussions demonstrate a prudent and sophisticated approach towards questions of neural coding. The work of PITTS is more difficult to judge, as much of his work has been destroyed. It seems that he was interested in a 'statistical physics' approach towards processes in neural systems, which undoubtedly would have been an innovation at that time. He was, however, not present in many different communities and, besides the MCCULLOCH/PITTS paper, not an influential person, especially within the biological community. SHANNON shows up in our analysis in two respects (besides being the 'founder' of information theory): First, he was critical towards an extended application of information theory, although his critique mainly focussed applications towards human systems, where semantics is inevitable. Second, he showed an interest in questions of biological information, which is shown in his later work on building 'intelligent artefacts' and his participations at biological conferences. WIENER and TURING, finally, do not show up in our analysis to a relevant degree. They supposedly failed to gain any involvement in the 'neural coding' and 'neural information processing' discussion, although WIENER can be considered as the founder of the brain-computer analogy. However, one has to keep in mind that we did not discuss the development of artificial intelligence, where TURING was a more important figure.¡

The conceptualization of information promoted several new fields of discussion: The first concerns the application of the information vocabulary on neural systems, exemplarily shown in the work of BARLOW and FITZHUG in the visual system in the 1950s. At the same time, a discussion on neural channel capacity emerged, promoted by MCCULLOCH, MACKAY and RAPOPORT. This discussion was entangled with the question, which 'code' a neuron may use. Together with the assumption of 'information capacity maximization' (proposed by MCCULLOCH), the field for new, temporal, codes was open. One example is the 'pattern code', that was found to be an attractive theoretical concept (e.g. by RAPOPORT and HARMON) and found support by experiments performed by WIERSMA. In the late 1950s and early 1960s, several conferences – among them the 1962 Leiden conference organized by GERARD – discussed this matter. In that time, the neural coding debate had his first peak – some important persons in this respect are BARLOW, BULLOCK, MOUNTCASTLE, SEGUNDO and UTTAL. Unlike the discussion about the genetic code (that happened at the same), no stabilized concept of a 'neural code' emerged, rather an explosion of candidate codes was visible – exemplified in the NRP work session on neural coding of 1968.

Another field of discussion can be grouped around the concept of 'neural information'. In the early 1950s, this discussion was already resumed in the field of molecular biology (QUASTER), where it lead to a disappointing result. The discussion of the concept of neural information – by BRAZIER, MACKAY and RAPOPORT and others – had a similar faith: a growing scepticism within the neuroscience community towards the useability of information theory within their field. The same scepticism also emerged towards the brain-computer analogy, that was further developed in the 1950s in the cybernetic (and artificial intelligence) community (e.g. by ASHBY and GEORGE). More and more, however, it was clear that this analogy should not be used to explain the brain (although for example the *Perceptron* has been introduced by ROSENBLATT for exactly that purpose), but to find new technology. This 'bionic inspiration' was first formulated by MCCULLOCH and VON NEUMANN and pushed forward especially by VON FOERSTER.

A recalcitrant problem was neuronal variability that inspired the search for neuronal noise sources (e.g. by FATT/KATZ, and later by VERVEEN/DERKSEN and CALVIN/STEVENS). It also led to a growing importance of statistical aspects in respect of neuronal connectivity. For the modeler community, a new challenge was born, as one had to assess, whether these aspects of randomness still allow a stable systems behavior. For the neurophysiologist, on the other hand, two problems emerged: First, a new need for statistical tools that allow the handling of variability emerged. This led in the 1960s to the development and consolidation of a statistical toolbox for spike train analysis (GERSTEIN, MOORE, PERKEL, STEIN). Second, a dispute about the consequences of neuronal variability for measuring neuronal activity emerged: One side argued, that only the mean response of a neuron can be considered as relevant – for the experimentalist *and* for the neuron (BURNS), whereas others refuse the notion of 'neuronal unreliability' (BARLOW, BULLOCK). Latter standpoint was successful in the sense that it was able to formulate a new role for neurons as 'feature detector', leading to the single neuron doctrine. In this field of discussion, also a new standpoint emerged that criticized the usage of the term 'noise' as a disturbing element. Rather, it was argued, that noise in biological systems might indeed have a functional role, although no consensus emerged, what this role could be.

In summary, the situation in the early 1970s led to the following specification of the information processing brain: The information vocabulary has been introduced and 'tested' within the context provided by the emerging neuroscience. Scientists that investigated explicitly the signal processing properties of neuronal systems started to use terms like 'neural code', 'neural information' or 'neural noise'. In that sense these concepts became part of the vocabulary of neuroscience. However, the 'test' of these terms also showed that precise definitions in the sense of information theory were not practicable within neuroscience. Alternative and generally accepted concepts were not found, rather an increasing number of proposals (especially for a 'neural code') made it necessary to use these concepts in a metaphorical sense and not as precisely defined scientific concepts. The turn from the 1960s to 1970s marks the end of a period, where a first attempt to integrate and conceptualize the information vocabulary within neuroscience has been made. However, at the end of the 1960s, a large variety of models as well as a mathematical toolbox for spike train analysis was available. This allowed a steady but slow growth of those fields of neuroscience where the development of theoretical concepts that describe the information processing properties of the brain was in the focus. This growths increased markedly in the mid 1980s, leading to a 'second phase' of interest in the theoretical investigation of the information processing brain. Although we did not investigate this point in detail, we suspect that the increased possibilities of computational modelling was a major driving force for this development. In this development, physicists and computer scientists that had a much larger affinity to the information vocabulary entered neuroscience. This also enforced the interest in finding more precise definitions for 'neural information', 'neural code' or 'neural noise'. Up to now, however, a 'success' in this respect has not been reached.

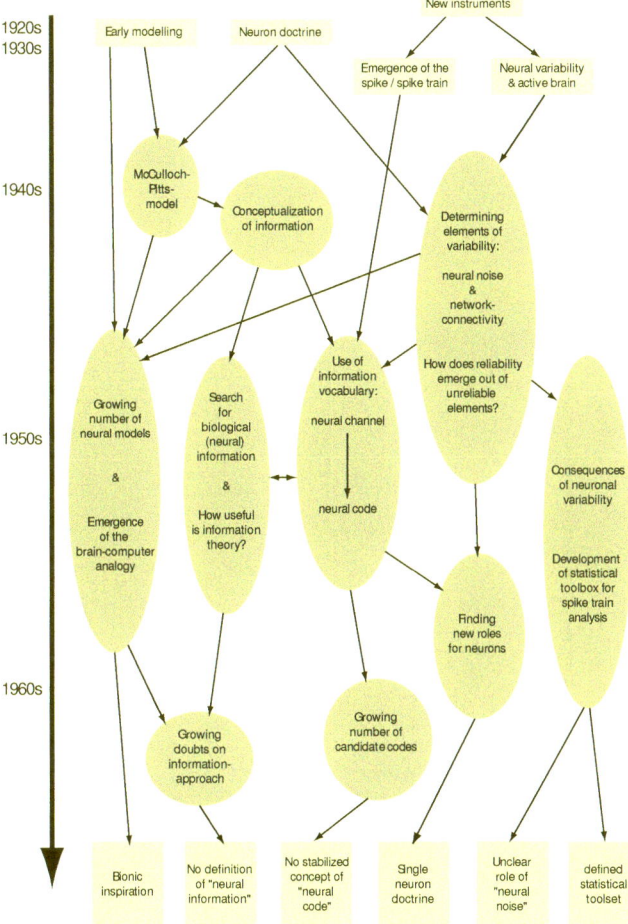

Figure 4.11: An overview of the main historical developments, that lead to the notion of the 'information processing brain' in the 1940s to 1960s.

Part II

Patterns in Neuronal Spike Trains

Chapter 5

Neural Coding and Computation

This chapter sets a general framework for the spike pattern discussion by defining the terms 'neural coding' and 'neural computation'. We start from the finding that the early neural coding debate did not solve the coding problem. We list the main difficulties that appear when the concepts 'code' and 'computation' are used to denote processed in neural systems and present a definition of both terms.

5.1 Neural Coding

5.1.1 What is a 'Neural Code'?

When referring to processes performed by neurons, the terms 'coding' and 'computation' are widely used in today's (theoretical) neuroscience [443, 651]. As our historical analysis suggests, we distinguish two phases where 'neural coding' and 'neural computation' became (more) intensively studied topics of neuroscience. The first phase (1940s to early 1970s) ended without obtaining any generally accepted concept of a 'neural code' or 'neural information processing'. In the second phase (starting in the mid 1980s), the situation has not changed fundamentally. Although the brain-computer analogy of the early days of the neural coding debate is no more considered as adequate for defining these terms precisely [544]:98, neural coding is still an important issue. On the other hand, some believe that the terms 'code' and 'computation' are inadequate for describing neural processes [296]:153,167; [517, 336]. Any such conclusion depends on the definition of the terms involved. We first focus on the term of a 'neural code'. It is undisputed that the meaning of 'code' in the sense of codification (a legal term) or in the sense of a computer code (a symbolic arrangement of data or instructions in a computer program) do not apply when using the term 'neural code'. As shown in chapter 3, the term 'neural code' appeared in neuroscience after the introduction of SHANNON's scheme of a communication system (see Fig. 2.1) [227], although the term usually has not been used in the precise definition of a (discrete) transducer proposed by SHANNON. In the mathematical notation of KOLMOGOROV [133], the communication scheme is translated into a chain of operations $\xi \to \eta \to \eta' \to \xi'$, where $\eta \to \eta'$ stands for a transmitting device, $\xi \to \eta$ for a coding operation and $\eta' \to \xi'$ for a decoding operation (ξ stands for a message and η stands for a signal). Any such communication process can only operate (and be observed) with finite precision. Whatever entities ξ and η are composed of, in the act ob observing them we attribute numbers or other types of symbols to them. In that sense, the most general definition of a code is that of a relation between sets of symbols [437]. We define:

Definition 5.1 *A countable set of symbols* $\mathbb{A} = \{a_1, a_2, \ldots\}$ *is called* alphabet. *A finite, arbitrarily arranged string of symbols of length* l $(a_i, a_{i+1}, \ldots, a_{i+l})$ *from an alphabet* $\mathbb{A}$ *is called* word.[1]

Definition 5.2 *For two alphabets* $\mathbb{A}, \mathbb{B}$, *a mapping* $f_c : \mathbb{A} \to \mathbb{B}$ *relating words of* $\mathbb{A}$ *with words of* $\mathbb{B}$ *is called* code relation. *If* f_c *is injective, the code relation is called* non-degenerative, *if* f_c *is bijective, the code relation is called* uniquely decodable.

Definition 5.3 *For two alphabets* $\mathbb{A}, \mathbb{B}$ *and a code relation* $f_c : \mathbb{A} \to \mathbb{B}$, *we call the set of all words of* $\mathbb{A}$ *that can serve as input for* f_c *the* code input set $\mathcal{I}$ *and the set of all words of* $\mathbb{B}$ *that can form the image set of* f_c *the* codeword set $\mathcal{C}$.

Definition 5.4 *For a code input set* $\mathcal{I}$, *a codeword set* $\mathcal{C}$ *and a code relation* f_c, *the triplet* $(f_c, \mathcal{I}, \mathcal{C})$ *is called* code.

This definition is very general and applies for any relationship between two entities described by two sets of symbols. The term 'symbol' refers here in most cases to measurement values that provide information about the state of the system. Thus, the alphabet is formed by rational numbers. The main point is that analyzing processes in a natural system in the framework of a code consists in determining f_c together with $\mathcal{I}$ and $\mathcal{C}$. This is the reason, why not only f_c (as for example proposed by [652]), but the whole tripled should be called code.[2] The question is now: how are f_c, $\mathcal{I}$, and $\mathcal{C}$ determined?

In linguistics and semiotics, three restrictions hold for a justified application of the term 'code': First, symbols are a special type of signs whose meaning can only be understood by incorporating the interpreter of the symbol and the conventions that attribute certain entities to a symbol.[3] Second, f_c is the result of in principle arbitrary and modifiable conventions [296]:167. Third, the rules used to relate phrases of $\mathcal{I}$ and $\mathcal{C}$ do not only have a syntactic, but also a semantic and pragmatic character [308]:57-61. The linguistic notion of a code assumes the existence of a well-known language that serves as a foundation for defining the

---

[1]Note, that the indices i of a_i within the word denotes the position of the symbol in the word and not the number of the symbol in the alphabet. The same symbol can appear several times in a word.

[2]We follow the proposal of UTTAL, according to which a code consists of sets of symbols used to represent messages (patterns of organization) and the set of rules that govern the selection and use of these symbols [249]:208. Some remarks to the use of the terms 'encoding' and 'decoding': Former mainly refers to sensory system (a stimulus is encoded), whereas latter is usually used in the sense of 'cracking the code'.

[3]Several different interpretations of the concept of symbol exist: In the typology of CHARLES SANDERS PEIRCE, signs are classified as icon, index and symbol, whereas the latter is characterized by a purely arbitrary attribution between the entity and the symbol that stands for this entity.

code by its users (i.e. the attribution of phrases from $\mathcal{I}$ to $\mathcal{C}$). From this perspective, single neurons or groups of neurons lack principal arbitrariness of choosing a specific code relation. Furthermore, the semantic and pragmatic character of neural signals as well as the existence of symbols, the 'symbol restriction', are (in the best case) unclear. Therefore, linguists and semiotics refuse the notion of a 'code' when referring to neural systems [308]:31.

However, it is not compulsory to give up the notion of a 'code' when one is unable to fulfill all restrictions formulated above. If we agree that information theory is 'allowed' to used the term code, then semantic and pragmatic constraints can be neglected. On the other hand, syntactic constraints are present due to the physical structure of the system performing the code relation that sets some limits upon the succession of states, e.g. due to the inertia of the system, and thus restricts the set of possible sequence of symbols (i.e. measurement values). The term 'symbol' is furthermore used much more 'liberal' in information theory, referring to the signs used to represent messages. In the following, we use this broader meaning of 'symbol'. The second restriction, the requirement of arbitrariness, cannot be neglected completely. When a communication system is constructed, f_c can indeed be chosen arbitrarily to some degree so that the system is able to fulfill a certain purpose – e.g. to maximize channel capacity or to hide encoded information (cryptography). If the restriction of arbitrariness would be given up completely, the use of the term 'code' becomes meaningless. Otherwise, one could say that a falling stone 'encodes' a certain height at time t in a certain speed at time t. The system that 'codes' for something must have a certain degree of freedom in respect to relate the input with its output. At that point it is important to mention that the physical process which performs the code relation (the code transformation, see below) is not arbitrary at all, but has to be as deterministically as possible in order to provide a reliable coding. The question is now: what degree of arbitrariness in determining f_c is allowed? The answer to this question is provided by analyzing the history of the system. The 'genetic code' serves as an example: Here, arbitrariness means that there is no chemical necessity that determined which nucleotide tripled codes for which amino acid. In genetics, the emergence of this relation is considered as a 'frozen accident'. But the physical process, that relates a specific nucleotide triplet to a specific amino acid is chemically determined by the tRNA molecule. The degree of arbitrariness is very low, as the system cannot change the relation any more after the 'frozen accident' happened. If the fact that the 'frozen accident' could have happened different, is considered as a sufficient degree of arbitrariness, the use of the term 'code' for the genetic code is justified. Also under this assumption the genetic code should not be mistaken for a code in the linguistic sense.

Is a certain degree of arbitrariness also fulfilled for a 'neural code'? In other words: is the system able to change the input-output-attribution over time? The answer to this question depends on the spatial and temporal scale of the analysis: If, for example, a single synapse is considered and the input string is a time series of the number of transmitter molecules secreted by the synapse in a small time unit and the output string is a series of values of the post-synaptic membrane potential, then the relation is fully determined up to a stochastic moment given by diffusion processes and channel gating. On a larger spatio-temporal scale, however, effects like adaption and long-term potentiation serve as a basis for arbitrariness in the sense that there is no temporally invariant mapping from a given stimulus intensity to a resulting spike train [312]. We propose that this degree of arbitrariness is sufficient not to exclude the notion of a 'neural code' *a priori*. To discuss the 'symbol restriction' in a neural code, we introduce the type-token distinction, that originates from the logician and philosopher CHARLES SANDERS PEIRCE [174]:

Definition 5.5 *The* type *of a symbol (word) denotes the abstract entity that represents the semantic content of the symbol (word). We use the term* represent *to denote the relation between a type and its semantic content.*

Definition 5.6 *The* token *of a symbol denotes the physical entity that stands for the symbol. We use the word* realize *to denote the relation between a token and a type. We call the tokens which realize the symbols of the code input set the* input tokens *and the tokens which realize the symbols of the codeword set the* output tokens.

Definition 5.7 *If input tokens and output tokens are given, we call the physical process t_c that transforms input tokens in output tokens the* code transformation.

In a technical coding context, the type-token relation is unambiguous, as the *code is constructed* on the type level in order to serve its purpose defined on the semantic level. Later, the tokens are chosen such that the realizations of phrases and codewords is reliable given the noise of the system. The 'semantic' aspect refers to the functional value of the phrases and codewords (see Fig. 5.1). When – as in neuronal systems – one does not construct the code but one analyzes whether a certain process can be understood as a code, the problem gets more complicated. One has to answer the question, whether tokens can be identified in neural processes to which symbols can be attributed. This problem is related to the fundamental challenge for explaining neural mechanisms: how to decompose the central nervous system into functionally meaningful parts [337]:49. This decomposition involves a spatial aspect (e.g. expressed by the localization debate) and a temporal aspect. Latter involves the search of certain entities that express significant moments in time within the processes of the system. We call these entities *events*, which are represented by symbols. Note, that the term 'event' can correspond to either a token (a specific event in space-time) or a type. In other words, there must be a natural coarse-graining of the parameter under investigation. One may certainly argue that the measurement itself produces a coarse-graining, as the measurement resolution is finite. Each single measurement point, however, stands probably not for a significant event of the system. One has to find a way to separate the significant events from all other phenomena that can be measured at some level of organization in the brain. These events are then related to symbols.

The spike is the 'classical event' that serves as a basis for analyzing neuronal processes in a coding framework [651] (see section 2.3.3). The 'all-or-none' character of spikes made them candidates for a symbolic description – e.g. by using the symbol '123' to indicate that a spike occurred 123 ms after a certain reference point in time. Three objections have been raised that question spikes as basic entities for discussing neural coding: A theoretical objection denying the possibility that spikes may serve as symbols; an empirical objection referring to other possible tokens that may serve as basis for defining symbols; and a practical objection stating that the determination of spike trains based on measurements (the spike sorting problem) is not sufficiently reliable. We shortly discuss the first two problems. For the third problem we refer to section 6.1.2.

The first objection claims, that spikes cannot be symbols due to two reasons [336]: First, spike times are continuous variables. Thus, the number of possible spike events is uncountable and one is not able to relate spike times with a countable alphabet. Second,

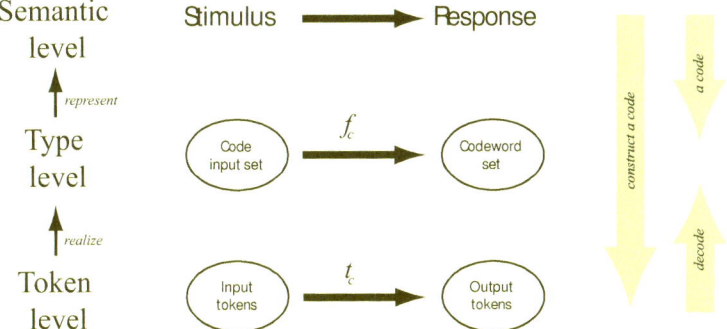

Figure 5.1: The several dimension of a code relation in natural systems: The semantic level defines the experiment in which a stimulus is presented to the system in order to get a certain response. The type level is the level where the definition of a code is applied. The physical processes that realize the code happen on the token level.

the assumption that single spikes are functionally significant is unwarranted. The first arguments reminds of the dispute between MCCULLOCH and RAPOPORT, whether a real-valued or a discrete mathematics should be used in order to model neural systems (see section 3.2.2). This distinction, however, misses the point. The problem is rather, if a coarse-graining on the temporal micro-scale is present or not. If it is present, it will show up also using a real-valued mathematics. MCCULLOCH and PITTS assumed an universal coarse graining when presenting their model in 1943 (the synaptic delay) – which is certainly a shortcoming. However, although synaptic delays or spike-widths are variable, they are not arbitrary small – a coarse graining is thus possible. Furthermore, measurements involve a finite sampling rate. The 'empirical world' in which measurements are performed and theories are tested is thus always a 'discrete' world. The argument, whereby the single spike is not functionally significant, is an empirical question – but does not affect the more fundamental problem whether one can relate a single spike with a symbol or not. Spikes might just not be the relevant tokens. Indeed, other events within spike trains have been proposed to be relevant, e.g. bursts, local firing rates etc. (see section 6.1.2). All these events, however, are characterized using the spike-times of the individual spikes. In that sense, the symbolic representation of the spike train as sequence of event-times is a prerequisite indeed for analyzing the problem. We thus consider the first objection as unsupported. The second objection, the possibility that other tokens than spikes – like local field potentials [408] or neurochemical aspects [343] – are more relevant for understanding a code relation is a relevant point. This aspect, however, has to be decided empirically and is not a fundamental argument against the role of spikes in a coding relation. We will in the following restrict ourselves to spikes as the fundamental event in order to discuss the neural coding problem.

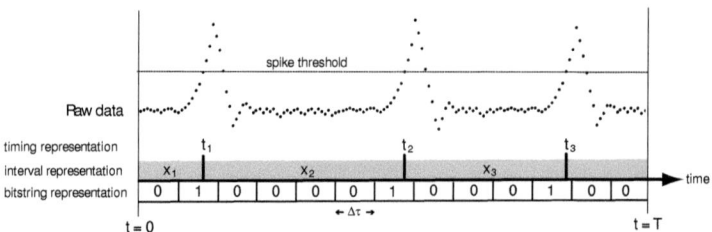

Figure 5.2: Different representations of spike trains: Raw data (in this picture, no real data are displayed) is given as voltage per sampling time unit (symbolized by dots). By defining a spike threshold, a spike train is obtained in the timing representation $\{t_1, t_2, t_3\}$ and the interval representation $\{x_1, x_2, x_3\}$. After choosing a bin-with $\Delta \tau$, one obtains the bitstring representation.

The recording of spikes leads to the notion of a 'spike train':

Definition 5.8 *A sequence of points in time* $t = (t_1, \ldots, t_L)$ *indicating the occurrence of spikes attributed to a single neuron of length L is called* spike train.

A spike train according to Def. 5.8 is in the *timing representation*. An equivalent representation of a spike train is given by the sequence of interspike intervals (ISI): $x = (x_1, \ldots, x_L)$, where $x_i = t_{i-1} - t_i$ (x_1 is the time interval from the beginning of the measurement to the first spike) – the *interval representation*. Usually, spike trains are analyzed in the timing or the interval representation. Another form of representation is obtained by transforming the spike train $t = \{t_1, \ldots, t_L\}$ into a bitstring – the *bitstring representation* – a standard procedure for applying methods of information theory [74]. For this transformation, the measurement interval $[0, T]$ is partitioned into n bins of width $\Delta \tau$ ($n \Delta \tau = T$). If at least one spike falls into the i-th bin, the letter '1' (and otherwise the letter '0') is written at the i-th position of the string. Usually, $\Delta \tau$ is chosen so that maximally one spike falls into one bin. This is achieved by setting $\Delta \tau = T_{\text{spike}} \approx 1$ ms, as the spike-widths is (in most cases) of the same order. The result is a string of the form $X_n = (x_1 \ldots x_n)$ (see Fig. 5.2). Another representation is the *local rate representation* of a spike train: If $[0, T]$ is partitioned into n bins of width $\Delta \tau$ ($n \Delta \tau = T$) where $\Delta \tau \gg 1$ ms, one calculates the firing rate r_i for each bin (in spikes/second) and obtains the desired representation: $r = \{r_1, \ldots r_n\}$. Spike trains serve as basis for defining further events of interest that consist of several spikes (like burst or spike patterns) or that are calculated out of spike trains (like local firing rates). If other events than spikes are of interest, their time of occurrence are calculated and a new symbol string is obtained that is object of the analysis. We conclude that the processes in neurons can be investigated from a coding perspective, because one is in principle able to identify symbols when describing the process, and because a sufficient degree of arbitrariness for the code relation is given. 'Decoding' a neural code needs the solution of three problems:

1. One has to identify entities in the input as well as the output of a process that stand for types which are in a code relation. In this respect, one has to decide, whether one

chooses an external view or an 'organism point-of-view' [394, 454, 555].[4] Which of the two perspectives one wants to adopts depends on the problem one wants to solve.

2. One has to identify a physical process that serves as a code transformation. The mere correlation between two classes of tokens is not sufficient. Here, the problem emerges whether the physical process identified should be understood as a code transformation or as some kind of computation (see section 5.2).

3. A neural code has a functional role [417]. Events that represent information which is not used by the organism are of no interest for establishing a code, as one expects that the nervous system encodes stimuli that have a significance for the organism [450]. A behaving animals is not a passive sensor, but seeks for sensory input (e.g. in olfaction [555]). Thus, one has to identify on the semantic level an appropriate stimulus-response-pair that may have significance for the organism.

Therefore, the neural coding problem cannot be solved 'by construction' (semantic level → type level → token level), but needs two parallel approaches (semantic level → type level and token level → type level, see 5.1). The final goal, describing the triplet $(f_c, \mathcal{I}, \mathcal{C})$, depends furthermore on the *spatial* and *temporal* scale considered as relevant. Sometimes, the problem of neural coding has been addressed when analyzing the information flow from stimulus to behavior [467, 533] – thus, the spatial scale would involve the whole organism. This problem cannot be practically handled. A coding problem in the sense that $\mathcal{I}$ stands for all possible 'stimulus phrases' and $\mathcal{C}$ stands for all possible 'behavioral response' would overload the code relation f_c and makes it impossible to handle as soon as the organism under investigation reaches a certain level of complexity. Furthermore, the processes involved would probably be more successfully analyzed in the framework of a 'computation' than of a 'code'. The proposal of JOHNSON [533] that the ultimative test of the correctness of a neural code would be to apply an electrical stimulation mimicking the sensory input such that a expected behavior is the result is inadequate, as it involves far more aspects than mere coding. Therefore, coding is experimentally analyzed on a much smaller spatial scale and involves in most cases sensory stimuli – physical entities characterized by spatial, temporal and quality (e.g. luminance) parameters – of one modality as input tokens and neural responses in the form of spikes or membrane potentials of a single neuron or small groups of neurons as output tokens. The code relation is a function that describes the spike train as the response of a neuron to a certain stimulus [347]. *Decoding* is understood as constructing an 'inverse filter' that operates on the spike train(s) and produces an estimate of the unknown stimulus waveform [468]:358.

The usual approach to solve this problem is based on a BAYESian scheme [621]: Different stimuli are presented many times to obtain a histogram of responses for each stimulus, leading to an estimation of the probability that a particular response occurred given that a particular stimulus occurred. Once this probability is known, BAYES' theorem is applied to derive the probability that a particular stimulus occurred given that a particular response occurred. In a similar way, in the framework of information theory, the mutual information

---

[4]The 'organism point-of-view' refers to the fact, that the neuron receiving a stimulus is not able to perform the averaging that is used when an external observer investigates a code relation. The neuron is rather confronted with the question of how much it could infer from a single observation. In this respect, measures like 'specific information' [456] and 'stimulus-specific information' [411] have been proposed.

between input and output is calculated [577]. For both approaches, a significant number of data is required in order to properly estimate the underlying probability distributions needed for information-theoretic based calculations [578]. This problem is intrinsic to any information theoretic approach. Therefore, these studies involve in most cases only a few cells: the sensory cells and the neurons that pool the information received from the sensory cells and transmit them to neurons of the central nervous system. Most studies of neural coding are thus studies on *sensory* coding: the transformation of a stimulus into spike trains. To conclude, we discuss in some more detail the temporal and the spatial aspect of neural coding. The first aspect leads to the 'rate code' vs. 'temporal code' distinction, the second aspects is discussed in the context of 'single neuron coding' vs. 'population coding'.

5.1.2 Rate Coding vs. Temporal Coding

If one assumes that the information processed by a neuronal system is present in the spike trains of the neurons involved, then only the timing of the spikes matters. The rate code vs. temporal code distinction is a matter of the relevant time scale. The rate coding hypothesis claims that the mean firing rate – the average number of spikes in some time interval (time scale: averaging interval) – carries the relevant information, whereas the temporal coding hypothesis claims that the precise placement of the spikes in time (time scale: jitter, see Def. 6.4) is (also) significant [450]. The averaging interval is usually of the order of 10-100 ms, the jitter interval is of the order of 1-10 ms.

> **Rate Code:** The hypothesis of a rate code states that it is sufficient to understand neural coding such that the spike train is an estimator of a time-varying firing probability [740]. In our framework, the symbols in $\mathcal{I}$ and/or $\mathcal{C}$ refer to firing rates measured over a certain time interval. The term 'firing rate' has two different meanings depending on the way of averaging [651]: The first possibility is to average the spike count over the whole length T of the spike train of a trial, the 'spike count rate'. The second possibility is to average over shorter time intervals, which leads to the 'time-dependent firing rate' that measures the probability of spike occurrence for a certain time. In practical applications, the post-stimulus time histogram obtained from several trials of the same experiment serves as an estimate of the time dependent firing rate. The convenient way to model a spike train in the rate code picture is to use a inhomogeneous POISSON process that includes absolute and relative refractory periods in order to take care of the specific biophysical properties of a neuron. The rate code hypothesis and the POISSON hypothesis (see section 6.2.2) are closely connected in the sense that rate coding is compatible which a POISSON distribution of the spike times within interval smaller than the time interval chosen for calculating the firing rate. The rate code hypothesis is often related to the 'feature detector' concept, where the time-dependent firing rate reflects the time course of the intensity of a certain feature of the stimulus. The sensory modality is preserved by combining a rate code with a 'labelled line' code, because no other means internal to the spike train can convey the modality. Rate codes are interpreted by neurons or populations of neurons that have sufficiently long integration times [468]. This requirement motivates a common counter argument against the rate code hypothesis: reaction times for stimuli can be very fast (in the visual system in the order of 150 ms [724]) and it is unclear how this can be achieved by a rate code that needs a certain averaging time for each neuron in the processing chain. It has been claimed [491] that even transmitting the simplest signals reliably using a

rate code requires either excessive long observation times, or excessive large numbers of redundant neurones.

Formally, we define a rate code as follows:

Definition 5.9 *Assume a spike train as output of a coding procedure, let T_{spike} be the time scale of the spike, and let T_{rate} be a time interval such that $T_{rate} \gg T_{spike}$. If symbols of $\mathcal{C}$ are attributed to events in time intervals such that the temporal arrangement of spikes within T_{rate} does not affect the attribution of the symbol, then we call this code a* rate code.

Temporal Code: Theoretical considerations on the increased information capacity of a system using the precise timing of spikes compared to rate coding were an important motivation for developing concepts of time coding (see section 3.2.2). Also later, this argument has been used repeatedly, e.g. by STREHLER in 1977 [242] and by SOFKY in 1996 [687]. Furthermore, time coding allows the possibility that a spike train contains more than one type of information [609] and may even allow to code for information of different sensory modality [160]. These arguments lead to an increasing interest in temporal coding since the 1980s in the community of empirical neuroscientists [480, 693] as well as in the modelling community [499, 718]. The concept of a 'temporal code' can mean two different things: Either is the temporal structure in the response a direct consequence of temporal structure in the input, or the temporal structure in the response that carries information results from the dynamics of individual neurons or the network. It has been suggested to call the first possibility 'temporal coding' and the second 'temporal encoding' [723]. It is undisputed that temporal coding is realized by neuronal systems, e.g. in the auditory system of barn owls and bats, and in the electrosensory organs of several electric fish [418, 419], indicating that some neurons are capable to reliably transmit the temporal structure of the input to succeeding structures. It is more controversial, however, if also temporal encoding (referred by [723] as 'true' temporal coding) is possible. A recent review paper [626] concluded based on data of primate visual and motor cortex that the evidence for a role for precisely timed spikes relative to other spike times (variance/jitter: 1-10 ms) is inconclusive. The data indicate that the signalling evident in neural signals is restricted to the spike count and the precise times of spikes relative to stimulus onset (response latency) [738]. Other investigators claimed, that it is not possible to recover information about the temporal pattern of synaptic inputs from the pattern of output spikes [669, 667]. This claim is based on a model of balanced excitation and inhibition, which is said to be consistent with the observed variability of cortical spike trains. There was, however, a debate whether the model would still allow temporal coding [686, 668]. Other investigators noticed that finding additional information in the temporal structure of the encoded stimulus does not mean that this information is used by the organism [625, 739].

Based on the idea, that the distinction between rate code and time code is a matter of time scales, we define a temporal code as follows:

Definition 5.10 *Assume a spike train as the output of a coding procedure and let T_{time} be a time interval such that $T_{rate} \gg T_{time} > T_{spike}$. If symbols of $\mathcal{C}$ are attributed to events in a spike train such that the temporal arrangement of spikes within T_{time} does not affect the attribution of the symbol, but the temporal arrangement of spikes within T_{rate} does affect this attribution, then we call this code a* temporal code.

This definition follows the intuition that a neuron may be coding for something or for nothing without any observable change in its firing rate [735]. We will define the concept of temporal structure more rigorously when introducing the concept of 'spike pattern'. Furthermore, synchronous firing and coincidence detection have a close relation to the concept of temporal coding as there is as well a (small) time scale involved that defines if spikes are considered as being synchronous or coincident. These aspects will be discussed in more detail in sections 6.4.1 and 6.4.2.

5.1.3 Single Neurons vs. Neuronal Assemblies

The single neuron vs. neuronal assemblies discussion deals with the spatial aspect of coding. As in the first discussion, the distinction between single neuron coding and population coding is to some degree arbitrary, as it is obvious that the functioning of a nervous systems is the result of the activity of many neurons. Thus, population coding is today an implicit theoretical assumption underlying most of the work done in neural coding. The distinction between single neuron coding and population coding is made upon the role one acknowledges to the single neuron. One can ask: Does the cortex function by pooling together large numbers of essentially independent neuronal signals as in an election, or does it work through the coordination of its elements as in a symphony [450]? The metaphor of an 'election' puts emphasis on the importance of single neurons.

Single Neurons: The discussion to what extent the activity of single neurons matter for the functioning of the whole system traces back to the mid 20th century (see section 3.2.1). In this picture, the neuron acts as a feature detector embedded in a network of labelled lines and, as one goes along the different pathways, the neurons become more and more specialized for a certain feature of the stimulus. Hypothetically, at the end of such a neuronal structure, a neuron would sit that represents specific objects (like a grandmother – thus the name 'grandmother cell hypothesis'). In the motor system, a similar picture holds, as so-called 'command neurons' would initiate certain types of movements by activating specific central pattern generators that provide the temporal structure of impulses needed to lead to a coordinated action of muscle groups in order to perform a behavior [549]. It is undisputed that there are 'important neurons' in neural systems – especially in systems of invertebrates – which fulfill the requirements formulated in the single neuron picture [382]. For more complex nervous systems, this framework has, however, been criticized: The number of neurons in the brain is too small in order to represent all possible sensory input an organism is confronted during its lifetime. Furthermore, this framework is not compatible with the redundancy of brain functions which shows up after lesions of certain brain structures [656].

The second alternative, catched by the metaphor of a 'symphony', became more attractive in recent years. An important reason of the development are improvements in multi electrode recording techniques. As described in section 3.2.3, this technique has already been introduced in the mid 20th century, but the lack of statistical tools and computer power to deal with multi electrode data combined with the 'neuron doctrine' formulated in the late 1960s led to a dominance of single neuron recordings. Today, multi electrode array recordings have become a standard tool in neuroscience [412]. They allow stable recordings for long periods (several hours [618]). This makes it possible to ask questions concerning the encoding of task-relevant information in awake and behaving animals [446], and even to use signals derived from multi electrode array recordings to control movements of a robot arm (e.g. [753] in rat motor cortex).

Neuronal Assemblies: In the 1980s and 1990s, the ensemble activity as well as circuit-oriented concepts became a major focus of research in neural coding [447, 461]. The relevant question in this framework is to find criteria that assign neurons to a certain assembly, which performs a 'population code'. In HEBB's original model, an assembly is differentiated by others when the neurons of the assembly show higher activation than the other neurons. Later, the concept of synchronously firing neurons that form assemblies has been introduced (this aspect we discuss in more detail in section 6.4.1). More generally, any set of neurons in which a coordinated timing of action potentials is observed that could code for messages can form a population [451]. Several new statistical concepts have been introduced to deal with such populations. A common example is the population vector used for computing movement directions in the primary motor cortex of primates [493]. Later, an information-theoretic analysis demonstrated that such interactions caused by correlated neuronal activity carry additional information about movement direction beyond the firing rates of independently acting neurons [601]. Such observations are necessary to be able to claim the existence of a population code. The main problem of the population coding hypothesis still is to find a reliable criterium that defines the clusters of neurons forming a population. One has to take into account that the interactions that define such a cluster is context-dependent and dynamic on several different time scales [735]. It is also a problem to estimate to what extent a certain population is able to represent more than one message. If assembles are defined by neurons with similar firing rates, co-activation of several assemblies in order to allow parallel processing would be impossible. Assemblies that use spatio-temporal patterns on the other hand could represent more than one message.

The single neuron vs. population coding discussion is orthogonal to the rate vs. temporal coding discussion. Single neuron as well as populations can use rate codes or temporal codes. We propose the following definition of a population code

Definition 5.11 *We call a criteria that groups a set of spike trains in a certain class in order to analyze a coding procedure a* population criteria *and the resulting set of neurons a* population.

Definition 5.12 *If there exists a population criterion and a time interval T_{pop} such that the population criteria can be applied for defining output tokens of a code relation leading to a symbolic labelling of populations, then the code is called* population code.

5.2 Neural Computation

A problematic connotation of the 'neural coding' metaphor is that one may understand neuronal processes as a mere transmission of information [296]:153, e.g. in the sense that the 'eye transmits the information it sees to the brain'. The use of the term 'code' does not necessarily presuppose such a view, as the code relation f_c is very broadly defined. Nevertheless the question emerges if neural coding is accompanied by 'neural computation'.

This leads to the question how 'neural computation' should be defined. One may be tempted to use the formal definition of computation, originally proposed by TURING [248] in order to understand 'neural computation'. However, neuroscientists generally agree, that TURING's definition of computation (for a formal definition we refer to [680]) is not an adequate definition for neural computation [434]. An alternative approach is the use of a more general framework. It has been proposed, to understand computation as any form of process in neuronal systems where information is transformed [450, 683] or destroyed [286]. This approach is certainly confronted with the problem to find a definition for transformation of information, such that not almost all natural systems count as computational information processors [315]. Furthermore, defining computation as transformation of information would probably lead to the conclusion, that every coding step along sensory pathways is also a computation step, which would make the two concept interchangeable.

A more precise approach for defining natural computation resides in the context of measurement and prediction, presented by STOOP et al. [695, 696]. In this approach, not individual strings but classes of strings should be considered as input. Computation is then defined by describing a measure of computation that expresses, how much, in a statistical sense, the computation simplifies inferring future from past results. Computation thus calculates the average over all problems the natural system is able to solve. In this picture, the computation is performed by maps that transfer input strings to output strings, measures the reduction in complexity [697] and averages this reduction over all possible input-output relations for the system under investigation. Using this approach, intermittent systems, for example, perform almost no computation. Maximal computation is performed by trivial maps (e.g. $f(x) \equiv 0$). The definition of natural computation in the sense of STOOP allows to relate computation to our definition of coding such that computation expresses a special property of the code relation f_c. Coding reflects the process of transforming words of $\mathcal{I}$ in words of $\mathcal{C}$, whereas computation reflects the change of information of the code relation in a statistical sense, by taking into account the average change of information when f_c is performed on all words of $\mathcal{I}$.

Definition 5.13 *If a code $(f_c, \mathcal{I}, \mathcal{C})$ is given, then* computation *refers to the change of information measured over all possible input words of the code relation.*

In this sense, every code in a natural system stands also for a computation performed by the physical process that transforms input tokens in output tokens. But the concepts are not interchangeable, as they refer to different perspectives on the problem. Two problems remain open: First, what is the basic unit that performs computation in nervous systems? Candidates are not only single neurons [383, 542], but also synapses [761], or groups of neurons like cortical columns. Certainly, in our framework, all of them could serve computational purposes as long as measurements are possible in order to construct a code. In that sense, it depends on the observer's standpoint and interest which process the observer wants to understand. Second, to what extent do long-term changes in the system, usually related to learning, affect our definition of computation (and coding)? Take the example of precise relative timings of spikes which may induce LTP [450]. This aspect will be discussed in some more detail in section 6.4.2.

Chapter 6

Defining Patterns

In this chapter, we introduce a general definition of a spike pattern along with the notions of the 'background' and the 'stability' of a pattern. These definitions are then related to the concepts of reliability of firing and neural noise. In a second step, we review experimental findings of spike patterns in different neural systems and major theories that claim a functional role of patterns. In a third step, we present a hypothesis, where reliability, noise and coding merge.

6.1 Characteristics of Patterns

6.1.1 Defining 'Spike Pattern'

Several important concepts in science – like 'system', 'complexity', and 'pattern' – are often used ambiguously. This ambiguity is expressed by a general and imprecise usage of these terms. This is not a problem *per se*, as the phenomenological description of a difficult problem needs such vague concepts. But their ambiguous use indicates an insufficient understanding of the object to which they refer. Precise definitions do not forbid to use the concepts in their vague sense, but provide an option for clarification, if necessary.[1] The concept 'pattern' falls into this category of terms. In its widest sense, the word stands for any kind or regularity that could be the object of a scientific investigation. In philosophy of science, a definition has been proposed by DANIEL DENNETT, according to which a 'real' pattern is an entity that can be described in a more efficient way [306]. The scientific task is then to find this simpler description. In this sense of 'pattern', one can consider 'spike patterns' as regularities in spike trains whose appearance give a specific information about the task the neuron is doing – e.g. the appearance of bursts in spike trains as an indicator for complex cells [420]. From this point of view, almost all studies in neuronal electrophysiology intend to find spike patterns and the term does not express a specific problem to solve.

---

[1] Precise definitions are not suited for all cases. Take the example of 'complexity', where a definition has been given by the complexity measure of KOLMOGOROV and CHAITIN, which is adequate for describing problems of computational complexity, but inadequate for natural systems [697].

If the concept of spike pattern is put in the discussion of neural coding or computation, one assumes that patterns should help to explain the coding or the computation performed by the system. For example, spike patterns may express the message being encoded, the signature of the biophysical spike generation mechanism or even changes in the noise that affect the system. In this context, the term 'spike pattern' should not be restricted to sequences of single spikes. Rather, any pattern of events, which can be defined using single spikes (like bursts or local firing rates) is a candidate for a spike pattern. The events that form a pattern are arranged along a time axis. They also can have a spatial character in the sense that the events are (or are constructed out of) spikes originating from different neurons. Practically, finding a pattern requires the creation of experiments in which regularities appear that can be separated from a background. One has to find an experimental procedure that can be repeated sufficiently often so that all relevant parameters remain sufficiently equal in each repetition.

Definition 6.1 *A* background *is a statistical hypothesis about the arrangement of events within sets of spike trains.*

Definition 6.2 *A* spike pattern *of length l is a repeated sequence of l events in a set of spike trains whose repeated appearance is improbable when assuming a certain background.*

A pattern is an abstraction on the type level, i.e. the sequence of events is a type (see Def. 5.5). Every pattern on the type level has as many counterparts on the token level, as the pattern repeats itself in the data. Because an experimentalist can only control the stimulus, but not the internal state of the system sufficiently precisely, some events within the sequence will not always appear when the whole sequence appears and the timing of the events will not always be the same either. This leads to the concepts of 'variation' and 'jitter'. Furthermore, there are *different kinds* of sequences with increasing constraints: The most liberal way to define a repeating sequence is by demanding that the *order* of appearance of the events that form the spike pattern stays the same [559]. Then, one can demand that the *relative time intervals* between the events of a repeating sequence remain fixed (proposed e.g. in [445]). Furthermore, one can demand that the *absolute time intervals* between the events remain fixed (e.g. [353, 444]). Finally, one can demand that the *timing of the events in respect to an external time frame* remains constant (e.g. [477]). We introduce the following notation:

$E_{i,j}$ An event E in a sequence ($i = 1, \ldots, l$ denotes its position in a sequence of length l, $j = 1, \ldots S$ denotes the number of the spike train from which the event emerges).

$p_{i,j}$ The probability that an event $E_{i,j}$ appears in the sequence.

$X_{i,j}$ The time interval between the events $E_{i,j}$ and $E_{i+1,j'}$ in a sequence.

$T_{i,j}$ The timing of the event $E_{i,j}$ by referring to an external time frame.

We remind that the term 'event' refers to tokens an types, whereas the term 'pattern' only refers to types. The event $E_{i,j}$ in a pattern then also denotes a type. If a sequence is repeated n times in the data and we want to refer to the specific event $E_{i,j}$ that appears in the k-th repetition of the sequence ($1 \leq k \leq n$), then we write $E_{i,j}(k)$ and the expression denotes a token. For $X_{i,j}$ and $T_{i,j}$, we proceed likewise.

Definition 6.3 *For n repetitions of a sequence of events in a data set and the set of all time intervals between the the events $E_{i,j}$ and $E_{i+1,j'}$ in the sequence $\{X_{i,j}(k)\}$, $k = 1, \ldots, n$, the variation $\Delta X_{i,j}$ of replication of $X_{i,j}$ is:*

$$\Delta X_{i,j} = \frac{\max\{X_{i,j}(k)\} - \min\{X_{i,j}(k)\}}{2}$$

Definition 6.4 *For n repetitions of a sequence of events in a data set and the set of all timings of the event $E_{i,j}$ $\{T_{i,j}(k)\}$, $k = 1, \ldots, n$, the jitter $\Delta T_{i,j}$ of $T_{i,j}$ is:*

$$\Delta T_{i,j} = \frac{\max\{T_{i,j}(k)\} - \min\{T_{i,j}(k)\}}{2}$$

Definition 6.5 *For probabilities of appearance of two succeeding events of a pattern $p_{i,j}$ and $p_{i+1,j'}$, the probability of appearance of the time interval between the two events $\bar{p}_{i,j}$ is*

$$\bar{p}_{i,j} = (1 - p_{i,j})(1 - p_{i+1,j'})$$

Variance, jitter and probability of appearance indicate the *noise level* of the system and the *precision* of sequence replication: If variance and jitter are high and $p_{i,j}$ is low, the noise level is high and the precision is low. Vice versa, the noise level is low and the precision is high. The concepts 'noise level' and 'precision' are, however, not used by us as quantitative, but as qualitative terms. Based on this notation and the previous definitions, four different types of spike patterns are defined:

Definition 6.6 *A order pattern of length l is a sequence of 2-tuples of the form:*

$$\{(p_{1,j}, E_{1,j}), \ldots (p_{l,j'}, E_{l,j'})\}$$

Definition 6.7 *A scaled pattern of length l is a sequence of 3-tuples of the form:*

$$\{\bar{p}_{1,j}, X_{1,j}, \Delta X_{1,j}), \cdots, (\bar{p}_{l,j'}, X_{l,j'}, \Delta X_{l,j'})\}$$

whereas for the n appearances of the sequence in the data there exist $n - 1$ numbers $r_k \in \mathbb{R}$, $k = 2, \ldots, n$ such that the following condition is fulfilled:

$$r_k \cdot X_{i,j}(k) = X_{i,j}(1) \quad \forall (i = 1, \ldots l; j = 1, \ldots S; k = 2, \ldots n)$$

Definition 6.8 *A interval pattern of length l is a sequence of 3-tuples of the form:*

$$\{(\bar{p}_{1,j}, X_{1,j}, \Delta X_{1,j}), \cdots, (\bar{p}_{l,j'}, X_{l,j'}, \Delta X_{l,j'})\}$$

Definition 6.9 *A timing pattern of length l is a sequence of 3-tuples of the form:*

$$\{(p_{1,j}, T_{1,j}, \Delta T_{1,j}), \ldots, (p_{l,j'}, T_{l,j'}, \Delta T_{l,j'})\}$$

Some additional remarks on notation: For $j = 1$ we speak of *single train patterns*, for $j \geq 2$ we speak of *multi train patterns*. A timing pattern of length 1 where the external time frame is given by a stimulus onset is also called *latency pattern*. A interval pattern of length $l > 1$ where all events originate from different spike trains is also called *T-pattern* [586]. A multi train interval pattern where $X_{i,j} = 0, \forall\, i = 1, \ldots l$ holds is also called *unitary event* [511]. A simple type of interval patterns is the frequent emergence of specific time scales (interval patterns of length 1), compared to a POISSON background. Furthermore, the interspike intervals of interval pattern must not necessarily be intervals of succeeding spikes, although we usually refer to interval patterns where no such 'additional' spikes are allowed, unless otherwise mentioned. Finally we note, that many pattern detection methods (see next chapter) only look for patterns where $p_{i,j} = 1$ holds.

6.1.2 Types of Events

In this section, we specify what types of events can occur in spike patterns. We list the events according to the degree of ambiguity when attributing tokens found in the data to the event of a pattern:

1. **Single spikes:** A single spike shows up as peak in the voltage trace of a measurement. The ambiguity of the event 'single spike' consists in discriminating between peaks originating from an action potential and peaks resulting from other electrical phenomena, in determining the time point when the spike occurred and in attributing peaks to action potentials originating from different neurons (spike sorting problem).

2. **Inequalities:** It has been proposed [47, 230] to consider inequalities one obtains when succeeding interspike intervals are compared as events that form patterns.[2] The ambiguity is to determine, when two succeeding ISI should be considered as equal.

3. **Bursts:** A group of succeeding spikes considerably closer than other spikes is called burst. The additional ambiguity relies in finding a criterion that allows to distinguish groups of spikes, that are classified as bursts, from other spikes.

4. **Local firing rates:** By assuming a time interval T_{rate} over which the firing rate is averaged, local firing rates are defined as events that could be part of a pattern [723]. The additional ambiguity is to find T_{rate}.

5. **Patterns:** Its also possible to consider a certain pattern as an event and look for patterns of patterns (*higher order pattern*), e.g. patterns of unitary events.

---

[2]Consider the following example: A short sequence of ISI is given by (2,5,2,2,7,8,4). The sequence of inequalities is obtained as follows: The first and the second ISI is compared. As 2 is smaller than 5, one writes '−'. Then the second and the third ISI are compared. 5 is bigger than 2, thus one writes '+'. The third and the fourth ISI are compared. They are equal, thus one writes "0". Proceeding in this way the ISI sequence is transformed to the sequence of inequalities (− + 0 − − +). A pattern could be (− +).

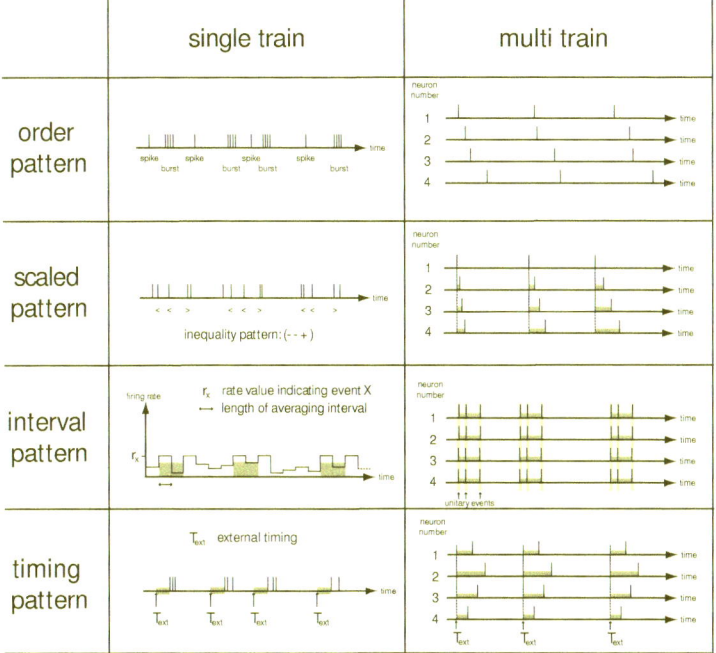

Figure 6.1: Examples of all possible types of spike patterns in the single and multi train case. Single train (from above): Order pattern between two types of events; inequality pattern (see 6.1.2); interval pattern of rate events; latency pattern. Multi train: order pattern; scaled interval pattern; interval pattern of unitary events (a higher order pattern); timing pattern.

In principle, several types of events could form a single pattern, for example an order pattern (spike of neuron 1, burst of neuron 2, spike of neuron 3). Such combined patterns, however, are supposedly only rarely considered in the literature (we found no such example). The most common patterns involve either the events 'single spike' or 'burst'. Patterns of local firing rates are less common and usually fall into the discussion of rate coding [723], although they are sometimes placed into the temporal coding debate [492].

Spikes and Spike Sorting: In intracellular recordings (sharp electrodes or patch clamp), the identification of spikes usually is less problematic than in extracellular recordings, where the voltage trace of an electrode contains more peaks of variable form and amplitude. Only in the optimal case, one type of unvarying peaks can clearly be distinguished from low amplitude noise. Otherwise, many different kinds of peaks are visible, which indicates, that more than one neuron has been measured. In this situation, the problem of spike sorting arises, i.e. attributing voltage waveforms to individual neurons. This problem gains special attention when dealing with multi electrode array recordings, as one is not able to optimize the recording quality to the same degree by optimal placing of the electrode as in a single electrode recording [497]. The classical approach to solve the spike sorting problem is template-based [2, 98]: Based on (usually) visual inspections, templates are chosen and the data are then processed by a computer to generate the spike trains of all neurons measured by he electrode. This approach cannot handle the problem that the waveforms of individual spikes of single neurons are variable. This variability is due to changes in the background activity correlated to the depth of anesthesia [539] and due to correlation of the waveform to the length of the precedent ISI [473] – like for example a broadening of the action potential during repetitive activation [436]. Furthermore, spikes occurring at the same time might not be distinguishable [568], which produces a bias against coincident spikes. Such effects certainly affect the pattern analysis. For example it has been shown that they can lead to apparent temporal ordering effects between neurons in the absence of any specific temporal relationship [641]. Thus, spike sorting, especially in multi train recordings, is still considered an important problem. The spike sorting problem is furthermore a good example of how theoretical expectations influence what is really measured – this corresponds to the classical argument of philosophy of science that observations are 'theory-laden' [320]. Novel proposals for the spike sorting problem are based on clustering techniques [640] and MARKOV analysis of the waveform variability [638].

Firing Classes and Bursts: The firing behavior of (cortical) neurons as a response to sustained current injections is generally classified into the categories of regular-spiking, intrinsically bursting, fast-spiking cells and chattering cells (a variant of a bursting firing type) [435, 436, 506, 602, 623]. These firing classes emerge as a result of the type and densities of ion channels and the dendritic geometry [587]. The majority of cortical cells, usually pyramidal or spiny stellate cells, are classified as regular-spiking. They adapt strongly during maintained stimuli. Fast-spiking cells can sustain very high firing frequencies with little or no adaptation. Morphologically, they are smooth or sparsely spiny, non-pyramidal cells and are likely to be GABAergic inhibitory interneurons. Intrinsically bursting cells are usually large layer 5 pyramidal cells. They generate bursts of action potentials in response to depolarization through intrinsic membrane mechanisms. Chattering cells are usually pyramidal cells and generate 20 to 70 Hz repetitive bursts firing in response to supra-threshold depolarizing current injections. The intra-burst frequency is high (350-700 Hz) and the action potentials have a short duration (< 0.55 ms). Beside this classification, model studies suggest that bursting may also result as a network effect [396].

Recognizing a burst as such is an easy task in some cases, as they appear as a stereotypical sequence of two to five fast spikes and of a duration of 10-40 ms riding upon a slow depolarizing envelope. These sequences are usually terminated by a profound afterhyperpolarization leading to a long ISI [543]. This appearance of two lengthscales (the ISI of spikes within bursts and the interval between bursts, the *interburst interval*) also shows up in the ISI histogram in a typical way (see section 7.2.1). There is, however, no general agreement on the exact size of the smaller time scale. In hippocampus, bursts are considered as a series of two or more spikes with ≤ 6 ms intervals [518], whereas in thalamic spike trains the interval is shorter (<4ms) [733]. In other cases, however, determining bursts is less easy as the variability of the ISI within the burst may overlap with the variability of the interburst intervals. For these cases it has been proposed to assume a POISSON process of equal mean firing rate like the original spike train as background hypothesis and to define a burst as a sequence of ISI that significantly deviates from the probability that such a sequence emerges from the POISSON process (POISSON surprise) [560]. Determining the significance is the ambiguity involved in burst classification. Bursts are a special kind of event as the burst itself has a much bigger variability compared with other events (like the spike waveform). In some applications it might thus be of relevance to indicate the length of the burst, the firing frequency within the burst duration (in this case one could attach a local firing rate to the burst and analyze the problem in the context of local firing rate patterns) and possible single spike patterns within the burst. The reason why bursts are an interesting class of events is their more profound effect on the postsynaptic cell. The presence of a burst is more reliably transmitted over synapses than the presence of a single spike. Thus, bursts possibly have an important role in reliable information transmission [574, 684]. In a model study it has also been shown that the timing and the number of spikes within a single burst are robust to noise [538]. Furthermore, the ISI within a burst may serve as an additional coding dimension. It has been shown that different firing frequencies within a burst are most likely to cause different postsynaptic cells to fire [529]. Thus, bursts provide a mechanism for selective communication between neurons. Finally, bursts may also be considered a basic element in synaptic modification in the context of long-term potentiation (LTP), as bursts tend to increase the magnitude of LTP [440].

6.1.3 The Stability of Patterns

If we assume that the pattern under investigation represents a certain content, e.g. the presence of a certain stimulus, then we have to take into account that a physical process in the neuron is responsible for the generation of this pattern. The reliability of the pattern generating process, however, is affected by intrinsic noise sources and noise that arises from network effects and synaptic transmission, which induce a stochastic component on the level of the neuronal dynamics [498] (see also the next section). Therefore, it can be expected that the neuron sometimes fails to fire the whole sequence of the events that form the pattern. This failure to generate the whole sequence might also be caused by the properties of the

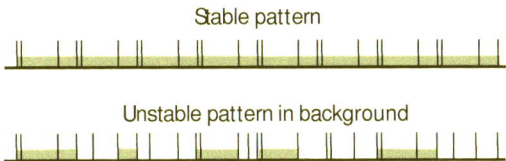

Figure 6.2: The stability of an interval pattern: a) Stable, no noise. b) Unstable, noisy background. Gray bars indicate ISI forming the pattern.

stimulus, for example when the temporal overlap of stimuli does not allow a neuron to finish the firing of the sequence [444]. The *stability* of a pattern concerns the question to what extent a neuron is not able to fire the whole pattern when confronted with a stimulus that usually generates this pattern [661]. If a pattern is unstable in that sense, one expects that not only the sequence of events that defines the patterns is present, but also subsequences. The so-called 'ghost doublets' [565] that have been detected in single spike trains when looking for interval patterns of length 3 are an example of such fragments of sequences. We exemplify the concept of pattern stability in Fig. 6.2: The first spike train describes an ideal case, in which the whole spike train consists of the repetition of an interval pattern of length 3. The second spike train describes the situation, in which the interval pattern and its fragments are immersed in a noisy background. To obtain a formal definition of the stability of a pattern, we restrict our discussion to single train interval patterns where $\tilde{p}_{i,j} \equiv 1, \forall i, j$ and discuss spike trains in the interval representation. For simplifying notation, we write X_i for $X_{i,j}$ and define:

Definition 6.10 *For interval pattern* $\{(X_1, \Delta X_1), \cdots, (X_l, \Delta X_l)\}$ *of length* l, *the set* $\mathcal{P}_l$ *that contains this sequence and all its subsequences is called* pattern group

$$\mathcal{P}_l = \{(X_i, \Delta X_i), \cdots, (X_{i+k}, \Delta X_{i+k})\},$$

where $k = 0, \ldots, l-1$ *and* $i = 1, \ldots, l-k$.

Example 6.1: If the interval pattern is given as $\{2,3,6,4,8\}$ (we assume $\Delta X_i \equiv 0$ to simplify notation), then the pattern group is $\{(2),(3),(6),(4),(8),(2,3),(3,6),(6,4),(4,8),$ $(2,3,6),(3,6,4),(6,4,8),\ (2,3,6,4),(3,6,4,8),(2,3,6,4,8)\}$. If the interval pattern is given as $\{2,2,2,2,2\}$, then the pattern group is $\{(2),(2,2),(2,2,2),(2,2,2,2),(2,2,2,2,2)\}$.

The size of $\mathcal{P}_l$ depends on the number of equivalent $(X_i, \Delta X_i)$ in the sequence. If $(X_i, \Delta X_i) \neq (X_{i'}, \Delta X_{i'})$ for all $i \neq i'$, then $\mathcal{P}_l$ has $\frac{l(l+1)}{2}$ elements. If $(X_i, \Delta X_i) \neq (X_{i'}, \Delta X_{i'})$ applies for certain i, i', the set is smaller, depending on the relative position of the equivalent $(X_i, \Delta X_i)$ within the sequence. We introduce the following notation:

$\tilde{p}_i$	The relative frequency of intervals $X_i \pm \Delta X_i$ within the data.
$N(X_i, \Delta X_i)$	The number of appearance of intervals $X_i \pm \Delta X_i$ within the data.
$P_l(m)$	The m-th element of $\mathcal{P}_l$
$N(P_l(m))$	The number of appearance of $P_l(m)$ within the data.
$N^{\text{exp}}(P_l(m))$	The expected number of appearance of $P_l(m)$ within the data.
$l_{P_n(m)}$	The length of $P_l(m)$

Although the expression $N^{\text{exp}}(P_l(m))$ could be approximated analytically, it is for practical applications more appropriate to shuffle the ISI of the spike train and to determine

$N^{\text{exp}}(P_l(m))$ based on the shuffled train. After having determined the number of appearances of each sequence of the pattern group for the original and the shuffled train, the pattern stability is defined as follows:

Definition 6.11 *For a pattern group $\mathcal{P}_l$ of an interval pattern, the stability $s_\mathcal{P}$ of the pattern is*

$$s_\mathcal{P} = \frac{\sum_m l_{P_l(m)} |N(P_l(m)) - N^{\text{exp}}(P_l(m))|}{2 \sum_l N((X_i, \Delta X_i))},$$

In this way $0 \leq s_\mathcal{P} \leq 1$. If the ISI of a sequence $\{a_1, \ldots, a_l\}$ only appear within the data as part of the whole sequence, then $s_\mathcal{P} = 1$. Such a pattern is maximally stable. If the sequence and its fragments appear as expected by chance, then $s_\mathcal{P} \approx 0$. There are two interpretations of this case: Either one assumes ISI-shuffling as the correct background (see next section) and one concludes that no pattern is present, or one assumes a POISSON background and concludes the presence of an unstable pattern.

6.2 The Background of a Pattern

6.2.1 The Significance of Patterns

We introduced the term 'background' (Def. 6.1) as a statistical concept that describes the 'unpatterned' firing of a cell. In this way, we implicitly attach a certain 'meaning' to the pattern as it may represent a stimulus feature, which is relevant for an animal [671]. From the organism point-of-view, a pattern may serve to predict the presence of a stimulus-feature. This is one way to understand the term 'significance'. There are, however, two notions of 'significance': First, a pattern can be significant in relation to a background model. Establishing this 'statistical significance' is a mathematical problem. Second, a pattern can be significant in the sense that it allows the establishment of a code relation by relating patterns with symbols. In this framework, a pattern is significant because it has an informational content that is important for the behavior of the organism. Describing this relation is an empirical problem. Both aspects of significance are, however, interrelated. A wrong background model can lead to the detection of 'wrong patterns' which, in later experiments, cannot be related to any behavioral significance. In this section, we deal with the statistical significance of a pattern.

6.2.2 The Poisson Hypothesis

In principle, the probability of an event occurring at any given time could depend on the entire history of preceding events. If this is not the case and the events are statistically independent, then they are the result of a POISSON process [443, 543, 651, 732] – a popular assumption for neuronal firing (especially for cortical neurons). An important reason for its popularity is its simplicity from a stochastic modelling point of view. In this framework, a spike train is seen as the result of a point process, which is defined as follows:

Definition 6.12 *Let (t_0, t_L) be a ordered set of points such that $0 = t_0 < t_1 < \ldots < t_L$. The series is generated by a POISSON process $\{P(t), t \geq 0\}$ with rate $\rho > 0$ if*

1. $P(0) = 0$

2. the random variables $P(t_k) - P(t_{k_1})$, $k = 1, \ldots, L$ are mutually independent

3. for any $0 \leq t_i < t_j$ and $k \in \mathbb{N}_0$, $P(t_j) - P(t_i)$ is a POISSON random variable with probability distribution

$$Pr\{P(t_j) - P(t_i) = k\} = \frac{(\rho(t_j - t_i))^k}{k!} e^{-\rho(t_j - t_i)}$$

If ρ is constant, the POISSON process is called homogeneous. If $\rho(t)$ depends on time, the POISSON process is called inhomogeneous.

The following two statistical measures are important for identifying a POISSON processes:

Definition 6.13 *If* $y = \{y_1, \ldots y_n\}$ *is a time series with mean* $\langle y \rangle$ *and variance* $\sigma(y)$, *then the* coefficient of variation c_v *and the* FANO factor F *are defined as*

$$c_v = \frac{\sqrt{\sigma(y)}}{\langle y \rangle} \qquad F = \frac{\sigma(y)}{\langle y \rangle}$$

If a spike train is given, then the following conditions must be fulfilled in order to claim, that a POISSON process is an appropriate statistical model for the train [432, 543, 692]:

1. The ISI distribution has an exponential decay.

2. For the coefficient of variation of the ISIs $c_v \sim 1$ holds.

3. For the FANO Factor of the spike count $F \sim 1$ holds

4. Consecutive ISIs are statistically independent.

The ISI distribution is usually approximated by the 1D histogram (see section 7.2.1). For calculating c_v and F, one has to check the presence of negative correlation on a short time scale and positive correlation on a long time scale. Otherwise, when choosing the counting window unfavorable, one may wrongly conclude that a POISSON process generated the spike train [380]. Tests for independence usually rely on the autocorrelation or on testing, whether a MARKOV-dependency can be found in the data [102, 673]. Such tests are difficult to perform if the firing rate is not stationary [523]. In this case it is difficult to distinguish between dependencies that contradict the POISSON hypothesis and non-stationary firing that still could be modelled by an inhomogeneous POISSON model using an appropriate rate function. A stochastic process is stationary, when its probability distribution does not change in time. As the sampling time of a process is always finite, stationarity can, strictly speaking, never be positively established. Only non-stationarity can be detected. Several tests for stationarity have been developed – many of them in the field of economics and financial markets, where large time series are available. In electrophysiology, however, it is much more difficult to obtain stationary time series that cover a time interval which is much longer than the longest characteristic time scale relevant for the evolution of the system. As the internal dynamics of the neural system cannot be fully controlled, one has to expect an increasing non-stationarity the longer the measurement lasts (e.g. due to effects of adaptation). One way to avoid this problem is to perform different trials and to concatenate the trains of each trial to one spike train. This new train can, for example, be checked for stationarity using the KOLMOGOROV-SMIRNOV test on the distributions obtained for the first half of the train and the second half of the train [690]. In general, a practical way for detecting non-stationarity is to measure the desired property (e.g. the mean ISI) for several segments of the data and to check, if they are sufficiently equal.

Is is well-known, that the assumption of independence does not hold for short ISI due to the absolute and relative refractory period in neuronal firing. The absolute refractory period is the time interval after occurrence of a spike during which it is not possible to induce a new spike. The relative refractory period is a time interval during which the probability of firing gradually increases to the 'normal value'. The length of both absolute and relative refractory period depends on the type of neuron (e.g. in retinal ganglion cells both periods are of the order of several milliseconds [389], whereas in neurons of the corticospinal tract they are of the order of one millisecond [622]). Various neuron models based on the POISSON model can take these effects into account and, furthermore, include a time-varying rate function. [389, 433, 443]. The POISSON *hypothesis* claims that a (inhomogeneous) POISSON model which includes refractoriness is a reasonably good description of a significant amount of spike train data [443, 247]. In the context of our pattern discussion, the hypothesis also claims that the only relevant events that could form patterns are local firing rates. Looking for patterns on a finer time scale would not make any sense.

But is the POISSON hypothesis true? The historical analysis demonstrated that this question was already investigated in the sixties to a considerable degree – but a consensus was not reached (see section 3.2.3). The majority of the findings indicated that the hypothesis is incorrect. Furthermore, seldom *all* four conditions that should be fulfilled in order to claim a POISSON process as an adequate statistical model had been tested. The fourth condition (MARKOV dependence) is – in the papers reviewed for this PhD thesis – almost never mentioned, when the validity of the POISSON hypothesis is claimed. On the other hand, studies, that looked for this dependence indeed found MARKOV dependencies up to the fourth order [209, 230]. Also, most of the newer studies usually use the first three criteria to claim the validity of the POISSON hypothesis. A study of BAIR et al., based on the spike trains of 212 cells in the area MT of macaque monkey, showed that about half of the cells fire in accordance to the POISSON hypothesis [377, 378]. This claim was based on the argument, that the distribution of the interspike intervals (and interburst intervals, if the cells where bursting cells) are of exponential type. A study of SOFKY and KOCH investigated this matter using data of 16 cells of V1 (15 non-bursting cells) and 409 neurons of MT (233 non-bursting cells) of macaque monkey [685]. As the firing rates were variable, the spike trains have been split into smaller trains with stationary rates. Those with comparable rates have been concatenated. By using the coefficient of variation criteria for the concatenated trains, it has been concluded that the POISSON hypothesis holds for almost all cells. Similar studies in the area MT of macaque monkey report a c_v even higher than 1 [405, 667].

Although often stated, such positive tests in favor of the POISSON hypothesis, are not incompatible with the existence of patterns in spike train. Only spike trains that fulfill *all* four criteria of the POISSON hypothesis cannot contain repeating sequences above a certain chance level. This can be easily shown by a simple thought experiment: Assume a spike train in interval representation generated by a POISSON process. From this spike train, n intervals that match the templates $(X_1 \pm \Delta X_1)$, $(X_2 \pm \Delta X_2)$ and $(X_3 \pm \Delta X_3)$ are extracted and arranged to n sequences of the form $((X_1, \Delta X_1), (X_2, \Delta X_2), (X_3, \Delta X_3))$. These sequences then are randomly reinserted into the train at random positions. The resulting train still fulfills the criteria 1 to 3 and only misses criterium 4.

6.2.3 Randomization Methods

The POISSON model serves as an important null hypothesis for several pattern detection methods. It is, however, not the only possible background. In this section we review the most common null hypotheses (ways of randomization) for pattern detection (see figure 6.3):

1. **Uniform randomization:** The assumption that every order or timing of events or time interval between events (latter only over a certain interval) is equally probable. A uniform background is common when looking for order patterns. Examples: [559].

2. **Homogenous Poisson randomization:** The assumption that the events are POISSON distributed. The homogeneous Poisson randomization changes the probability distribution of the original measurement unless the original data were itself POISSON distributed. Examples: [353, 560, 586, 722].

3. **Inhomogeneous Poisson randomization:** Like method 2), but taking changes in firing rate into account. This is done by either using a rate function based on knowledge of the stimulus dynamics (e.g. periodic) or by calculating the firing rate and then by performing homogeneous Poisson randomization only for events that belong to an interval with quasi-stationary firing rate. Examples: [353, 722]

4. **ISI-shuffling:** A common way of randomization for interval patterns: The order of succeeding intervals is randomly rearranged but the intervals themselves are not changed. This preserves the 1D ISI histogram. In principle, large changes in firing rates can be considered by shuffling time segments with quasi-stationary rates separately (**inhomogeneous ISI-shuffling**). Examples: [444, 725]

5. **Spike-jitter randomization:** This randomization assumes a jitter ΔT and redistributes the timing t_i of each spike i within the interval $[t_i - \Delta T, t_i + \Delta T]$ uniformly. Examples: [348, 442].

6. **Spike shuffling:** This randomization is applied for multi train patterns and allows to keep the post stimulus time histogram (PSTH) unchanged: For each bin used to generate the PSTH, the membership of a spike to a train is randomly rearranged. Examples: [562, 639, 725].

7. **Spike-count match randomization:** This randomization is a more restrictive variant of the spike shuffled background: The shuffling of spikes within a bin across the spike trains is restricted, such that the spike counts of each trial before and after shuffling remains comparable, and by forbidding shuffles that would generate ISI shorter than the empirically derived refractory period. Examples: [379, 627].

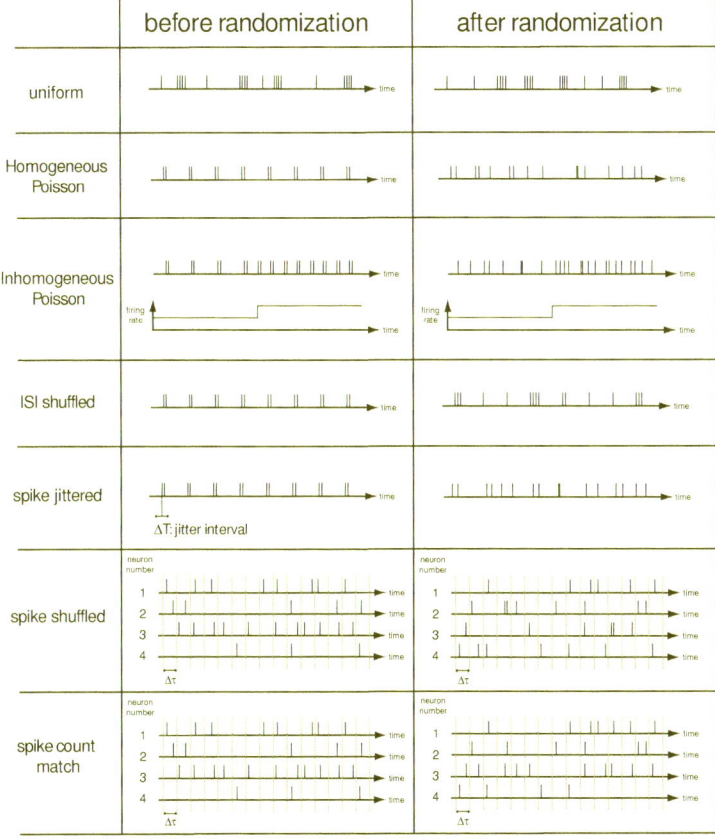

Figure 6.3: Schematic overview of different methods for randomization (see text).

The background models differ in the degree of randomness they impose on the original data. This affects the significance of found patterns when compared with the statistical null: In the studies mentioned above, all backgrounds except the last one [379, 627] have led to the conclusion that patterns have been found. Therefore, when comparing such studies one

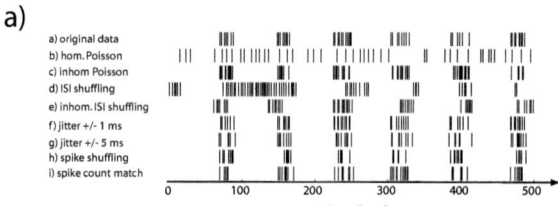

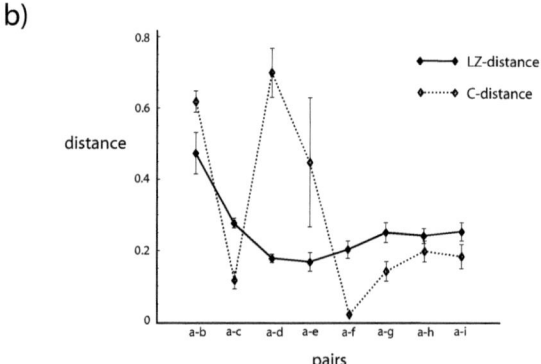

Figure 6.4: The degree of randomization introduced by shuffling procedures shown at spike trains of simple cells of V1 of macaque monkey: a) original spike train and shuffled spike trains. b) Mean LZ-distance and C-distance (see section 7.1.3) between original and shuffled spike trains.

has to take into account which way of randomization has been used. Also the use of pattern detection methods should include an assessment of the degree of randomness introduced by the shuffling procedures (see example 6.2).

Example 6.2: To demonstrate the degree of randomness introduced by the different shuffling procedures, we use a set of 10 trial spike trains obtained from simple cells stimulated with drifting ratings (see section 8.1 for experimental details). We applied all (except the first) procedures to the spike trains and calculated the LZ-distance and the C-distance (see section 7.6) between the original spike train and the shuffled train. The LZ-distance serves as an estimator for the randomness imposed on interval patterns. The C-distance serves as an estimator in respect to timing patterns. The larger the distance, the more random is the shuffled train compared to the original one. Figure 6.4 displays mean and variance of the distances between the pairs of spike trains for the different randomization procedures. It is obvious that both the randomization procedure and the criterion (distance measure) used to its evaluation matter: The homogeneous POISSON randomization (a–b) is probably an inappropriate randomization procedure for timing and interval patterns, as the degree of randomness introduced is significantly higher compared with the other procedures. The same argument holds for both variants of ISI shuffling (a–d and a–e) when timing patterns are object of analysis. Finally, the rather complex spike count match randomization procedure does not seem to impose a significantly different degree of randomization compared with most other procedures and can thus be replaced by alternative and simpler methods.

6.3 Noise and Reliability

6.3.1 What is Noise in Neuronal Systems?

From the perspective of the observer of a system, two classes of noise are distinguished [536]: *Measurement noise* (or additive noise) refers to the corruption of observations by errors that are independent of the dynamics of the system. Measurement noise is a unwanted noise component and one tries to minimize it by optimizing the experimental design and measurement instruments, so that no relevant information about the system is lost. *Dynamical Noise*, in contrast, is a result of a feedback process by which the system is perturbed by a small random amount at each time step. This noise results due to stochastic elements in signal generation. In spike trains, dynamical noise may affect the probability of spike appearance as well as spike timing. There is, however, a different perspective on the problem of neural noise, as the distinction between 'signal' and 'noise' is a distinction usually made from the observer's perspective. What appears as noise for the observer, may have importance for the neuron. In other words, from the 'neuron perspective', the important question is whether noise can have any functional role for a neuron. This perspective has already been formulated in the mid 20th century as we have shown in the historical analysis (see section 3.2.2).

To assess the influence of noise, one distinguishes between internal and external noise. By *internal noise*, we refer to the variability of the spike generating processes (channel gating, ion fluxes etc.) that are attributed to the neuron under investigation. *External noise* is attributed to the variability of (at least) three aspects: First, the (possible) noisy character of the input which emerges from the activity of all neurons projecting to the neuron under investigation. Second, the activity of glia cells that are known to influence neuronal firing [476]. Third, fluctuations in the extracellular chemical environment of the neuron under investigation. The influence of these external noise sources is difficult to measure. Especially the second and third aspect are seldom investigated, whereas the first aspect is analyzed unter the general notion of 'background activity' (see below). It is an open question whether internal or external noise sources have a 'purely' random basis (i.e. referring to quantum effects) or express a higher-dimensional deterministic part of the dynamics. One can suspect that especially external noise sources are of the latter case [642, 730].

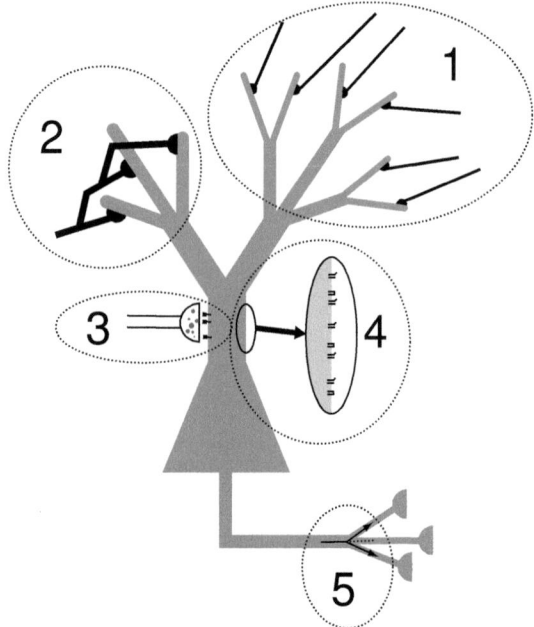

Figure 6.5: Schematic overview of neuronal noise sources: 1) Background activity. 2) Input delay variability. 3) Synaptic noise. 4) Channel noise as an example of membrane noise. 5) Conductance failure at axonal branching points as an example of conductance variability.

6.3.2 Noise Sources in Neurons

In the historical part we have demonstrated that the 'noisiness' of neurons became a controversial topic as soon as the information vocabulary was applied in neuroscience. Later, several neuronal sources of noise have been investigated. The underlying goal of these investigations is to estimate to what extent these noise sources change the timing of spikes, suppress spikes that 'should' occur due to the stimulus characteristics or generate spontaneous spikes. Today, a rather complete picture of neuronal noise sources is available (see Fig. 6.5). We list these sources according to the pathway of incoming signals and concentrate our discussion on cortical neurons:

1. **Background activity:** Background activity is a component of noise in cortical neurons for *in vivo* conditions. Although also glia cell activity and fluctuations in the

chemical *millieu* of the cell fall into this category, usually only the input a cortical neuron obtains (in humans $\sim 10^5$ synapses contact one neuron) is taken as the main component of the background activity. The number of spikes per time unit a cortical neuron receives, depends on a number of factors like cortical area and layer. One estimates that a neuron receives several hundred inputs for each spike it emits [666]:70.

2. **Input delay variability:** Whenever a presynaptic neuron contacts a postsynaptic neuron with several synapses, the action potential of the presynaptic neuron will not instantly affect the postsynaptic neuron, but the impact will be spread over some time due to the delays in axonal transmission [561].

3. **Synaptic noise:** Stochastic elements in synaptic transmission are one of the first neuronal noise sources that have been studied in more detail [79, 131]. Synaptic noise subdivides into the following aspects [385, 728]:

    - Spontaneous release of neurotransmitter
    - Random nature of neurotransmitter diffusion
    - Chemical reactions in the synaptic cleft
    - Varying number of released vesicles during stimulation
    - Varying size of released vesicles.
    - Varying number of postsynaptic receptors activated by transmitter release

4. **Membrane noise:** Membrane noise is a collective term for several aspects that affect the cell membrane. The most important component is called *channel noise*, the randomly opening and closing of (usually voltage gated) ion channels [71, 755]. A minor noise source in membranes, called *Johnson noise*, is caused by thermal effects. Certainly, also small lesions in the membrane are a noise source.

5. **Conductance variability:** The current flow within the neuron as well as axonal action potential spread may be affected by the internal cellular structure (cytoskeleton etc.). This aspect has been, however, rarely investigated (we found no reference on this matter). One aspect of conductance variability is caused by axonal branching points, as sometimes the action potential does not reach some of the synaptic terminals of the axon and is unable to induce a reaction at the postsynaptic site [417].

Experiments to investigate the significance of each noise source are not easy to conduct. It is obvious that there is a large difference between the *in vitro* and the *in vivo* condition, as in latter case the background activity will play a major role [523]. Usually, the term 'spontaneous activity' is used when referring to the spiking activity of a single neuron driven by background noise [603]. The role of the background noise for the variability of neuronal firing will be discussed in the next section. Concerning the other noise sources, estimations have been conducted in empirical as well as model studies using detailed neuron simulations. Already one of the first (empirical) studies on this issue (1968) concluded that synaptic noise mainly accounts for the observed firing variability [64]. A model study which investigated the magnitudes of the voltage standard deviations of thermal noise, channel noise and synaptic noise in a space-clamped somatic membrane patch confirmed this finding

[591]: Thermal noise and Na$^+$ channel noise were one magnitude lower (2×10^{-2} mV and 6×10^{-2} mV) compared to K$^+$ channel noise and synaptic noise (5×10^{-1} mV and 9×10^{-1} mV). Which aspect of synaptic noise is the most relevant? This question has, for example, been analyzed in hippocampal neurons. Four potential causes for synaptic unreliability have been investigated: Threshold fluctuations at the postsynaptic site, conduction failures at the presynaptic site, temperature, and probabilistic transmitter release. The study concluded that the probabilistic release mechanisms at low capacity synapses are the main cause of unreliability of synaptic transmission [361].

The dominant role of synaptic noise does not mean that membrane noise is unable to induce spontaneous spikes or a failure of firing a spike. Several model studies suggest, that 'additional' or 'missing' spikes also can occur due to membrane noise [424, 659, 755, 756]. This is so because what matters is not the total number of ion channels – this number is high and fluctuations would usually have been averaged out – but the number of channels that are open near threshold – which is considerably smaller. Moreover, the membrane patches involved in spike generation or transmission, such as nodes of RANVIER, also contain only small numbers of channels, so that fluctuations in channel gating cause fluctuations in firing. Counteracting channel noise needs an increase in the number of ion channels. This implies a large energetic investment for the cell, not only because of the costs of creating the channel but also because large ionic currents imply a greater accumulation of ions and thus increase the demand of ion pumps. As the energetic costs per action potential are considerable,[3] channel noise is probably the result of a trade-off between the energetic costs assigned with higher channel numbers and the reliability of firing. The role of input delay variability has been investigated in experimental as well as in model studies [410, 597]. These studies concluded that the standard deviation of the output jitter in spike timing is linearly related to the standard deviation of the input jitter with a constant smaller than one. Furthermore, anatomical studies (reviewed in [561]) suggest a interbouton-distance of around 15 μm, implying a delay of only a few μs. Therefore, input delay variability has probably no decisive role as a noise source.

6.3.3 The Reliability of Neurons

The historical analysis demonstrated that neuronal '(un)reliability' has been analyzed as soon as an experimental setup that provided stable measurements of spikes were available (1930s, see section 2.3.3). It later became an important issue when the question emerged how a reliable system behavior could emerge out of unreliable elements (see section 3.2.5). In other words, if we discuss the reliability (or variability) of neuronal firing, we are judging the performance of the neuron from an observer-point-of-view. The basic idea is to expose a neuron to identical stimulus conditions and to see, whether identical responses result. Sometimes, the term 'variability' of firing is used to classify *single* spike trains in the sense that 'variable' trains are POISSON trains (e.g. in [523]). We, however, suggest not to use the terms 'variability' or 'reliability' for this type of problem (i.e. the question, what degree

---
[3]In a study on photo-receptors and interneurons in the fly compound eye [553], the energy consumptions in terms of ATP molecules has been determined. It has been estimated, that the cell needs 10^4 ATP molecules to transmit a bit in a chemical synapse and $10^6 - 10^7$ ATP molecules for graded signals in an interneuron or a photoreceptor. These numbers are five to eight orders of magnitude higher than the thermodynamic minimum required for these operations.

of randomness a spike train has). These concepts should refer to the problem, whether a neuron is able to maintain a certain firing behavior in different trials. For this problem, different measures of reliability have been proposed:

1. The simplest measure or reliability is the **spike count** in a certain time window, respectively the FANO factor of the spike count [277] (alternatively, the coefficient of variation of the spike count is used as a measure of variability).

2. Another possible measure is to evaluate, if a certain **firing rate pattern** is preserved. The simplest such pattern is to compare the peak firing rates [654].

3. A common measure refers to the reliable **spike timing** of some particular spikes, which is usually measured based on the post stimulus time histogram (PSTH). This measure needs a criterion to identify the 'important spikes' whose reliable preservation of spike timing is considered as relevant. Usually, this is done by smoothing the PSTH by some filter function and to define a threshold value to determine the important spikes. The reliability is then measured as the fraction of spikes that lie in the 'important peaks' of the PSTH and the precision is defined as the mean standard deviation of the spike timings for all spikes that are part of these 'important peaks' [588] (a similar approach where the spikes of different trials are projected to one train has been proposed in [725]). An alternative measure only concerns the timing of the first spike (reliability of response latency) [616]. If the spike timing of all spikes is considered, distance measures are applied (see section 7.6).

4. Finally, the reliable presence of certain **interval patterns** could serve as a measure for reliability. In this respect, distance measures that take the temporal order of spikes into account could serve as a measure. We will present this approach in more detail in section 7.6. (a comparable approach is the use of the entropy of spike trains [453]).

These measures of reliability fall into the context of the neural coding discussion: Can a neuron maintain a certain encoding of (sensory) information? This addresses the general problem whether a neuron is able to transmit information reliably. An alternative concept of reliability refers to neural computation. Here, the question is: Can a neuron always perform the same computation when the set of inputs the neuron is confronted with corresponds to the same type of computation? To test this variant of reliability, one has to find a set of different inputs that nevertheless leads to the same output (in one of the four meanings above). To analyze to what extent a certain input affects the reliability of computation, the following test procedure has been proposed [442]: Choose a time interval, bin the time axis using this interval and jitter the spikes within the borders of each bin. As long as the choice of the partitioning interval does not change the reliability of the output of the system, one is within the class of inputs that generates the same output. Note that reliability in this context also can mean that a neuron always generates a *random* output for a large class of inputs. Such a neuron could be considered as a 'reliable noise generator'. In other words, the concept of reliability is dependent on the context one is interested in — neural coding or neural computation. The determination of the reliability of neurons in experiments has to consider several practical problems:

1. One will expect a difference in reliability under *in vitro* or *in vivo* conditions. It is generally assumed that this difference results from the higher background activity, and is not due to changes in internal noise sources in the neuron [523].

2. The design of the experiment is also important. Measuring the reliability of response for directly injected current inputs will lead to different results than measuring the reliability for stimuli the neuron receives via the sensory pathway. There is the possibility that variability emerges due to overlooked sources of variability in the experimental design. For example, a study showed that eye movements cause a large component of response variance in alert monkeys [515]. It is thus crucial that eye position during data collection is carefully monitored, when doing such studies in the visual system. Besides, one has also to take into account the firing rates one obtains, as for high firing rates, spike timing reliability could be a result of the refractory period of neurons [389].

3. The kind of stimulus used will have an impact on the result of reliability estimations. First, the stronger the stimulus the more reliable the neuron reacts (for example strong contrast when using visual stimuli) [644, 749].[4] Second, constant stimuli lead to imprecise timing, while aperiodic stimuli yield precise spike timing. This has been shown in model studies [402, 516] as well as in several experiments (H1 neurons in the visual systems of flies [453][5], in rat cortical neurons [588] and in *Aplysia* neurons [48]).

4. The results of reliability measurements differ when measured in different individuals. For example in fly motion-sensitive neurons it has been shown that the inter-individual differences are large, they differ by a factor up to ∼3.5 [749]. Also in *Aplysia*, the differences in variability between animals are substantially larger than the trial-to-trial differences in one animal [758].

The fact that reliability can refer to different measures and that there are several practical difficulties to consider when measuring neuronal reliability might explain the nonuniform picture that emerges when different experimental results are compared. For example, using the spike count reliability as a measure, one study carried out in the striate cortex of macaque monkey reported a coefficient of variation of 0.35 ± 0.2 [211]. A study measuring simple cells in cat striate cortex obtained a FANO factor of ∼1.2 [448]. Another study in cat LGN cells obtained a FANO factor of ∼0.3 [575]. Using the spike timing as measure of reliability (standard deviation of mean spike timing), often a high reliability has been found: A study in the retina of salamander and rabbits showed a high reliability, as long as the contrast of the visual stimulus (checkerboard stimulus) was not too low [390]. Studies in cat LGN cells showed a spike timing reliability in the order of 1 ms [575, 645]. Also in higher areas of the visual pathway a markedly reproducible reliability is visible: In the area MT of monkey, it is of the order of only a few (∼3) ms [376, 409]. A review for spike timing reliability in the mammalian visual system showed that the reliability is generally in the order of 1 to 10 ms [375]. The reason why newer studies found higher reliability (especially when considering spike timing reliability) lies probably in the stimulus used. Several experimental and model studies emphasize the importance of fast membrane potential fluctuations for

---
[4]There are exceptions: A study analyzing the response variability in macaque monkey retinal ganglion cells showed that the variability was independent from the amplitude of the stimulus [438].

[5]There is yet some disagreement with this conclusion, see [750].

spike timing reliability [384, 548, 624, 662]. Such fluctuations result when the stimulus has a more 'natural' appearance (aperiodic and not constant, see point 3 in the listing above).[6]

6.3.4 Can Neurons Benefit from Noise?

We now change the perspective and ask the question if the neuron could benefit from the noise it is imposed to. This aspect will play an important role in the hypothesis on neuronal coding and computation proposed in section 6.6. Here, we discuss the influence of background noise on the reliability of neuronal firing. As mentioned, the background activity is made responsible for the major part of neuronal firing variability *in vivo* [369, 374, 388]. The background, however, is not irregular in the sense that it can be averaged away by some pooling process. Rather, it represents a spatially and temporally coherent state of the cortical network that continuously fluctuates and plays a crucial role in regulating the response strength of cortical neurons to sensory inputs [373, 422, 479]. Furthermore, model studies suggest that background noise might also enhance the precision of spike timing [763]. Several other studies showed that the background noise a neuron receives correlates with stimulus properties and is thus potentially an additional information source for the neuron [745, 749]. Finally, background noise may have furthermore supportive functions in sensory systems for several specific purposes (enhancement of contrast invariance of orientation tuning [365], improved encoding of the stimulus waveform in the auditory system [569]). Such findings have been recently connected to a concept developed in physics: stochastic resonance.

> **Stochastic Resonance:** The phenomenon of stochastic resonance has been found in noise driven, periodically modulated bistable systems. When the intensity of the external noise is properly tuned to the internal parameters of the bistable system, then the external noise and the periodic driving mechanism interact by pumping power from the whole noise spectrum into a single mode which is coherent with the signal. This produces a well defined peak in the power spectrum of the output signal and the corresponding signal to noise ratio has a maximum. For larger noise amplitudes the signal is increasingly corrupted. In the past years, this concept of stochastic resonance has gained much attention in biology. Several authors suppose that stochastic resonance plays an important role in the nervous system [728, 754]. This speculation is supported by physiological experiments and model studies: The phenomenon of stochastic resonance has for example been demonstrated for voltage-dependent ion channels [391] and crayfish mechanoreceptors [463]. Model studies showed the phenomenon of stochastic resonance in pattern detection experiments using holographic neurons [702] or in experiments using the FITZHUGH-NAGUMO model [589]. They show how neurons may use noise in order to enhance their sensitivity to detect external stimuli. One model study gave more insight into the detailed process of stochastic resonance on the level of the membrane. In a model study with HODGKIN-HUXLEY-type neurons, the influence of noise on the detection and timing precision of neural signals has been investigated. It has been shown, that the timing precision is improved by internal noise sources only for deterministically subthreshold stimuli [634]. In a detailed model of a morphologically reconstructed neocortical pyramidal neuron, it furthermore has been shown, that the

---

[6] Some authors claim that the increased spike timing reliability could be caused by a 'resonance' when the input contains a frequency equal to the 'natural' firing rate of the neuron [525, 526] — a proposal, which we will not discuss further.

introduction of voltage fluctuations increased responsiveness. This again is considered as a possible example of stochastic resonance [522].

6.4 Functional Aspects of Spike Patterns

In this section, we discuss some main theoretical frameworks, which may explain the presence and functional role of patterns. Early investigations (1960s) distinguished two causes for temporal patterns found in neuronal spike trains: Either neurons follow patterned time cues of stimuli, or patterns result from intrinsic timing cues (e.g. pacemaker cells) [54]. This basic distinction that either aspects inherent to neurons (like refractoriness) or aspects of the stimulus cause the presence of patterns, is still made [400]. It, however, disregards the role of the neuronal network in which the neuron is embedded as a specific connectivity may favor certain types of patterns and exclude others [704]. Thus, patterns may result from three general causes: The temporal structure of the stimulus, the biophysical properties of neurons, or the network connectivity. This classification of potential causes does not imply a functional classification. For example, stimulus aspects may cause patterns which are functionally irrelevant, as there are no neurons able to use the pattern for any functional purpose. We discuss two thematic fields: First, we review two theoretical frameworks that explicate, why neurons produce patterns: the synfire chain hypothesis and the temporal correlation hypothesis. Second, we sketch two proposals that illustrate, how neurons could make use of patterns: Coincidence-detection and spike-time dependent plasticity.

6.4.1 Causing Patterns: Network, Oscillation, Synchronization

The very early neural models have been based on sophisticated network structures that implement logical functions (see section 3.2.4). In such structures, e.g. in the delay line model of STREHLER [243], patterns obtain a clear functional relevance as causers of 'neuronal switching', such that only specific patterns cause a firing of the postsynaptic cells. The assumption of a sophisticated connectivity on the micro-scale has been challenged already in the mid 20th century and was gradually replaced by a 'statistical perspective' on network connectivity. Today, it is well known that there is a specific connectivity within different areas of the brain (e.g. in the visual system [362]). Although the anatomically established connectivity is not identical to the functional connectivity, an intricate mutual relationship between anatomical and functional connectivity exists. For example, using methods from graph theory, it has been shown that distinct anatomical motifs are associated with particular classes of functional dynamics [688]. Also on the micro-scale, the hypothesis of a 'random connectivity' between neurons has been sophisticated by investigating the intralayer connectivity and neuronal micro-circuits [462].

Local Connectivity in Cortex: In humans, neocortex covers an area of around 2600 cm^2 and has a thickness of about 2-4 mm, The total number of neurons in neocortex is estimated as 28×10^9, the number of glia cells is comparable (only in neocortex!). The number of synapses is estimated as 10^{12}. Two basic types of neurons can be distinguished: those whose dendrites bear spines (stellate and pyramidal cells, these are excitatory cells) and those whose dendrites are smooth (smooth cells, these are inhibitory cells). Occasionally, sparsely spiny cells have been described, but these neurons form a very small subclass of cortical neurons. The overall proportions of these types remain approximately constant between different areas: 80 % spiny cells (70% of which are pyramidal cells) and 20% smooth cells. The basic structural unit of neocortex is called *minicolumn* with a diameter of 50 to 60 μm and a total number of 80-100 neurons (in visual cortex, the number is 2.5 times larger). About 80 minicolumns form the ontogenetic unit in the developing neocortex: the *column*. The diameter of a column varies between 300-600 μm [460, 611].

Traditionally, six neocortical layers have been distinguished by neuroanatomists. This has been subject to some lumping and splitting: Layers 2 and 3 are often lumped together, whereas (in visual cortex), layer 4 has been subdivided in two to four sublayers. Concerning connectivity on the micro-scale, the inter-layer and intra-layer connectivity are of interest. The inter-layer connectivity between excitatory cell types is summarized as follows: The input originating from thalamic relay nuclei mainly reaches layer 4. Layer 4 mainly projects to layer 2/3. Neurons in layers 2/3 project to layer 5, to other layer 2/3 neurons in distant cortical areas and to layer 4 neurons. Neurons in layer 5 project to neighboring layer 5 neurons that project back to layers 2/3, to layer 6 neurons, that project back to the thalamus and to layer 4, and finally to subcortical areas. Layer 5 is thus expected to be the main output area [416, 459, 636, 719].

Furthermore, there is a specific intra-layer-connectivity, which has been extensively analyzed, e.g. in the barrel cortex of rats. The most information is available for layers 4, 5 and 2/3. In layer 4 (rat barrel cortex), around 70 % of the excitatory cell are spiny stellate cells and 15 % are pyramidal cells. These cells are highly connected within a single barrel: 91% of the axons and 97% of the dendrites of neurons remain within the home barrel, and also the analysis of functional connections showed, that synaptic transmission appears to be restricted to a single barrel. This motivates the claim that a barrel acts as a single processing unit. Pairwise testing of neurons with distances below 200 μm show, that one fifth of the pairs were directly connected. The probability of connectivity of a single excitatory neuron in a barrel is estimated in the order of 0.31-0.36. It is furthermore claimed, that L4 excitatory neurons mediate a reliable and strong feed-forward excitation of thalamic signals within L4 and to L2/3, so that layer 4 acts as an amplifier of thalamo-cortical signals. [475, 474, 580, 636]. The layers 2/3 receive their input mainly from layer 4. The (functional) connectivity of the pyramidal cells has been estimated as 0.09. More than 70 % of all synapses on a layer 2/3 pyramidal cell originate from other pyramidal cells within a radius of 600 μm [519, 619, 691]. Layer 5 is the major cortical output layer. The two major layer 5 pyramidal cell subtypes are intrinsically burst spiking and regular spiking. It has been estimated that each pyramidal neuron is connected to at least 10% of the neighboring pyramidal cells by fast excitatory glutamate synapses. It has been proposed, that (tufted) layer 5 pyramidal neurons form local networks of around 300 μm diameter consisting of a few hundred neurons, which are extensively interconnected with reciprocal feedback. Statistical analysis of the synaptic innervation suggest that the network is not randomly arranged, the degree of specificity of synaptic connections seems to be high [474, 594, 595, 664].

There is an ongoing debate, how precisely the connectivity pattern within a micro-circuit replicates [689] and to what extent neocortical micro-circuits discovered in primary sensory areas generalize to the whole cortex [459]. Furthermore, it is not clear, how fast the network connectivity can change, e.g. due to learning. Recent studies showed rather discrepant results [631]: Two studies in mice revealed different surviving times of spines. In one study [727], even the stablest pool (60% of spines) had a limited lifetime of about 100 days (barrel cortex), whereas in a second study [512], about 96% of spines remained stable for over a month, translating to a half-life of more than 13 months. This implies that synapses could persist throughout a mouse's lifetime (visual cortex). It furthermore remains open if the time-scale of changes on the anatomical level is the same as the time scale of changes on the functional level. The question whether the structure of a specific micro-circuit prohibits certain spike patterns is probably hard to answer. Model studies suggest, that already the positioning of a single synapse along the distal-proximal axes may matter. Earlier studies claimed that the size of the individual somatic EPSP is independent of the dendritic input location [585, 584]. More recent model studies, however, argue that this independence is only valid for the *in vitro* case and that *in vivo*, distal synapses become weaker than proximal synapses [579]. This results from a several-fold increase in dendritic membrane conductance due to the activity of many synapses *in vivo*. They concluded that the dendritic location (proximal versus distal) of even a single excitatory synapse activated in the presence of massive background synaptic bombardment can have a noticeable effect on axonal spike output (shifting or disappearing of spikes) [577].

Due to the difficult questions that emerge when the micro-structure of the neuronal network is related to the occurrence of specific temporal patterns, statistical approaches are used. They assume a locally random connectivity based on a certain probability of connectivity, but require some ordered connectivity pattern on the level of groups of neurons. A famous hypothesis of this type is the *synfire hypothesis*.[7] The focus of the synfire hypothesis is the transmission of information by putting emphasis on synchronous firing events, but it doesnot provide a scheme of how this information is processed.

The Synfire Hypothesis: The synfire hypothesis claims the existence of networks of groups of cells with converging connections between these groups (synfire chains) [350, 352, 358]. Each link in a synfire chain consists of a set of synchronously firing neurons. Each set excites synchronous firing in the next set, which in turn excite the next set of neurons synchronously etc. The prerequisites of the hypothesis are: 1) The cells in the sending node are indeed firing synchronously. 2) The synapses must be strong enough to ensure synchronous firing of the cells in the receiving node. 3) A mechanism exists that prevents an accumulation of a small jitter in firing times of the synchronous volleys. Synfire chains imply the existence of multi train patterns of single spikes that stretch over time intervals of up to hundreds of milliseconds. Several claims about the existence of such patterns have been made (see section 6.5). Some of these studies explicitly tested whether the detected multi train patterns are consistent with the synfire chain hypothesis. In one study, two alternatives have been investigated [639]: Either the patterns are the result of slow dendritic processes [381] or they emerge as the result of superposition of periods of regular firing across different neurons. The first alternative has been considered as unlikely, as it is hard to imagine that a neuron is able to 'remember' a given spike for a delay of hundreds of milliseconds and to fire a second, precisely ($\pm$ 1 ms) timed spike, while (between these two spikes) the same cell fires other spikes. The second alternative leads to the expectation, that mostly single spike patterns should be detected, which was not the case in the study. Furthermore, there was no excess of intervals in the range of 15-100 ms: the corresponding periods of the observed *in vitro* oscillations, making it unlikely that such oscillations were present. The authors concluded that the patterns found are best explained by the

---

[7]In 1953, GEORGE H. BISHOP and MARGARET CHALRE proposed a model of propagation of neuronal excitation that may be called a forerunner of the synfire model of ABELES, see [34]: Figure 3.

synfire hypothesis. A second, *in vitro* study in rat V1 investigated spontaneous firing and found synchronous spikes correlated among networks of layer 5 pyramidal cells [592]. These networks consisted in 4 or more cells. Sets of neurons are also sequentially activated several times. These findings are compatible with the synfire hypothesis. In a further study in cat LGN it has been shown that neighboring geniculate neurons with overlapping receptive fields of the same type (on-center or off-center) often fire spikes, that are synchronized within 1 ms [363]. They project to a common cortical target, where synchronous spikes are more effective in evoking a postsynaptic response, which is again compatible with the synfire chain model. Finally, the plausibility of the synfire hypothesis has been tested in several model studies. One study came to the conclusion, that a near-synchronous wave of activity can propagate stably along a synfire chain [520]. A second model demonstrated how a synfire chain of excitatory neurons in combination with two globally inhibitory interneurons that provide delayed feed-forward and fast feedback inhibition recognizes order patterns (a certain sequence of neurons that fired) [530].

Generally, synchronization as well as oscillations in neuronal firing are phenomena, that are considered important for temporal representation and long-term consolidation of information in the brain (see [413] for a recent review). First, we briefly discuss oscillations. **Oscillations** express themselves via periodically appearing interval patterns in single spike trains. For the general case, we can define oscillations as follows

Definition 6.14 *A single spike train which is described as the repetition of a interval pattern* $\{(\bar{p}_{1,j}, X_{1,j}, \Delta X_{1,j}), \cdots, (\bar{p}_{l,j}, X_{l,j}, \Delta X_{l,j})\}$ *of length l is said to reflect an* oscillation *of periodicity l.*

A more detailed introduction into oscillations within the framework of dynamical systems theory is provided in section 6.6.1. It is well-known that several types of neurons are capable to generate intrinsic oscillations [576]. Oscillations appear in several functional units of the nervous system: the olfactory system, the visual system, the hippocampus and in the somatosensory and motor cortices [507] (some further examples will be provided in section 6.5). Their functional role is usually associated with the possibility that oscillations may serve as a timing framework necessary for ensembles of neurons to perform specific functions [414, 478, 717]. A study in rat cortex for example demonstrated, that oscillations induce sinusoidal subthreshold oscillations, and thus periodically recurring phases of enhanced excitability of the neuron [746]. Weak and short-lasting EPSP evoke discharges only if hey are coincident within a few milliseconds with these active membrane responses. Thus, oscillations seem to impose a precise temporal window for the integration of synaptic inputs, favoring coincident detection. But the detection of oscillations does not imply *per se* that they have a function. One example, where oscillations are probably not functionally relevant, are oscillatory patterns in the visual system due to the phase locking of neuronal responses to the refresh rate of computer display monitors, on which visual stimuli are displayed. Such oscillations are a well-known artefact in visual experiments [483]. In the cat visual system, for example, it has been shown that almost all LGN cells and substantial fractions of simple and complex cells are sensible for a refresh rate of 60 Hz [757]. It is also known that retinal and LGN neurons of cats can lock to the flicker frequencies from fluorescent tubes (100 Hz), which result of the 50 Hz alternating current current [472]. Another study in V1 of macaque monkey demonstrated that more than 50% of all measured neurons synchronized to the refresh rate of 60 Hz [604]. This type of oscillations, however, does not seem to have perceptual consequences [734][8] – but this temporal artefact has a significant influence when analyzing single and multi train patterns in the visual system, especially when cross-correlation techniques are used [757].

Synchronization is always a multi train phenomenon and shows up as a succession of unitary events. The basic problem when claiming the occurrence of synchronization is the same as in pattern detection in general: One has to determine a time interval within which spikes are called 'synchronous' [734] and one has to compare the empirically derived number of unitary events with the number obtained after a certain type of randomization [503]. Synchronized oscillations show up as unitary events of a certain periodicity and are thus a higher order pattern. Alike oscillations, synchronization in neuronal firing has been found in the visual system (retina, LGN, V1 [734]), the olfactory system – where (as shown in bees) the blocking of synchronization impairs the discrimination of molecularly similar odors [713] and (as shown in locust) neurons capable to read out the synchronized oscillatory firing of groups have been found [582] – and the motor system, where (as shown in macaque monkey) synchronization ($\Delta X_{i,j}$ = 5 ms) is associated with distinct phases in the planning and execution of voluntary movements [650]. Basically two possibilities concerning the functional role of synchronization are discusses: It might allow a reliable information transfer and offers an additional coding dimension for increased information transfer. The first possibility is discussed within the synfire chain model so that we focus here only the second aspect. Generally, as neuronal responses may synchronize without a rise in neuronal firing rate, an additional coding dimension is provided [452]. In vision, this might solve the 'bottleneck problem': As the number of possible patterns of unitary events largely exceeds the number of ganglion cells, this might allow the retina to compress a large number of distinct visual messages into a small number of optic nerve fibers [606]. Experimental studies have shown that (in salamander retina) nearby neurons fire up to 20-fold more often synchronously than expected, given the fact that they share the same input ([607], $\Delta X_{i,j}$ = 20 ms) and that (in cat LGN) considerably more information can be extracted from two cells if temporal correlations between them are considered [441] (the average increase is 20 % for strongly correlated pairs of neurons). The additional coding dimension is also expressed by the idea that transiently synchronized groups of neurons reflect a read-out of a certain spike pattern. A model study demonstrated, that the pattern of synaptic connections can be set such that synchronization within a group of neurons occurs only for selected spatiotemporal spike patterns [524]. However, also synchronization can appear without having any functional role. For example, synchronization may emerge as a mere coincidence when measuring two pacemaker neurons or may result from a common input source [179, 734]. Difficulties may also arise with the classical test used for detecting synchrony, the cross-correlogram. Based on this test, a peak at zero time is taken as evidence for synchronous firing of two neurons. This condition, however, is not sufficient. A study using model LGN neurons demonstrated that also in the absence of any synaptic or spike synchronizing interaction between two geniculate neurons, slow covariations in their resting potentials can generate a cross-correlogram with a center peak. This can erroneously be taken as evidence for a fast spike timing synchronization [407].

---

[8]One the other hand, these findings may be a neurophysiological basis for the discomfort reported by individuals working in environments illuminated by fluorescent tubes or for more severe effects like photosensitive epilepsy, headaches and eyestrain [537, 581].

The Temporal Correlation Hypothesis: A theoretical framework where both oscillations and synchronization gain a functional interpretation is the temporal correlation hypothesis. The hypothesis addresses the following problem: If an assembly of neurons codes for a single perceptual object then a mechanism should bind the different sensorial modalities of the object together ('binding problem'). The basic idea (first proposed 1974 [161]) is that the discharges of neurons undergo a temporal patterning and become synchronized if they participate in the encoding of related information [747]. The following predictions derive from this hypothesis [677]: 1) Spatially segregated neurons should exhibit synchronized response periods if they are activated by stimuli that are grouped into a single perceptual object. 2) Individual cells should be able to rapidly change the partners of synchronous firing, if stimulus configurations change and require new associations. 3) If more than one object is present, several assemblies separated by different synchronization patterns should emerge. Several studies claimed to have found support for these predictions: In the primary visual cortex (cat, monkey), neuronal oscillations in the gamma frequency band (40-60 Hz) have been found that occur in synchrony from spatially separated neurons (same column, across columns, and between the two hemispheres), that the synchronization is influenced by global stimulus properties, and that set of cells with overlapping but differently tuned receptive fields split into two independently synchronized assemblies [465, 470, 471, 508]. These and other findings (reviewed in [469, 505, 637, 678, 679]) made the temporal correlation hypothesis a widely debated proposal.

Severe critique was raised against the temporal correlation hypothesis. It appeared difficult to replicate the results reported by the SINGER and ECKHORN group. Either oscillations were not found or, if they were, they appeared to be rhythmic responses not correlated to the stimulus [759, 484, 726], but rather resulting from spontaneous oscillations of a subpopulation of retinal ganglion cells [501]. Recently, a careful experimental study in area MT that tested whether neuronal synchrony constitutes a general mechanism of visual feature binding did not find evidence for the temporal correlation hypothesis [632]. Furthermore it has been criticized that the time necessary to establish an oscillatory group (tenths of milliseconds) is not compatible with the fast reaction times observed [726] (this critique has later been rejected, see [679]) and that oscillations, which are not a response to a visual feature, will seriously disrupt the system, if oscillatory activity is important for the establishment of synchrony [503]. In a detailed critical review in 1999, the temporal correlation hypothesis has been classified as incomplete because it describes the signature of binding without detailing how binding is computed [666]. Moreover, while the theory is proposed for early stages of cortical processing, both neurological evidence and the perceptual facts of binding suggest that it must be a higher-level process. Finally, the architecture of the cerebral cortex probably lacks the mechanisms needed to decode synchronous spikes and to treat them as a special code. These problems may explain why a considerable shift in argumentation of the supporters of the temporal correlation hypothesis has occurred [633]: At the onset of the theory, SINGER insisted particularly on the occurrence of oscillations. Later, in the mid 1990s, the reference to oscillations gradually vanished, giving a growing weight to the occurrence of synchronization. In the same period, SINGER began to develop the hypothesis, that oscillations, besides being a prerequisite, might also be a consequence of synchronization. These, and other, shifts in argumentation demonstrate the controversial character of the temporal correlation hypothesis.

6.4.2 Using Patterns: LTP and Coincidence-Detection

Whoever claims a functional role of patterns has to show how neurons can use them. There are three perspectives on this matter that either focus the synapse, the neuron, or the network. The synaptic perspective refers to the effect of patterns on a single synapse. This aspect was already analyzed in the early 1950s and is one of the first examples indicating how temporal structure might be important for neuronal information processing (see section 3.2.2). Later, this discussion led to the concepts of long-term potentiation and long-term depression. The neuronal perspective concerns the effect of patterns on single neurons. The major theoretical concept in this respect is the proposal that neurons act as coincidence-detectors: They fire with high probability if they receive unitary events as input.[9] The network perspective concerns the effects of patterns on a whole neuronal network. The conceptual and methodological difficulties related with this issue are, however, considerable and we will not discuss it in more detail.

We first discuss the effect of patterns on single synapses. The heterogeneity in release probability of transmitter after arrival of a spike at a synapse is remarkable. Even synapses within an ostensibly homogenous population, such as those arising from a single presynaptic axon and terminating on a single postsynaptic target, have different release probabilities [760]. Changes of the release probability of single synapses over timescales of milliseconds up to seconds are discussed in the context of short-term plasticity [443, 543]. This is also a widely discussed field that we are not able to review in detail. An interesting question, however, is whether the release probabilities of synapses form a continuous spectrum, or if classes of synapses in terms of their release probability can be formed. For the latter possibility, evidence in several systems has been found: In a study [521] in NMDA-receptor mediated synapses of hippocampal rat cells *in vitro*, it has been shown, that two classes of synapses can be distinguished: ~15% of the synapses have a strong release probability (0.37 ± 0.04), the rest has a low release probability (0.06 ± 0.01). A second study in rat hippocampus (glutamate synapses) claimed the existence of two classes of synapses of comparable size in terms or release probability [655]. In the auditory cortex (mice *in vitro*), two such classes have been found, too [371]: In this study 37 synaptic connections in 35 pairs of pyramidal neurons of layer 2/3 (two were reciprocally connected) have been investigated. In 24 connections, transmission (defined as a detectable postsynaptic event) occurred with a low probability (0.13 ± 0.02), in the other 13 connections, the probability was high (0.68 ± 0.02). Generally, more and more indications show that the synaptic strength is not described by a continuum of efficacy, but rather expresses itself in discrete states [610]. Based on these findings one may postulate two classes of synapses, 'reliable' and 'unreliable' ones. What does this mean for the effect of a pattern on the temporal structure of the postsynaptic excitation? 'Unreliable' synapses will certainly fail to transform the pattern into a equally patterned postsynaptic stimulation. But even 'reliable' synapses may achieve this task only for short patterns – a problem that is enhanced when taking conductance failures at axonal branching points into account. To what extent a neuron may use this fact for a 'filtering' of patterns is an open question [417]. Therefore, one must assume that a pattern of single spikes

---
[9]An alternative proposal is that intracellular signalling pathways allow a storage of spatio-temporal sequences of synaptic excitation, so that each individual neuron can recognize recurrent patterns that have excited the neuron in the past [381]. This interesting theoretical proposal is, however, difficult to analyze empirically and we found no experimental paper on this issue.

will usually not be transformed into a postsynaptic stimulation of equal temporal structure. For any theory that relates patterns with an information-bearing symbol that has to be 'transmitted' reliably, this is a severe problem. One possible way out of this problem is to consider burst patterns, as bursts are more reliably transmitted over a single synapse (see section 6.1.2). For the neuromuscular junction, this issue has been analyzed in detail. It is well-known that a burst-input to the neuromuscular junction in *Aplysia*, crayfish, and frog leads to a stronger excitation of the muscle [731, 270, 271]. For *Aplysia*, it has been shown that the mean release of acetylcholine is sensitive to the temporal pattern of firing, even to patterns of time scales much faster than the time scale on which the release is averaged [403, 404, 743]. For cortical neurons, a study in superior cervical ganglion cells of the rat suggests that burst-patterns may determine the type of neurotransmitter which is released (acetylcholine vs. a non-cholinergic neurotransmitter – two substances which regulate the stimulation of catecholamine biosynthesis in the postsynaptic neuron) [528]. Newer studies on the influence of patterns on synapses of nerve cells of the central nervous system fall into the discussion of spike-time dependent plasticity.

Spike-Time Dependent Plasticity: The basic idea of spike-time dependent plasticity is, that a certain pattern of spikes arriving at a synaptic site changes the strength of the synapse. The first concepts studied in this respect are long-term potentiation (LTP) and long-term depression (LTD). They have become very popular in the last decades, as they provide a possible physiological basis for HEBB's postulate [397], whereby a neuronal connection is strengthened if pre- and postsynaptic neurons are active together. Later, the concept of short-term plasticity was introduced, although the distinction between short and long-term plasticity is somewhat arbitrary [543]. The reason why LTP and LTD are linked with the spike pattern discussion is that – to induce LTP in this case – synaptic activation should occur during postsynaptic depolarization. This coincidence of synaptic stimulation and postsynaptic membrane depolarization may be induced by succeeding spikes – but also possibly by action potential back-propagating into dendrites [593]. Therefore, a time window for sequentially arriving spikes exists so that the synaptic changes can take place [449, 500]. It has been shown using repetitive pair-pulsed stimulations, that through a 'delay-line' mechanism, temporal information coded in the timing of individual spikes can be converted and stored as spatially distributed patterns of persistent synaptic modifications in a neural network [393]. There are, however, cell-type specific temporal windows for such synaptic modifications [392].

Early LTP-studies have been undertaken in the hippocampus and used stimuli well beyond the normal physiological firing rates of hippocampal neurons (e.g. [653]). Later studies showed the induction of LTP also for stimulation at frequencies and patterns comparable to those exhibited by hippocampal neurons in behaving animals [552]. Today, LTP and LTD are considered as a general class of synaptic phenomena. Neurons can vary in terms of the specific form of LTP and LTD they express and it has thus been claimed that it is no longer productive to debate the generic question of whether LTP and LTD are synaptic mechanisms for memory [590]. Rather, LTP and LTD are experimental phenomena that demonstrate the repertoire of long-lasting modifications of which individual synapses are capable – but it is very difficult to prove that these modifications subserve essential functional roles. It is clear, however, that these mechanisms could enable neurons to convey an array of different signals to the neural circuit in which it operates [346].

If synchrony is claimed to be important in neural coding, it must make a difference for a neuron if they receive many coincident spikes or not. It is not disputed that such coincidence-detection plays an important role in sensory systems that are able to precisely encode temporal properties of the stimulus. Examples are localization systems for electric fields or sonor waves (e.g. electric fish, bats) [419]. It is more controversial if cortical neurons could also act as coincidence detectors. This discussion started in the early 1980s based on a model of ABELES that related the incoming excitatory and inhibitory postsynaptic potentials to the intracellular membrane potential fluctuations and to the firing rate of a single neuron [357]. From this model, the strength of the synapses was assessed in two ways: the ability of several synchronous presynaptic spikes to initiate a postsynaptic spike (synchronous gain of the synapse) and the ability of several asynchronous presynaptic spikes to add a spike to the output spike train (asynchronous gain). It was found that for the conditions prevailing in the brain's cortex, the synchronous gain is almost always higher. Out of this, it has been concluded that cortical neurons could act as a coincidence detector. Later, several theoretical arguments have been raised in favor of coincidence-detection when compared to the concept of integrate-and-fire neurons [546]: First, neuronal systems utilizing coincidence detection can process information much faster. Considering the adaptive value of processing speed, this provides a teleological argument in favor of coincidence detection. Second, neurons operating as integrators will integrate all incoming activity including noise and potentially misleading signals. Systematic influences are transmitted to other neurons and potentially accumulate along processing pathways, as the information is integrated. A system operating in the coincidence mode could discard a substantial fraction of erroneous postsynaptic potentials. Third, in an network employing coincidence detection, the contribution of individual postsynaptic potentials is much higher because fewer inputs are needed to drive a cell if they arrive in synchronous volleys. Thus, the size of functionally effective neuronal populations can be smaller.

A major point of critique against the coincidence-detector model was, that the membrane constant must be at least one order of magnitude smaller ($\sim$1 ms instead of > 10 ms as measured in cortical neurons) in order that coincident spikes can have an effect as a result of their arrival within a time window of $\sim$1 ms [669]. Model studies, however, claimed that the time window for synaptic integration in passive dendritic trees can be much smaller than the membrane time constant [545], and that the temporal synchronization of excitatory inputs can indeed increase the firing rate of a neuron if the total number of synchronized spikes is not too high [387]. By using another neuronal model it has further been demonstrated, that voltage-gated Na^+ and K^+ conductances endow cortical neurons with an enhanced sensitivity to rapid depolarization that arise from synchronous excitatory synaptic inputs. Thus, basic mechanisms responsible for action potential generation seem to enhance the sensitivity of cortical neurons to coincident synaptic inputs [372]. Although it remains open whether coincidence detection is a phenomenon of which neurons generally make use, there is evidence to claim a role of coincidence detection in the *in vivo* condition.

6.5 Spike Patterns: Examples

In this section we review interval and timing patterns of single spikes, which are hypothesized to be important in neuronal information processing. We focus on sensory systems.

Pattern in the motor system are usually discussed in the context of central pattern generators [415] that cannot be discussed additionally. Several scientist were very active in investigating the functional role of spike patterns. To gain a short overview, we introduce the most important protagonists of the last 20 years (alphabetical order): The Israeli scientist MOSHE ABELES introduced the synfire hypothesis, the coincidence detection model (see sections 6.4.1 and 6.4.2) and was involved in the development of several methods for spike pattern detection, especially for multi train patterns in order to find experimental support for the synfire hypothesis (see section 7.1.2). BARRY CONNORS, MICHAEL GUTNICK and DAVID MCCORMICK investigated neuronal firing in order to find intrinsic firing classes (see section 6.1.2). GEORGE GERSTEIN pioneered the use of statistical methods for spike train analysis (see also section 3.2.3 and introduced the term 'favored pattern', as well as several methods for pattern detection (see section 7.1.2). WILLIAM R. KLEMM and C.J. SHERRY investigated the serial dependence of interspike intervals by looking at patterns of increasing or decreasing interval sequences (order patterns) and claimed that the nervous system processes information on a moment-to-moment basis in terms of 'bytes' of short sequences of spikes with specific patterns of relative interspike duration [541, 672, 540]. REMY LESTIENNE and BERNARD L. STREHLER found interval patterns up to length 5 in a variety of data sets, and postulated that the mammalian brain uses precise patterns of discharges to represent and store specific information [714, 566] – a strong claim that was attenuated in later studies [563, 562].

In the **visual system**, the investigation of spike patterns was either related to the discussion of firing reliability (see section 6.3) or to their potential role in neural information processing. Only the latter aspect will be discussed here. In the vertebrate *retina* (retinal ganglion cells), multi train interval patterns of single spikes are usually investigated [605] – in most cases synchronized firing ($X_{i,j} = 0$) although the allowed interval variance $\Delta X_{i,j}$ can be quite high. This is justified with the low spontaneous firing rate of these cells. One study distinguishes three types of synchronized firing according to the difference in variance [406]: Narrow correlations ($\Delta X_{i,j} \leq 1$ ms), medium correlations ($1 < \Delta X_{i,j} \leq 10$ ms), and broad correlations ($10 < \Delta X_{i,j} \leq 50$ ms). Narrow correlations are postulated to emerge from electrical junctions between neighboring ganglion cells, whereas medium correlations seem to emerge from shared excitation from amacrine cells via electrical junctions. It is believed that the firing patterns in the optic nerve are strongly shaped by electrical coupling in the inner retina. It has been proposed that transient groups of synchronously firing neurons code for visual information. This is supported by a study in salamander retina [660] where it has been estimated that more than 50 % of all spikes recorded from the retina emerged from synchronous firing ($\Delta X_{i,j} \leq 25$ ms) of several ganglion cells – much more than expected by chance. But also other types of patterns may be important: A study focussing on timing patterns in single spike trains of cat retinal ganglion and LGN cells showed that these cells can exhibit a stimulus-dependent jitter ($\Delta T_{i,j} = 5$ ms) [643]. In the *primary visual area*, different kinds of patterns have been investigated. A series of experiments in V1 (and the inferior temporal cortex) of macaque monkey was based on rate and timing (latency) patterns [649, 647]. The stimuli were static WALSH patterns, so that temporal modulations in the response cannot result from temporal modulations of the stimulus. The studies showed that modulations in the rate profile (derived from the PSTH) as well as latency modulations were correlated to changing stimuli and could not be ascribed to changes in the spike count alone. It has been proposed that these modulations serve as representations in a code relation, as the spike count alone underestimates the amount of information contained in the spike train by at least 50% [648]. Single and multi train interval patterns were also found by several investigators: A study in primary visual and medial prefrontal cortex of mice (*in vitro*) claimed the existence of multi train interval patterns of lengths 2 to 9 using different background models (ISI shuffling, spike shuffling) [527]. Another study in cat striate cortex found single spike interval patterns of several lengths in 16 of 27 spike trains, using ISI shuffling as background model [445]. Both studies did, however, not evaluate whether the appearance of patterns is related to changes in stimuli or behavior. Single train interval patterns of length 3 have also been found in the area 18 of the visual cortex of monkeys [714, 566]. More detailed calculations have been performed for data obtained in the fly visual system, especially from the motion-sensitive neuron H1 [651]. One study demonstrated, that two close spikes carry far more than twice the information carried by a single spike [401]. Another study claimed, that the neuron transmits information about the visual stimulus at rates of up to 90 bits/sec – within a factor of 2 for the physical limit set by the entropy of the spike train itself [715].

In the **auditory system** (cochlear ganglion cells, auditory brain stem nuclei, auditory cortical areas), patterns that represent the temporal structure of a stimulus (e.g. phase locking to sinusoidal tones) have been extensively analyzed. We do not intend to review these studies (see [551] for an overview). Far less studies have been made on spike patterns, which are not directly reflecting the temporal structure of auditory stimuli. Studies found multi train interval patterns of length 2-5 in the spontaneous firing of the auditory thalamus of cats [744] and rats [721]. In neither case, the functional significance of this finding has been investigated. Newer studies focussed on patterns that seem to code for sound localization. Several types of patterns have been found (in the auditory cortex of cats) whose occurrence is correlated to a certain orientation of the sound source in space: Single train interval patterns [609], multi train interval patterns [489] and timing (latency) patterns [488]. It is, however, suggested, that the majority of the location-related information which is present in spike patterns relies in the timing of the first spike [488].

Olfaction is a chemical sense, differing from senses which process physical input like photon density (vision) or air pressure / particle velocity (audition). A major distinction is the synthetic property of olfaction, i.e. the ability to assign a specific identity to a great number of component mixtures [555, 554]. In humans, there are as many as 1000 different types of receptors present on the cell surfaces of the olfactory receptor neurons, allowing the discrimination of many thousands of different odors [481]. Therefore, a specific odor is not recognized by translating the activation of a single receptor into spiking activity in a single neuron. Rather, it is assumed that odor identity is encoded in the activity of many cells in the output neurons of the olfactory system (in vertebrates: the mitral cells of the olfactory bulb). Thus olfactory coding is a typical example of a population code. A well-studied system in this respect is the antennal lobe of the locust. One study demonstrated, that individual odors are related to single train interval patterns, which remain stable for repeated presentations (separated by seconds to minutes) of the same odor [556]. Older studies emphasize the important role of coherent oscillations of many neurons (expressed in oscillations of the local field potential) for olfaction encoding [557]. This led to the hypothesis that odors are represented by spatially and temporally distributed ensembles of coherently firing neurons. Alternatively, oscillations may provide the background for establishing a rank order of action potentials produced by neurons participating in coding [752]. In this scheme, the order of recruitment of neurons in an oscillating assembly is stimulus-specific. In the vertebrate olfactory system, the situation is less clear. Although single train interval patterns of length 2 (background: inhomogeneous POISSON) have been detected in the olfactory bulb of rats [562], no correlations between patterns and type of stimulus have been found. Oscillations are often found in the vertebrate olfactory system – usually reflected in the local field potential, or in subthreshold oscillations of the membrane potential of mitral cells in the frequency range of 10-50 Hz [455]. As mentioned earlier (see section 6.4.1), these oscillations seem to trigger the precise occurrence of action potentials generated in response to EPSPs and thus play a role for ensuring spike-time reliability. Newer studies for the vertebrate system also support that the identity and concentration of odors may be represented by spatio-temporal sequences of neuronal group activation [689].

Many examples of spike patterns have been found in the **hippocampus**. Early studies focussed scaled patterns (inequality pattern, see footnote 3 in this chapter), claiming that such patterns show up more frequently than expected (uniform background) during REM sleep, behavioral arousals and learning [47, 596]. Multi train interval patterns have been found in both awake and sleeping rat hippocampus [614]. Furthermore, it has been shown that spike sequences observed during wheel running were 'replayed' at a faster time-scale during single sharp-wave bursts of slow-wave sleep. It was hypothesized that the endogenously expressed spike sequences during sleep reflect reactivation of the circuitry modified by previous experience. Also order patterns have been investigated: A study demonstrated, that correlated activity of rat hippocampal cells during sleep reflects the activity of those cells during earlier spatial exploration [681]. Also the order in which the cells fired during spatial explorations is preserved. An example of a timing pattern, finally, is the well-known phenomenon of phase precession for spatial behavior in hippocampal cells [583]. There, the time frame is given by the theta rhythm (5-10 Hz) oscillation of extracellular current and the pattern consists of the phase of spikes relative to this time frame.

The search for patterns was also extended to **somatosensory systems** and higher cortical areas. An example of a timing pattern emerges in the pain system. Recordings in thalamic neurons responding to noxious stimulation showed that timing patterns emerged (time frame: stimulus onset) that change for different stimulus-intensities [75]. Also in the haptic sense, the existence of spike patterns has been claimed [164, 612]. These single train interval patterns, however, reflect in their ISI the stimulus period (the MEISSNER or quickly adapting afferents). If neurons of somesthetic cortex are measured, these patterns become noisier in the sense that the period of the stimulus is still visible in the autocorrelogram, but no more in the ISI histogram. Single train interval patterns have been found in the recordings of the spontaneous activity of neurons in the *substantia nigra zona compacta*, the *nucleus periventricularis hypothalami*, the *locus coeruleus*, the *nucleus raphe magnus* and the *nucleus hypothalamus posterior* of rats *in vivo* [421, 550]. The studies, however, made no attempt to investigate the significance of these patterns. Finally, several studies looked for the occurrence of multi train interval patterns in relation to behavioral tasks. In a study on the frontal cortex of monkey, where the activity of up to 10 single units was recorded in parallel, excess of patterns was found in 30-60% of the cases ($\Delta X_{i,j} = 1 - 3$ ms) [351]. Different patterns emerged for different behavior of the monkey. A similar result emerged in a study in the premotor and prefrontal cortex [639]. Here, multi train interval patterns ($l = 2$) have been found in 24 of 25 recording sessions, whose occurrence was correlated to behavior. A study in the temporal cortex of rats found multi train interval patterns in a Go/NoGo reaction task, some of which were specific for one condition [721]. All these studies, however, did not make any attempt to causally relate the occurrence of patterns with a certain behavior.

6.6 The Stoop-Hypothesis

We have provided a general definition of the term 'spike pattern' that covers all relevant ways of its use in the neuroscience literature. We have furthermore reviewed the basic theoretical frameworks (neural coding and computation) in which the spike pattern discussion is placed, as well as the main theories and mechanisms, where spike patterns may gain a functional role (synfire chains, temporal correlation hypothesis, coincidence detection, spike-timing dependent plasticity), and we have outlined the main problems associated with the functional use of patterns (pattern stability, neuronal noise, reliability of firing). Finally, we have shown that many findings indicate the presence of patterns in various neuronal systems and circumstances, although their functional significance remained untested or unclear. This shows that a coherent picture of neural coding, neuronal computation, and the role of spike patterns within this context is missing. Theoretical proposals like the temporal correlation hypothesis are controversial or have difficulties to unify seemingly contradictory concepts like noise and precision in firing. We suggest that connecting the degree of precision in neuronal firing necessary such that fast and reliable responses of the organism are possible with the various sources of noise in the system is a major challenge for current theories on neural coding and computation. In the following, we present a hypothesis, developed by RUEDI STOOP and co-workers in a series of publications [704]-[712] in the context of neocortical neuron-neuron interaction (pyramidal neurons). The hypothesis combines noise and precision of firing in an uniform framework of neuronal coding and computation. It postulates that neurons should be described as noise-driven limit cycles, where changes in the background-activity are encoded in specific firing patterns. Existing support for the hypothesis based on model and *in vitro* studies is mentioned in the following description of the hypothesis. At the end of this section, we provide some specific predictions that are connected to the pattern detection problem an that are analyzed further in this thesis.

6.6.1 Neurons as Limit Cycles

The Stoop-hypothesis is based on the well-established finding, that cortical neurons receive synaptic connections from numerous other neurons, whereas many of those synapses are on distal sites. Under *in vivo* conditions, distal synapses are weaker than proximal synapses (see section 6.4.1). Furthermore, many incoming spikes (according to conservative estimations several hundred, see section 6.4.1) are necessary in order to induce a output spike in the target neuron. In an initial step, we assume that the system is in a quasi-stationary state. This means that the neuron receives in average the same (high) number of inputs. Assuming a GAUSSian central limit behavior (other distributions that allow for a well-defined average are also suitable), an almost constant current I results, that drives the neuron [708, 711].[10]

---

[10]The following quantitative data should be available to obtain a realistic estimation of the driving current that a neuron receives under quasi-static conditions: a) The number of synaptic connections a specific target neuron has with its neighborhood. b) Mean and distribution of incoming spikes per time unit. c) The

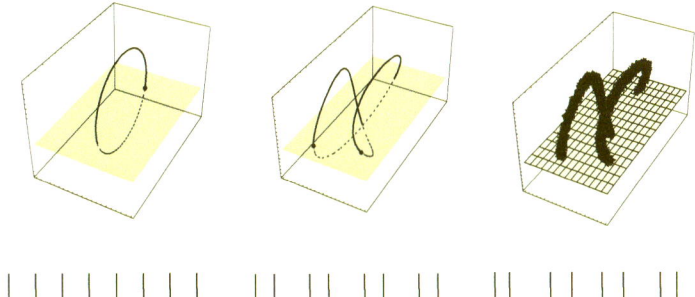

Figure 6.6: Schematic diagram of limit cycle firing and resulting spike train: a) Firing of periodicity 1. b) Firing of periodicity 2. c) Firing of periodicity 2 with noise. The gray plain indicates the POINCARÉ section.

The majority of cortical neurons are regular or fast spiking (see section 6.5), i.e. they fire regularly for constant driving currents. The standard neuronal models have this property, too. Even networks of neuron models driven by constant external stimuli show stable periodic firing patterns (multi train patterns) [531].

Mathematically, this behavior is described as a limit cycle – spatially extended regions in the state space of stable behavior. In other words, the general modelling framework that is used in this hypothesis derives from dynamical systems theory. In the following, we provide a short introduction to this matter (for a detailed introduction, see [368, 635, 694]). In this framework, neuronal firing is captured by trajectories in a state space whose dimension is defined by the number of variables determining the system behavior. If, for example, a neuron spikes regularly in periodicity 1, this shows up as a closed trajectory (= a periodic orbit) of periodicity 1 in state space (see Fig. 6.6). The state of the system is represented by a point moving along the closed orbit so that after each 'round', the neuron spikes. A POINCARÉ section rectangular to the orbit results in a single point – just the result one obtains when the spike train in the interval representation is embedded in dimension 2 (see definition 7.2). In this way, the notion of oscillations of periodicity l (Def. 6.14) gets a straightforward interpretation as orbits in state space of periodicity l. Note the difference between the terms 'period' and 'periodicity': The former denotes the duration until the point travelling along a periodic orbit has finished one cycle (the length of an ISI), the latter denotes the number of loops of the limit cycle. If the limit cycle is attracting, small noise present in the system shows up as a perturbation of the limit cycle: The trajectory

---

distribution of synaptic release probabilities. d) The spatial distribution of the synapses along the proximal-distal axis. e) A function that estimates the amount of synaptic depolarization at the axonal hillnock that results from an EPSP generated at a specific postsynaptic site along the proximal-distal axes. Up to know, these parameters can only be defined completely in model studies, whereas for biological neurons, only incomplete knowledge, rough estimations, or controversial theories are available.

representing the system behavior 'jiggles' around the limit cycle, which shows up in the POINCARÉ section as a cloud of points. Stronger noise means stronger perturbations of the limit cycle, which can lead to a qualitative change of the system behavior.

6.6.2 Locking – The Coupling of Limit Cycles

The most simple generic system that describes the interplay between periodic, quasi-periodic and chaotic movements in state space is the 1-dimensional circle map [368]. In its general case, the circle map is a two-parametric function of the form $f_{\Omega K}(\phi) = \phi + \Omega - K \cdot g(\phi)$ mod 1 (the meaning of Ω, K and $g(\phi)$ will be described below). The circle map displays a characteristic and well-analyzed behavior within the (Ω, K) parameter space, which will provide us with a tool to further analyze the interplay of two coupled neurons. Here, we show in a first step how a coupling (excitatory or inhibitory) of two neurons operating on limit cycles is related theoretically and experimentally to the generic circle map. In a second step, we introduce the phenomenon of phase locking, which will later (next section) provide us with a coding scheme.

Let us assume that two neurons, each displaying an attracting limit cycle behavior, are relatively strongly coupled (unidirectionally). The coupling parameter is denoted by K and it can be rescaled such that $0 \leq K \leq 1$ ($K = 0$ means no coupling and $K = 1$ means critical coupling). We assume (for the beginning) that K remains unchanged. The coupling implicates that the firing of one neuron perturbs the firing of the other, whereas the effect of the perturbation is dependent on the phase ϕ of the perturbing spike in relation to the unperturbed oscillation. The temporal succession of phases is captured by the phase return function $f_{\Omega K}(\phi)$ that has the mathematical form of a circle map [708]: $f_{\Omega K}(\phi): \phi_2 = \phi_1 + \Omega - g_K(\phi_1)$ mod 1, where the parameter Ω is the ratio of the intrinsic period of the targeting neuron divided by the intrinsic period of the targeted neuron (called winding number for $g_K(\phi_1) \equiv 0$, respectively $K = 0$), ϕ_2 is the phase of the next perturbation, if the last perturbation arrived at phase ϕ_1, and $g_K(\phi)$ is the phase response function. The latter measures the lengthening/shortening effect to the unperturbed interspike interval as a function of the phase ϕ_1 and depends on the coupling strength K. From an experiment, the phase return function is derived using the equation $X_p + x_2 = x_1 + X_s$ (see Fig. 6.7), where X_s is the ISI between successive perturbations (resulting from the driving current I_1), X_p is the ISI of the perturbed system (a function of ϕ), x_1 is the interval between the spike of the unperturbed train and the time, when the perturbation has been applied, and x_2 is the interval between a spike of the perturbed train and the next time of the perturbation. Dividing this equation by X_u, the ISI of the unperturbed neuron (driven by current I_2), one obtains the form of the equation:

$$\phi_2 = \phi_1 + \Omega - g_K(\phi_1) \mod 1 \quad \text{where} \quad \Omega = \frac{X_s}{X_u} \quad \text{and} \quad g_K(\phi_1) = \frac{X_p(\phi_1)}{X_u}(K)$$

The relative change of the interspike interval for perturbations of fixed strength K applied at variable phases is measured experimentally. This has been performed for rat pyramidal neurons *in vitro* [658, 709], indicating a linear relationship for excitatory and inhibitory connection. This means, that $g_K(\phi_1)$ is of the form $K \cdot X_p(\phi_1)/X_u$ and the equation has indeed the desired form of a general circle map.

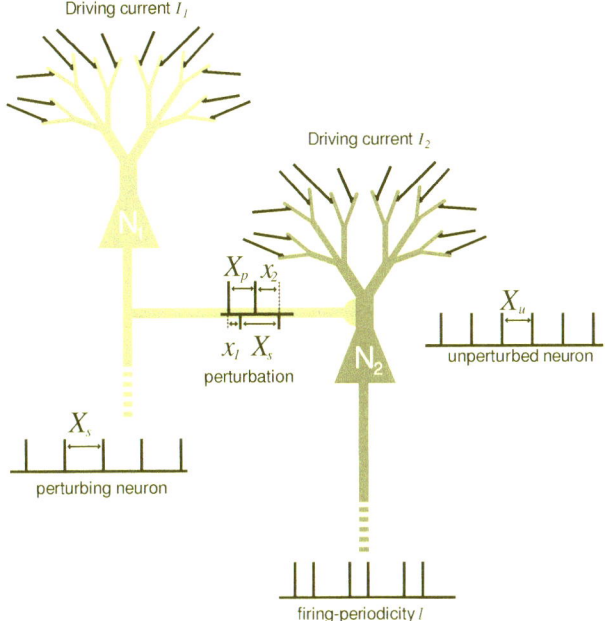

Figure 6.7: A general scheme describing locking between two neurons (see text).

A main objective of dynamical system analysis is to provide conditions for the parameters that determine system behavior for which the behavior is stable. As the circle map describes the interplay of two interacting oscillations with different frequencies ω_1 and ω_2, one wants to know, whether this interplay can lead to a periodic behavior of the whole system for an appropriate choice of Ω and K. This problem has been analyzed extensively by ARNOL'D [16]. From the point-of-view of topology, the circle map can be interpreted as the POINCARÉ-map of a trajectory whose movement on the surface of a torus is described by ω_1 and ω_2. If the ratio ω_2/ω_1 is rational ($\omega_2/\omega_1 = p/q$, where $p, q \in \mathbb{N}$, p and q are coprime, and $p \leq q$) the trajectory closes after q rotations (i.e. is a periodic orbit of periodicity q). If the ratio is real the trajectory fills up the surface of the torus densely and the orbit is quasi-periodic. ARNOL'D has shown, that for $K \neq 0$ there exist intervals within Ω, where ω_1 and ω_2 synchronize so that the system behavior becomes periodic. This phenomenon of synchronization is called *phase locking* – a generic phenomenon of coupled oscillators which has already been described in 1665 by CHRISTIAAN HUYGENS [125]. For each rational number p/q such an interval exists. For increasing K, the intervals increase and form a characteristic pattern of regions where phase locking is possible: the ARNOL'D tongues (see Fig. 6.9) [368]:479-491. For $K = 0$, the probability that the system develops a periodic behavior is zero, as the LEBESQUE measure of the rational numbers within the real interval [0,1] is zero. For $K > 0$, the probability is nonzero and increases up to 1 for $K = 1$. For $K > 1$, the situation becomes more complex, as the ARNOL'D tongues start to overlap, which leads to competing periodicities and a chaotic system behavior between the tongues. In excitatory and inhibitory coupled neurons (rat) *in vitro*, such an ARNOL'D tongue structure in the (Ω, K) parameter space has been found [707, 709, 710, 712]. As the emergence of ARNOL'D tongue structure is a defining property of coupled oscillators, this finding demonstrates that the firing behavior of pyramidal neurons is represented by limit cycles.

6.6.3 Emergence of a Coding Scheme

Let us remind the main problems one has to solve when the notion of a code is used in biological systems (see section 5.1.1): One has to identify entities that may serve as types in a code relation, a physical process that serves as code transformation, and a functional role for the code. Here, we introduce the notion of a 'code' in the framework of dynamical systems theory and present a coding scheme that emerges within the context described in the previous section. A dynamical system is represented by differential equations. The equations reflect the physical processes that govern the dynamics of the system. This approach lies in the tradition of RASHEVSKY (see section 3.2.6) and assumes real-valued mathematics as a correct basis for describing the system under investigation. In this way, the type level of the general coding scheme (Fig. 6.2 will be divided into two sub-levels, a real-valued (i.e. the analytical description of the system) and a rational-valued one (i.e. the numerical description of the system, see Fig. 5.1). In the following, in order to simplify notation, we assume that the system is described by a single variable $y(t)$. The goal of system reconstruction is to obtain the differential equation that governs the dynamics of $y(t)$: $\dot{y}(t) = f(y(t), \{a\})$, where $\{a\}$ stands for a set of parameters that determines the systems dynamics. System reconstruction is done by measuring $y(t)$:

Definition 6.15 *A measurement of a real-valued variable of a dynamical system $y(t)$ is a function $M(y) : \mathbb{R} \to \mathbb{Q}$ with*

$$M(y(t)) = y^*(t) \quad \text{such that} \quad y^*(t) \in [y(t), y(t+\delta t)) \quad \wedge \quad y^* \in \mathbb{Q}$$

whereas δt is called sampling rate.

For capturing the dynamics of the system, the measurement is performed during a certain time interval. The coarse graining imposed on the measured variable by the sampling rate of the measurement process generates symbol strings called time series:

Definition 6.16 *A time series* $\mathbf{y} = \{y_1, \ldots, y_L\}$ *that emerges by measuring a real-valued variable of a dynamical system $y(t)$ during the interval $[T_1, T_2]$ with sampling rate δt is the symbol string*

$$\{y^*(t_1), \ldots y^*(t_L)\} \quad \text{with} \quad t_i = \delta t(i-1) + T_1, \quad i = 1 \ldots L \quad \text{and} \quad L = \frac{T_2 - T_1}{\delta t}$$

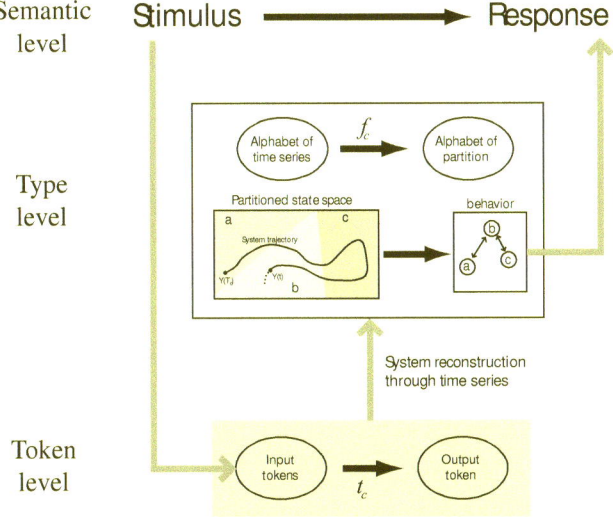

Figure 6.8: Scheme describing the coding problem in the framework of dynamical systems theory.

Time series are symbol strings where the symbols (= numbers) indicate the value of a certain parameter at a certain time interval. Based on time series, experiments with appropriate parameter setting as well as prior knowledge on the physics of the system (i.e. assumptions about the physical laws that govern the dynamics) allow to obtain (in the optimal case) the differential equations that govern the dynamics of the system. These equations reflect the dimensionality of the state space (e.g. if $y(t)$ describes the displacement of a simple 1D oscillator with a single parameter (spring constant), then the state space is two-dimensional). If an analytical solution is not possible (which is the usual case when analyzing complex systems), the embedding theorem is used for state space reconstruction [536, 635]. If we assume, that such a reconstruction has been successfully performed: What is a 'code' in such a framework? Note, that the level of description of the system changed: What is referred to 'input tokens' and 'output tokens' on the type level is now unified in a complete description of the system. What remains to be 'coded'? The answer to this question is: the behavior of the system, i.e. the succession of states. We explain this point in more detail: The *state* of a system is a point in the state space and the dynamics of the system is reflected by a trajectory in the state space. There now might be special regions in the state space, e.g. in the sense that the system stays for a significant amount of time in that region. Then the state space can be partitioned so that 'special regions' are separated from each other and each part of the partition is labelled with a symbol. Whenever a trajectory enters such a symbolically labelled part of the state space, the symbol is reported and a new and simpler symbol string is generated that describes the dynamics of the system. This is the basic idea of encoding a dynamical system [708]. In practical applications, encoding a dynamical system means to transform the time series (a symbol string) into a symbol string that emerges from the labelling of the partitioned state space. This string we call *state series*. We again assume that a single variable $y(t)$ describes the behavior of the system and denote the state of the system with $Y(t)$. Then, the generation of the state series is defined as follows:

Definition 6.17 *For a time series* y *where* $M(y(t)) = y^*(t)$ *determines the state* $Y(t)$, t_0 *is the beginning of the measurement and* $\Pi = \{\pi_i\}$, $i \in \mathbb{N}$ *is a partition of a state space labelled by symbols of the alphabet* $\mathbb{A}$, *a* state series *is the symbol string obtained by the following inductive rule:*

- **Anchorage:** *For* $y^*(t_0)$, *find* π_i *such that* $Y(t_0) \in \pi_i$ *and report the symbol* a_i *that labels* π_i

- **Induction step:**

    1. *For the time* $k\delta t$, *let* $Y(k\delta t) \in \pi_i$. *If for* $(k+1)\delta t$ *the condition* $Y((k+1)\delta t) \in \pi_i$ *is still fulfilled, repeat step 1, otherwise go to step 2.*
    2. *Find* $\pi_j \neq \pi_i$ *such that* $Y((k+1)\delta t) \in \pi_j$ *and add the symbol* a_j *to the string that labels* π_j. *Go back to step 1, using* π_j *instead of* π_i *in the condition.*

The state series allows the connection between the type level and the semantic level of our coding scheme: A specific behavior of a system as a reaction to a certain stimulus is expressed in the succession of states the system passes through in order to perform this behavior. This succession of states need not always be the same, but should rather be described as a set of similar state sequences. Take the *Drosophila* courtship as an example: Here the courting male goes through repeating successions of different elementary acts like tapping, licking etc. It has been proposed to consider closed orbits of such elementary acts as a definition of behavior [700]. These elementary acts can be related to the encoding of the dynamical system and the state sequences then reflect these closed periodic orbits. In summary, the coding problem within the framework of dynamical systems theory is described as follows (see also Fig. 6.8):

- The physical processes of the system happen on the token level such that a certain input is interpreted as defining a certain initial state of the system, which then evolves unter the influence of t_c in a succession of states.

- On the numerical type level, this succession of states is captured in time series by measuring the relevant system variables of the system and by choosing an appropriate measurement resolution (this determines $\mathcal{I}$).

- The time series is used in order to reconstruct the system analytically (if possible) or by using the embedding theorem.

- The reconstruction allows to find a partitioning of the state space by choosing an appropriate symbol alphabet for labelling the partition (this determines $\mathcal{C}$).

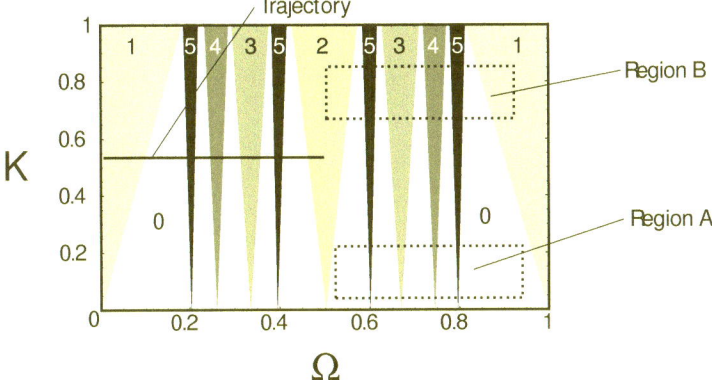

Figure 6.9: Scheme describing ARNOL'D coding in the (Ω, K) parameter space. Only areas of locking up to periodicity 5 are shown and labelled. Note, that real ARNOL'D tongues (from periodicity 2 on) have curved boundaries. For further explanations, see text.

- The code relation f_c then describes how the time series describing the state of the system is translated into a state series.
- Several trials of the experiment (settings of initial conditions) lead to a set of state series, from which a behavior is obtained.

We apply this general scheme to the model system we sketched in the previous section. Here, the state space is given by the (Ω, K) space. In this space, the ARNOL'D tongues reflect particular regions of stability that offer themselves for a symbolic labelling in the sense that regions reflecting the same periodicity of the limit cycle obtain the same symbol. The higher the periodicity of the limit cycle, the smaller the associated tongue. As the precision of observing the system is limited, only small periodicities lead to noteworthy regions of stability in state space. We assume periodicities up to 5 (this is also the highest periodicity found in the *in vitro* condition [710]) as realistic and label the regions accordingly with '1' to '5' (using '0' for the unstable state space regions, see Fig. 6.9). Trajectories in the (Ω, K) space emerge due to changes of Ω or K. The former parameter changes when the driving of the limit cycle change, the latter parameter changes when the coupling between the limit cycles changes. If we assume for example that K remains fixed and we increase Ω from 0 to 0.5, then the trajectory crosses the ARNOL'D tongues and leads to the symbol string (1,0,5,0,4,0,3,0,5,0,2), which is the encoded trajectory.

What are the properties of this code? Several indicators of efficiency and optimality can be observed [708]: Just by looking at the general coding scheme (Fig. 6.9), we see that, for a reasonable high coupling, the partition elements corresponding to low periodicity cover a large part of the state space, which means that the firing behavior of coupled cells converges fast to a periodic firing type (this has indeed been shown for the *in vitro* condition [709]); and that the smaller the periodicity, the larger the associated area in the (Ω, K) space. The simplest firing mode (periodicity 1) corresponds to the largest partition element in the state space, which means that the periodicity-1 firing is most common. As (for an equal background activity) periodicity-1 firing is also the 'shortest codeword' (the sequence of the pattern only covers one ISI), the coding has a HUFFMAN-optimality property – the most common 'codeword' is also the less resource-intensive (in terms of the time span to 'recognize' the 'codeword') and is associated to the most stable Arnol'd tongue [437].[11] If we consider the 'biological' background of Ω, further interesting indicators can be observed. We remind the relation $\Omega = X_s/X_u$. As the general noise level determines X_s (of the perturbing neuron) and X_u of the perturbed neuron, a homogeneous change of the network activity affecting the noise level for both neurons in the same way does not change the state. Only the emergence of a local gradient moves the trajectory and changes the state sequence. This is plausible, as local changes should induce such differences. Also LTP or LTD can have a role in changing the background activity a neuron perceives, if a substantial number of (distal) synapses that provide the background changed their synaptic strength due to this effect. Learning in this picture means for example, that the perturbing neuron gets a larger current inflow, increases the cycle time of its limit cycle and thus changes the periodicity of the output spike train of the perturbed neuron – which results in a change of the state sequence ('something has been learned'). As the effect of synaptic plasticity on the single neuron-neuron coupling is small (the EPSP increases only in the order of 50%) it might be more reasonable to speculate, that synaptic plasticity is as a *statistically* relevant phenomenon than changes the sensibility of the neuron for its noisy background. But also a homogeneous increase in network activity can have an influence: Then, the system is operating in a faster mode.

Changes in the parameter K, on the other hand, also can change the state sequence one obtains. Consider the example, that the parameters K and Ω are bounded such that only trajectories in the region A are possible (see Fig. 6.9). No state sequence that encodes trajectories within this area can contain the symbols 2 and 1. However, if K changes such that the trajectories are contained in region B, this is possible, which could be related a change of the global state of the system. The biological interpretation of K is the coupling of neurons. This coupling can, however, consist of two types: a synaptic type and a coincidence-firing type. First, the synaptic type means, that there must be a synapse sufficiently strong to provide a reasonable coupling. Assuming the correctness of the postulate of two classes of synapses in terms of release probability (see section 6.4.2), only the (fewer) 'strong' synapses would probably provide a sufficient coupling. Synaptic plasticity would then be a mean to increase K. One has, however, take into account that even the strong synapses will fail quite often to induce an EPSP. Burst-firing could then be a further way to obtain the desired coupling and to change K by longer or shorter bursts. The coincidence-type of coupling finally provides a solution of the problem, that the synfire hypothesis is only a theory that describes the stable propagation of activity - but not its role in a code relation. In this picture, the perturbed neuron is not coupled to a single neuron, but to a group of synchronously firing neurons. Chances in K are then expressed by the number of coincident

---

[11]Note, that for very small K, this type of optimality is not fulfilled, as the unstable regions in state space (the symbol '0') cover the largest part of the state space.

spikes. In that way, the proposed coding scheme integrates aspects considered in the rate coding picture (the firing rate determines the background) and the time coding picture.

To summarize, neurons on a limit-cycle solution are able to combine noise (background firing) and precise timing in the following sense: The noise can be thought of as the driving source for the neuron, if it follows a central limit theorem behavior closely enough. Superimposed on noise, the neuron emits and receives precisely timed firing events. The ARNOL'D tongue structure that emerges from the locking of two neurons provides a coding scheme, where analog information (e.g. given by the rate function of a group of neurons providing the synaptic background) is converted into an essentially digital one, namely the periodicity of firing. Let I_1 and I_2 be two currents that drive neuron one and two. Then the coding scheme can be described as [699]:

$$\text{currents } \{I_1, I_2\} \quad \to \quad \text{periodicity } l$$

In this way, the amount of a current driving a neuron is coded with reference to the current driving the target neuron. As the outcome, the encoded information is of analog and the coded information is of digital type. Thus, from a technical point-of-view, the coding scheme is an analog to digital converter.

6.6.4 The Role of Patterns

What is the role of patterns within the above sketched framework? Three different aspects have to be considered: First, how do the periodicities show up in the *in vivo* condition? Second, how can the different findings about the existence of various kinds of patterns be related to this scheme? Third, what are the consequences of the need of a noisy background for the types of neuronal firing one expects? The first aspect concerns the question, why the periodicities do not show up as regular spike trains in the *in vivo* condition, although experiments performed under the *in vitro* condition have indeed led to regularly patterned spike trains [658, 709]. The reason is that the assumption of quasi-stationary of the noisy input is not fulfilled in the *in vivo* condition, as one is less able to maintain quasi-stationarity by choosing appropriate experimental conditions. However, the system can still be locked, if the noisy input changes in a correlated way (i.e. the ratio Ω remains constant), but this does not show up in the spike train. If one would know, how the driving changes, one would be able to rescale the spike train and the periodicity would be visible again. Indeed, experimental and model studies, where the change in driving is known by the experimenter and one is able to do this rescaling, demonstrates the plausibility of this scenario: If HODGKING-HUXLEY neuron coupled via α-type synapses are driven under non-quasi-stationary conditions, locking remains, expect for the case when the modulation substantially interferes with the neuron's own firing frequency [699]. This has an important consequence for the design of neurophysiological experiments: Because of the strong response and suppressed adaptation, this is the preferred experimental situation. However, it just might miss the 'normal' working condition of the brain. When detailed compartment models are used to test, whether locking remains under non-quasi-stationary conditions, the result is again positive [698]. Finally, in the *in vitro* condition, where the background driving of both neuron has been modulated to some degree such that Ω remains unchanged, again locking is preserved (unpublished results of a diploma thesis of T. GEIGES performed under

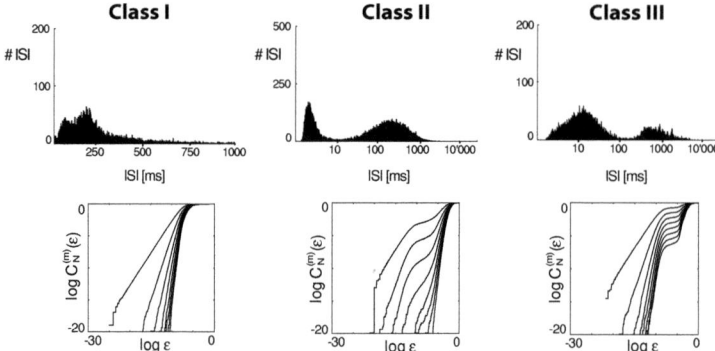

Figure 6.10: Classes of neuronal firing: I: noisy firing. II: Pattern in random background. III: Pattern firing. (figures derived from [428]).

the supervision of R. STOOP). Recent investigations on neural background activity (see section 6.3) indicate correlated activity that may also provide the conditions that locking is preserved. The fact, that one does usually not see the firing periodicity in single neuronal spike trains does therefore not rule out the hypothesis. Concerning the second aspects, to almost all types of patterns presented in our general overview (see section 6.1.1) a role within this framework can be attributed:

- **Order patterns:** They may reflect a stable network structure in noisy condition.

- **Scaled patterns:** They may represent firing periodicities for a changing background, where the change is slow compared to the mean ISI of the spike train under consideration.

- **Interval patterns:** They may represent firing periodicities for a quasi-static background (single train interval pattern). Multi train interval pattern indicate quasi-static firing conditions for a group of locked neurons.

- **Timing patterns:** They may reflect a strong perturbation signal.

The third aspect, finally, implies the existence of reliable noise generators: neurons, that are able to maintain uncorrelated (POISSON) firing for several stimulus conditions, whereas only the firing rate changes. This raises the question of the physiological requirements that such neurons can act as 'noise generators'. Possible causes are a specific dendritic arborization that 'randomizes' a correlated input, or a low number or specific distribution of ion channels that amplifies channel noise. Taking all three aspects together, one expects to find three classes of firing in cortical neurons, that can be sustained for various stimulus-conditions: I) The class of randomly firing neurons, i.e. the 'noise generators', which essentially provide the energy source for the network activity. II) The class of neurons where simple patterns are injected into a random or incompatible background, indicating that these neurons are unable to cope with a certain class of stimuli. III) The class of neurons that preferentially fire in patterns, i.e. those neurons that perform the information processing step. Earlier investigations of V1 data indeed suggested the existence of three classes of neuronal firing [705]. In Figure 6.10, the classes are illustrated by three cat V1 neurons using correlation integral based pattern discovery (see section 7.3). Whereas bimodal ISI histograms emerge in all cases, the corresponding log-log plots indicate clear differences in the associated firing behaviors. Neurons of class I show straight-line correlation plots whose slope fails to saturate. Neurons from class II show a dependence of the slope-ratio on the embedding dimension (see example 7.3 in section 7.3 and [428]), indicating that patterns of length 2 and 3 are present. The class III neuron's behavior is compatible with the earlier finding [704] that members of this class are generally associated with two (exceptionally: one) clearly positive LYAPUNOV exponents, and with fractal dimensions that saturate as a function of the embedding dimension. These indicators hint at unstable periodic orbits generating these responses, and imply that the data are essentially deterministic in nature. This classification has further the implication, that the POISSON hypothesis for neuronal firing (see section 6.2.2) is only valid for one class of neurons – which is indeed compatible with the majority of the early findings on neuronal firing statistics (see section 3.2.3). Note, that class II neurons are not intrinsically unable to show a class III behavior. Rather, their kind of firing expresses the fact that they are not able to cope with a certain class of stimulus, but they may show a class III behavior for a different class of stimuli.

6.6.5 The Integral Framework

We conclude this section by providing an integral framework that unifies the coding scheme described in this chapter with the variety of facts on network-connectivity (in particular inter-layer connectivity, see section 6.4.1), patterns found in sensory input to cortical layers (see section 6.5), and finally LTP and coincidence detection (see section 6.4.2). The integral framework is described in Fig. 6.11. We demonstrate using this scheme, how different aspects of stimuli can be combined and new combinations can be learned. To keep the general description simple, we do not make detailed descriptions on what exactly the stimulus consists of. We further assume, that we are dealing with visual and auditory input. We assume three stimuli, A, B and C. A and B may stand for different visual aspects of a single object, C may stand for a acoustic feature of this object, whose association to the object has to be learned. Input on the feature A emerges from the thalamus and includes (as shown in section 6.5) multi train patterns (unitary events). This input is probably a mixture of multi train interval patterns, that show up as unitary events and reflect the information about A (see section 6.5) and uncorrelated spikes reflecting the activation of the specific patch in the visual field. The input is sent to layer 4, where it activates a group of 'noise generators' through distal synapses, such that patterns of unitary events in the retinal input do not show up in the output. The uncorrelated firing of these neuron groups drive two groups of limit cycle firing neurons: One group is again in layer 4 – which is plausible as the connectivity between layer 4 neurons is high. This group also receives thalamic input on the

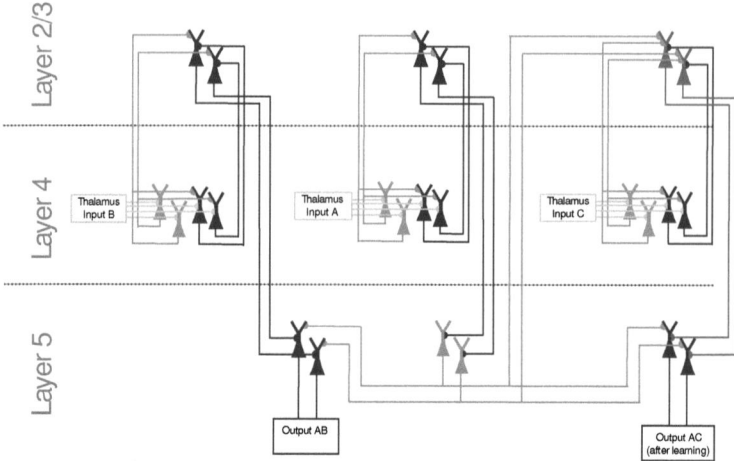

Figure 6.11: STOOP-hypothesis: integral framework. Class I neurons are in bright grey, class III neurons are dark and class II neurons are in mid grey color. For further explanation see text

stimulus A on proximal synapses. The proximal input allows spiking only if the number of coincident spikes emerging from the input is high enough, thus this group filters out those spikes that are attributed to A from noisy spikes that may also be part of the input. The second target group of the 'noisy' neurons lies in layer 2/3, where another group of limit cycle neurons is situated. The noisy input on distal synapses drives the neuron on a limit cycle solution that is associated with the general activity level of stimulus A, whereas these neurons are perturbed by those neurons of layer 4, which filtered out the specific coincident spikes associated with stimulus A and transmit them according to the synfire model. The output of these neurons thus reflects a pattern coding for the activity level of the important aspects of stimulus A. The same process also holds for stimulus B. For stimulus C, however, we assume that this is not the case: Although the processes in layer 4 are comparable, the 'noisy' input of layer 4 to the group of layer 2/3 neurons is considered to be too weak to induce limit cycle firing. The perturbation the neurons of the group obtain does therefore not lead to a phase-locked output as described by Fig. 6.7. Thus, these neurons are class II neurons, where only short patterns are present in a random background.

In the following, we describe, how the stimulus features A and B are combined (which can be considered as a solution of the binding problem) and how the binding of the features A and C can be learned: The A output is sent to a group of noisy neurons in layer 5, which produce again a uncorrelated output reflecting the activity of the important aspects of stimulus A. This output is sent to three different groups of neurons: The first group of neurons receives this input on distal synapses as a limit cycle driving current. This group is perturbed by the pattern reflecting stimulus B, such that the emerging pattern reflects the combination of both aspects. The second group lies in layer 2/3 and targets those neurons, that are unable to cope with stimulus C. When the stimuli A and C are correlated, we expect effects of LTP such that the synapses get stronger and the neurons changes from a class II to a class III firing. This new output is sent to neurons in layer 5, where it perturbs the limit cycle firing of another group of neuron, also driven by the noisy output representing the important features of stimulus A. In this way, the system has learned to associate stimuli A and C. This scheme is certainly still rather simple, but it could serve as a reference for a model study or to derive new experiments (see chapter 8). As we were not able to perform such experiments, we now derive the following experimental predictions, that can be tested using the data set that was available to us:

- Concerning the general firing statistics, we expect that only a minority of neurons shows POISSON firing (the 'noisy' neurons), whereas the majority of the neurons show a deterministic behavior, which is reflected in long tail distributions.

- We expect different reliability on different stages of the information flow: The thalamic input should display high reliability both in respect of timing and patterns. Higher cortical areas should loose their timing precision but keep their pattern precision.

- We expect different clustering of neurons in terms of their pattern distance: Before the stimulus, the groups are unstable, whereas during stimuli, we expect stable groups.

- We should be able to find the three classes of neuronal firing, whereas classes I and III should be larger than class II, if the set of different stimuli used is large.

- We should see differences between the stability of patterns that reflect stimulus timing not used by the system ('temporal coding' according to section 5.1.2) compared to patterns that reflect computation ('temporal encoding'): Latter should be more stable, as former are not relevant for the system.

Chapter 7

Detecting Patterns

In this chapter we outline the pattern detection problem and introduce several methods for solving it. Conceptually, we differ between pattern discovery, pattern recognition and pattern quantification. We review common methods for pattern detection in spike trains. As limiting the bias for pattern detection is one of the main requirements for any method applied in this respect, we introduce several methods that fulfill this requirement and compare them to classical ones.

7.1 The Problem of Pattern Detection

7.1.1 Outlining the Problem

There is a close relation between the method used for finding patterns in data and the pattern one actually finds. Prior knowledge is introduced in several ways in the task of pattern detection: by the experimental setup (e.g. which stimuli are considered as 'important'), the measurement (e.g. by choosing a certain measurement resolution), uncertainties in determining the events that are considered important (e.g. spikes or bursts, see section 6.1.2), the time span before and after the stimulus that is considered relevant, the method used for pattern detection, and the background selected to confirm the existence of a pattern [439, 490]. As consequence, patterns can remain unobserved due to an observer's bias. Any method for pattern detection should minimize the biases it introduces and should be aware of the remaining ones.[1] When one looks for patterns in a certain set of data, different concepts come into play: pattern discovery, pattern recognition and pattern quantification. The

---

[1] Data obtained from neuronal systems furthermore suffer in many cases from a sampling problem: Only a very small fraction of, mostly large, neurons can actually be measured in order to obtain spike data. Multi electrode array recordings and spike trains obtained using voltage-sensitive dyes may diminish this problem to a certain degree, but even today's technology is confronted with the problem that only a small part of a neuronal network of interest can actually be measured. This sampling problem gains importance when looking at multi train patterns [353]: Since only a limited number of neurons is examined, any multi train pattern found may be part of a larger pattern which involves neurons that are not under investigation. It is thus possible, that different patterns are associated with the same subpattern.

term *pattern discovery* is used, when no assumptions about the probability of appearance $p_{i,j}$ of events forming a certain pattern, the variation of inter-event intervals $\Delta X_{i,j}$ or jitter $\Delta T_{i,j}$ are made. The term *pattern recognition* is used, when such assumptions are indeed made – e.g. by predefining a template.[2] The term *pattern quantification* is used, when the number of patterns is determined and compared with the number that is present in the data after choosing a certain randomization method. Certainly, also pattern discovery and pattern recognition include the comparison with randomized data, although the conclusion whether patterns are present or not may not rely on a quantification but, for example, on a comparison between two graphical displays. For some questions (e.g. determining the reliability of neuronal firing), whole spike trains are compared. This leads to the idea of a *distance measure* that is used to quantify the difference (or similarity) between spike trains.

7.1.2 Pattern Detection Methods: Overview

Any pattern detection method relies on an appropriate description of the spike data. Usually, the spike trains are analyzed in the timing, the interval, or the bitstring representation (see section 5.1.1). After choosing an event type which is considered as important for the problem under investigation, the spike trains are translated into symbol strings of the particular events that conserve the timing (or only the order) of the events. These strings are object of the analysis. If patterns of joint spike activity in a set of simultaneously recorded spike trains are subject to investigation, then the data set may also consist of bitstrings *across* the trials. This means, that for a certain bin τ_i one checks for each train $j = 1, \ldots, S$ if a spike appears in this bin at position j. If this is the case, one writes '1', otherwise '0'. The result of this procedure are sequences of bitstrings of length S representing the joint-spike activity [510, 514].

The first statistical methods used for pattern discovery were histogram-based (see section 3.2.3). This approach relies on the counting of events, or sequences of events, available in the data. We will discuss and classify histogram-based methods in more detail in the following section. A second method for pattern discovery relies on the correlation integral [428, 429]. This method will be discussed in more detail in section 7.3. A further approach for pattern discovery relies on the concept of the ϵ-machine [671, 670]. As input, single spike trains transformed into bitstrings are used. The method finds the minimal hidden MARKOV model that is able to generate the bitstring. Pattern recognition methods rely on prior information about the potential pattern condensed in a *template*, which predefines the sequence and the matching interval. Template methods are widely used in spike train analysis [353, 444, 720], and also provide a way for pattern quantification. We will discuss them in more detail in section 7.4. When multi train patterns are in the focus of the analysis, the number of possible patterns becomes very large – especially when interval or timing patterns are considered. To overcome this problem, usually only order patterns [559] or synchronous firing are considered. The restriction to unitary events is common, as several theoretical frameworks are available to discuss the obtained results (the synfire hypothesis or the temporal correlation hypothesis, see section 6.4.1). Several methods are available to deal with such patterns. Usually, the data are transferred to bitstrings that represent the joint spiking activity (see above) and the

---

[2]We note, that the term 'pattern recognition' is also widely used in the field of neural networks. Here the term denotes the classification-ability of neural networks [558].

assumption of independent firing is taken as null hypothesis [511, 514] (for a refinement of the method that takes non-stationarity into account see [509, 510], for an older approach using the joint firing rate see [93]). Recently, several sophisticated methods that take higher order interactions into account (conditional inference [513], log-linear models [600], information-geometric measures [615]) have been proposed. These methods, however, face the practical problem that many possibilities have to be taken into account what makes the calculation computationally expensive. Furthermore, a large amount of individual trials is required, that often cannot be obtained in neuronal experiments. One of the few methods able to deal with multi train interval patterns is the *T-pattern algorithm*, a method originally proposed for the analysis of behavioral data [586].

T-Patterns: The T-pattern algorithm spots multi train interval patterns of single spikes (see Def. 6.8), whereas all spikes of single patterns emerge from *different* neurons. The algorithm starts with a pairwise comparison of spike trains. Let $\{t_1,\ldots,t_n\}$ and $\{t_1^*,\ldots,t_n^*\}$ be such a pair. For a spike t_i a so-called 'critical interval' $[t_i + t_a, t_i + t_b]$, where $t_a < t_b$, is determined, so that no other spike of the first train is in the interval $[t_i, t_a]$ and the number of spikes within the critical interval of the second spike train is significantly higher than expected (POISSON null hypothesis, see below). The borders t_a and t_b are chosen so that there is a single spike of the second train that fulfills $t_i + t_a \leq t_j^* \leq t_i + t_b$ and that the probability that this happens per chance is below a certain significance level. The expected number of spikes in the critical interval is based on the assumption that the spikes of both trains emerge from two non-correlated POISSON processes. To obtain longer patterns, the patterns resulting from all pairwise comparisons of spike trains are considered singular events (hierarchical approach). The procedure is repeated, until a predefined hierarchical level is reached. The algorithm delivers all patterns that are significant in the above defined sense and the number of repetitions of each such pattern. A method for finding multi train interval patterns that is related to the T-pattern algorithm is based on FISHER's exact probability test [486] – a computationally expensive method as soon as longer patterns are considered. The built-in way of randomization may be problematic for some applications. Tests performed by us show that POISSON randomization leads to higher number of patterns in spike trains compared with ISI shuffling. When the T-pattern algorithm has been applied to data sets (rat olfactory bulb, see section 8.1.1) and the minimal number of pattern occurrence has been set high (50), the number of different patterns found in the ISI shuffled set was in the order of 50%–80% of the number found in the unshuffled set. When the minimal number of occurrence has been set low (5), still 25%–30% of the patterns remained in the shuffled set. As we showed in section 6.2.3, the homogeneous POISSON randomization is probably uneligible for interval patterns as the degree of randomness introduced is too high and disregards the known statistical dependencies of succeeding spikes. It is therefore recommendable not to use the built-in randomization procedure in the Theme algorithm in order to analyze T-patterns, but to randomize the data separately using ISI shuffling, is interval patterns are subject of analysis.

7.1.3 Spike Train Distance Measures: Overview

Spike train distance measures allow to quantify the similarity of neuronal firing. This allows the solution of two problems of spike train analysis. First, it provides a tool for classifying neurons. In this way, information on the functional connectivity of the probed neuronal network can be gained from multi-electrode array recordings, or the distance may serve as a population criterium (see definition 5.11). Second, it provides a measure of the reliability of neuronal firing by calculating the mean distance of spike trains obtained in trials of the same experiment. The larger the mean distance, the less reliable the neuron's firing is. The similarity of neuronal firing is quantified by using measures usually called 'distances', although the fulfillment of the axioms of a metric (a distance is always ≥ 0, symmetric and satisfies the triangle inequality) is seldom checked. We nevertheless speak of spike train distance measures. A variety of distances has been proposed (see Figure 7.1):

- **Spike count distance:** This measure considers spike trains as similar if they have a similar number of spikes (or, more generally, events). If L is the number of spikes in the first train t and L^* is the number of spikes in the second train t^*, then the distance is defined as $d(t, t^*) = \frac{|L-L^*|}{\max\{L, L^*\}}$. Temporal structure is not taken into account.

- **Firing rate distance:** For the firing rate distance, the measurement interval over which a spike train has been sampled is partitioned into segments of width T_{rate}. For each segment, the local firing rate is calculated and the spike train is transformed in sequences of rate values. The distance of two such sequences $\{r_1, \ldots r_n\}$ and $\{r'_1, \ldots r'_n\}$ is defined as $\sum_i (r_i - r'_i)^2$. One may also calculate the difference of the two (continuous) approximate firing rate functions obtained by the shifting window approach. The time window has to be predefined.

- **Cost-function distances:** A family of spike train distance measures is based on the so-called 'metric space analysis' [742, 741]. The basic idea is to define a cost-function that determines the costs of transforming one spike train into another by moving, deleting and inserting spikes. The cost function has to be predefined.

- **Correlation distances:** Correlation distances refer to the correlation of spike trains, i.e. expressed by the hight of the highest peak in the pairwise cross-correlogram of two spike trains [748] or by synchronized activity measured by the so-called gravitational clustering approach [495, 496, 573]. Another correlation distance is based on a convolution of a spike train with an exponential or gaussian kernel. The distance is then calculated either as the integral over the difference of the convolved trains [737] or as the normalized dot product of the convolved trains [663]. The latter correlation distance – in the following referred as C-distance – is calculated as $d(t, t^*) = 1 - \frac{f(t) \cdot f(t^*)}{\|f(t)\| \|f(t^*)\|}$, where $f(t)$ denotes the convolved train. This distance measure requires prior information by determining the width of the Gaussian kernel.

- **Information distances:** Information distances are applied on spike trains in the bitstring representation. Here, several different concepts exist, relying on the Kullback-Leibler distance [532] or the Kolmogorov complexity [386]. The latter proposal emerges from DNA sequence analysis [570] and basically compares the lengths of the shortest descriptions of two bitstrings representing spike trains. Another distance proposes to use the average change in the bitstring representations of a given spike train for increasing bin-with [628]. The change is measured as the mean of the number of symbol-switches per bit-coded spike train. In this measure, spike trains with similar average change are considered as close. We will introduce a novel information distance, the LEMPEL-ZIV-distance (section 7.6).

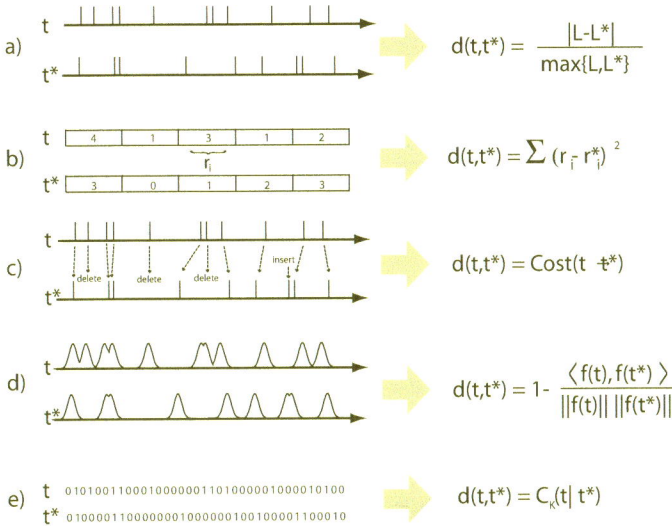

Figure 7.1: Spike train distance measures: a) Spike count distance ($L, L*$: number of spikes in train $t, t*$), b) rate distance (r_i: local firing rate at position i), c) cost-function distance, d) correlation distance, e) information distance ($C_K(t|t*)$: KOLMOGOROV complexity of train $t*$ given train t).

The use of distance measures for clustering spike trains will be discussed in section 7.6. Distance measures are also applied for the creation of *recurrence plots* [466]. This visualization tool has been applied for neuronal data to identify different modes of operation of a neuron (e.g. random firing vs. ordered firing) [534]. The basic idea is to measure the distance between an ISI sequence of length l starting at the i-th position in the whole ISI series of length L, and a sequence of the same length starting at the j-th position. If the distance is beyond a certain threshold, a black dot is plotted at the position (i, j). The procedure is repeated for all sequence-pairs $i, j = 1, \ldots, L$.

7.2 Histogram-based Pattern Detection

7.2.1 Classes of Histograms

Histograms display the result of a counting procedure. This procedure is usually applied to either the timing or the interval representation of spike trains. In the following, we classify histograms according to the spike train representation used and the number of spike trains involved (see Table 7.2.1). We start with some preliminary remarks on *binning*: The counting requires a binning of the time axes. Let T be the time interval covering the whole measurement (timing representation of the spike train) or the length of the longest interval (interval representation of a spike train). Binning means, that a bin-size $\Delta\tau$ is chosen such that $n\Delta\tau = T$ (n is the number of bins). In principle, it's also possible to use bins of variable size. Take the example of a timing representation of a spike train $\{t_1, \ldots, t_k, \ldots, t_l, \ldots t_L\}$: The number of spikes $t_k, \ldots t_l$ for which $(i-1)\Delta\tau \leq t_k \leq \ldots \leq t_l < i\Delta\tau$ represent the value of the i-th bin. The choice of $\Delta\tau$ depends on the problem under investigation.

Histograms of single spike trains, timing representation: These are the conceptually most simple histograms. Several possibilities emerge. If the histogram ranges over the whole measurement, then the histogram obtained is similar to the local rate representation of a spike train (see section 5.1.1). In case of a periodic stimulus, or some intrinsic periodicity present within the single spike train, a histogram over this period can be obtained – the *phase histogram*. It serves as a tool to analyze phase locking. One can also determine a time interval T^* and count the spike times from the first spike of the train t_1 until $t_1 + T^*$. If this is repeated for all spikes, one obtains the *auto-correlogram*. It serves as an estimation of the autocorrelation function.

Histograms of multi spike trains, timing representation: A histogram over several spike trains originating from many trials of the same experiment is called *post-stimulus time histogram* (PSTH) or peri-stimulus time histogram. It serves as an estimation of the firing rate function. Also the phase histogram can be obtained from many spike trains in the same way as for single spike trains. If, however, a periodic sequence of stimulus-occurring times serves as a reference to count the spike times of two spike trains *simultaneously*, one obtains several variants of *joint-peri-stimulus histograms*: The original proposal was the joint peri-stimulus-time scatter diagram [99] which can also be displayed in the same way as a 2D interval histogram (see below). An improvement of the joint peri-stimulus-time scatter diagram is called joint peri-stimulus diagram [359, 494]. Furthermore, a similar construction as the auto-correlogram is possible for two different spike trains, where the spike-times of one train serve as reference for counting the spike-times of other spike trains. The resulting histogram is called *cross-correlogram*. It estimates the cross-correlation function and serves as a tool to investigate the mutual dependence of spike trains and to derive their underlying (functional, not anatomical [608]) connectivity (e.g. shared input vs. synaptic connection) [179]. If this procedure is applied for two spike trains originating from the same neuron (different trials of the same experiment), it serves as a tool to differ between stimulus-driven and intrinsic oscillations. Intrinsic oscillations show up as periodic structures in the auto-correlogram but disappear in the cross-correlogram of different trials, whereas these structures remain if the oscillations emerge from the stimulus (*shift-predictor*). Another application of the cross-correlogram is the *information train* technique [487]. The basic idea of this approach is, to first generate the cross-correlogram and then to build a spike train using only those spikes contributing to the main peaks of the cross-correlogram. The characteristics of the new train is its time-locked relationship with the original train. The cross-correlogram technique can also be extended to three spike trains. This *joint impulse configuration scatter diagram* displays the relative timing of nerve impulses in 3 simultaneously monitored neurons [354, 175]. The time intervals between impulses are plotted on triangular coordinates. This technique allows to analyze various connectivity patterns of three neurons and serves as a method to detect interval patterns of length 2 [355].

	timing representation	interval representation
single train	Timing of trial: $\simeq$ *local rate* Periodic timing: *phase histogram* Timing relative to spikes: *auto-correlogram*	Interval histograms of *order n* and *dimension m*. *Scaled histograms* *Local trend histogram*
multi train	Timing of many trials, one neuron: *post-stimulus time histogram* Timing of stimulus-occurrence, two neurons: *joint peri-stimulus histogram* Timing relative to spikes of different trials, one neuron: *shift-predictor*. Timing relative to spikes of two neurons: *cross-correlogram* Timing relative to spikes of three neurons: *joint-impulse configuration scatter diagram*	Cross-interval histograms of *order n* and *dimension m*

Table 7.1: Overview of histograms of single/multi spike trains in timing/interval representation.

Histograms of single and multi spike trains, interval representation: In this representation, histograms of different *order* and different *dimension* are distinguished:

Definition 7.1 *For a spike train in the interval representation* $\{x_1, \ldots x_L\}$, *a histogram that displays the result of counting the occurrences of the n-th order intervals*

$$\left\{\sum_{k=1}^{n} x_k, \ldots, \sum_{k=L-n}^{L} x_k\right\}$$

in the spike train is called a n-th order histogram.

Creating a set of ISI that contains all n-th order intervals up to a certain n_{max} allows the search for interval patterns where additional spikes are allowed. The 1D histogram of all intervals between spikes up to a certain maximal time interval is called *cumulative histogram* [714]). The set of all n-th order intervals can also serve as basis for other methods for pattern discovery, e.g. the correlation integral based method (see next section). For defining m-dimensional histograms, we introduce the procedure of embedding, a standard procedure in nonlinear dynamics [458, 536, 635, 657, 716]:

Definition 7.2 *For a spike train in the interval representation* $\{x_1, \ldots x_L\}$, *the points* $\xi_k^{(m)}$

$$\xi_k^{(m)} = \{x_k, x_{k+1}, \ldots, x_{k+(m-1)}\}, \quad k = 1, \ldots, L - m + 1$$

are called embedded points *and m is called the* embedding dimension.

Definition 7.3 *For a spike train in the interval representation* $\{x_1, \ldots x_L\}$, *a histogram that displays the result of counting the occurrences of intervals embedded in dimension m*

$$\{\xi_1^{(m)}, \ldots, \xi_{L-m+1}^{(m)}\}$$

is called a m-dimensional histogram *(mD histogram)*.

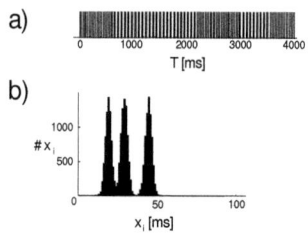
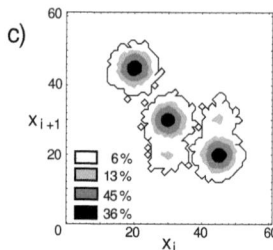

Figure 7.2: Example of a 2D interval histogram: a) Spike train of a noisy, unstable neural oscillator. b) 1D histogram of the spike train. c) 2D histogram. Gray-scales indicate the percentage of points lying in the corresponding region of the plot.

Example 7.1: Consider the ISI series $\{x_1, x_2, x_3, x_4, x_5\}$. The second order interval histogram counts the intervals $\{x_1 + x_2, x_2 + x_3, x_3 + x_4, x_4 + x_5\}$. The 2D histogram counts the sequences $\{(x_1, x_2), (x_2, x_3), (x_3, x_4), (x_4, x_5)\}$. A second-order 2D histogram counts the sequences $\{(x_1 + x_2, x_2 + x_3), (x_2 + x_3, x_3 + x_4), (x_3 + x_4, x_4 + x_5)\}$.

The most common interval histogram is the first order 1D histogram [356], on which the TUCKWELL classification scheme is based (see below). The first order 2D histogram – also called joint ISI histogram, scatter diagram, scatter plot, or return plot [201, 674, 675] – serves as a simple tool for pattern detection (see below). First order 3D histogram are also used occasionally, although the graphical display is more complex [464]. Histograms of higher dimensions can not any more be displayed. The correlation integral analysis offers an alternative to deal with higher dimensions (see next section). Rarely used are the *scaled histogram* (a histogram of the intervals between every 2^m-th spike, where m is a positive integer [201]) and the *local trend histogram* (a histogram that counts the sequences of differences of succeeding ISI of length 3 $(x_{i+1} - x_i, x_{i+1} - x_{i+2}, x_{i+2} - x_{i+3})$ [482]). In principle, it is also possible to consider histograms of intervals between spikes of different spike trains. Due to practicability, this approach is usually restricted to the comparison of two spike trains. The first-order histogram of intervals between the spikes of two spike trains is called *cross interval histogram* [179]. The sum of all cross interval histograms (usually up to a certain upper bound) corresponds to the cross-correlogram.

2D Interval Histograms: The 2D interval histogram of variable order is a convenient tool for interval pattern detection. It is able to take the serial order of intervals into account, can easily be displayed, allows to estimate the length of the pattern, and a (preliminary) quantification when the histogram is displayed as a contour plot [97]. To introduce the 2D histogram in more detail, we use model data generated by a noisy (σ of additive gaussian noise: 2 ms) unstable neuronal oscillator that switches between periodicity-1 (ISI: 30 ms) and periodicity-2 firing (ISI-sequence: (20,45) ms). In the first order 2D histogram, five different point clusters emerge, that are indicated by their centers of gravity (30,30), (20,45), (45,20), (30,20) and (45,30). The differences in point-density are displayed by the contour plot. The gray scalings represent the percentage of points that lie in the corresponding region compared to the total number of points displayed in the plot (Fig. 7.2). By systematically increasing the order of the 2D-histogram, information about the length of patterns is obtained: Assume that the ISI sequence $\{x_1,\ldots x_l\}$ is repeatedly present in a spike train. In the first-order 2D histogram, $l-1$ clusters with centers of gravity of the form (x_i, x_{i+1}) for $i = 1,\ldots,l-1$ will be present. In the second order 2D-histogram, $l-2$ clusters with centers of gravity of the form (x_i, x_{i+2}) will be present. Proceeding in this way, the number of clusters diminishes up to the l-th order 2D-histogram where only the cluster with center of gravity (x_1, x_l) remains. Thus, this procedure allows to estimate the length of a pattern. Figure 7.3 demonstrates this by using a spike train obtained by a POISSON spike generator with refractory period. In the spike train, the interval pattern $\{(15,2),(27,2),(39,2)\}$ has been inserted (see Def. 6.8). The first order histogram displays the two clusters with centers of gravity $(15,27)$ and $(27,39)$. The second order histogram displays only one cluster with center of gravity $(42,66)$. In the third-order histogram, no large cluster is visible.

To demonstrate the potential for pattern detection using the 2D histogram on real spike data, we display the result from four data files derived from extracellular field potential measurements of neurons from the primary visual cortex of macaque monkey (data description see section 8.1). Three files (Fig. 7.4 a,c,d) derive from complex cells stimulated with drifting gratings (drift frequency 6.25 Hz), one file (Fig. 7.4 b) derives from a simple cell stimulated with drifting gratings of frequency 12.5 Hz. In Fig. 7.4.a, a regular structure of point clusters is visible (~40% of the points belong to these clusters), whose centers of gravity are roughly described as $(9n_1, 9n_2)$, $n_1, n_2 \in \mathbb{N}$ — indicating a high frequency periodic firing, which is corroborated by calculating the FOURIER-spectrum of the spike train (data not shown). The large clusters in Fig 7.4.b result from the drift frequency of the grating, which is reproduced by the firing of the simple cell — a robust indicator for the identification of simple cells [682]. The 2D histogram reveals fine-structure within the large clusters — indicating periodic firing of

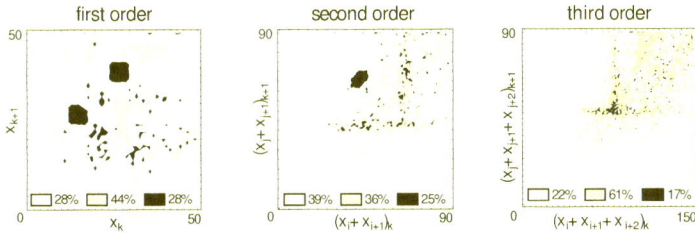

Figure 7.3: Using the order of 2D histograms for pattern length estimation: The first order 2D histogram of a spike train with a pattern of length 3 leads to two clusters, the second-order histogram to one cluster and in the third order histogram the large cluster disappears. The gray-scale code shows the percentage of points that lie in the corresponding region.

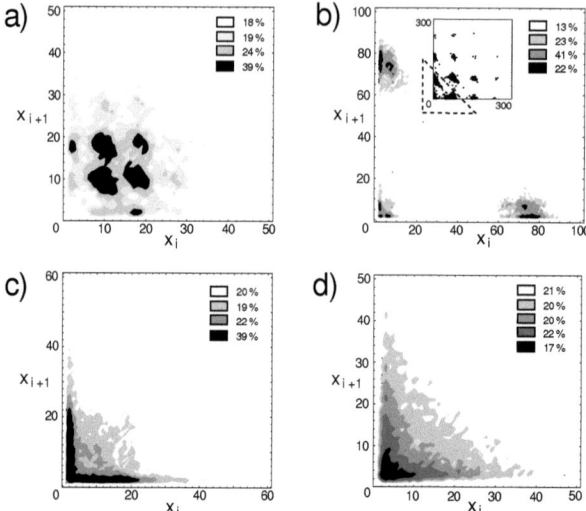

Figure 7.4: 2D histograms for neuronal data (macaque V1): a) Point clusters indicating high frequency oscillation. b) Fine-structure in point clusters resulting from drifting grating stimulation of simple cells indicating high frequency oscillation. c) High probability of a specific (symmetric) firing pattern d) Firing of POISSON type cell shows no specific structures.

high frequency. In Fig. 7.4.c, ∼20% of all ISI pairs are described by the expression (a, b) and (b, a), where $a \sim 1$ ms and $b \in (0, 20]$ ms – indicating the presence of patterns in the train (compared to a POISSON background). Fig. 7.4.d displays the 2D histogram of a spike train that shows an exponential decay in the 1D-histogram and a c_v of 1.02. As the 2D histogram does not indicate a serial dependence of the ISI, one may conclude that a POISSON process is a valid statistical model of the spike train.

7.2.2 1D Interval Histograms

The 1D interval histogram is (beside the PSTH) the most common histogram in neuroscience. We therefore briefly discuss the TUCKWELL classification scheme for 1D histograms [732] and introduce a measure to distinguish exponential histograms from long-tail histograms. The TUCKWELL classification scheme distinguishes 10 classes (see also Fig. 7.5):

1. **Exponential distribution:** Indicating very large EPSPs with POISSONian arriving times being transformed on a one-to-one basis to action potentials.

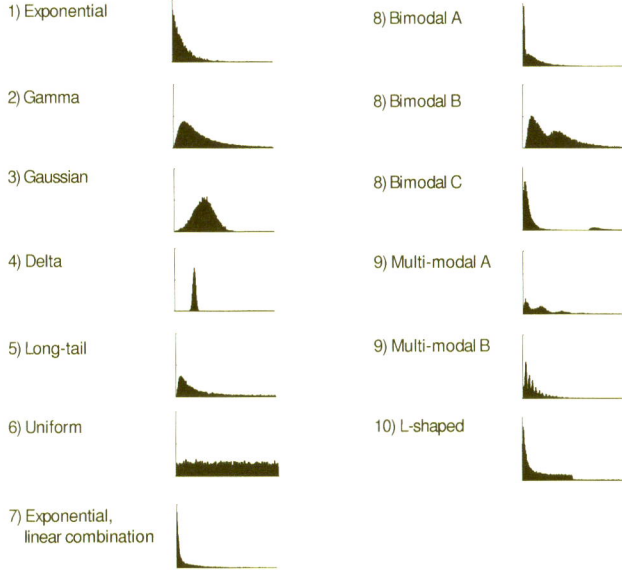

Figure 7.5: The TUCKWELL classification scheme. The histograms were obtained using model data.

2. **Gamma-like distribution:** Indicating an absolute and relative refractory period in neuronal firing. Probably only a few EPSPs occurring in a short time interval are sufficient to make the neuron firing, whereas inhibition is unlikely.

3. **Gaussian distribution:** Indicating that a large number of EPSPs is necessary to induce spiking, whereas significant inhibition is unlikely.

4. **Delta-like distribution:** Indicating intrinsic oscillatory firing behavior (e.g. pacemaker cells).

5. **Long-tail distribution:** Indicating a significant amount of inhibition. For the distinction between exponential distributions and long-tail distributions see below.

6. **Uniform distribution:** Indicating a cell in different firing states such that the distribution is the linear combination of several densities.

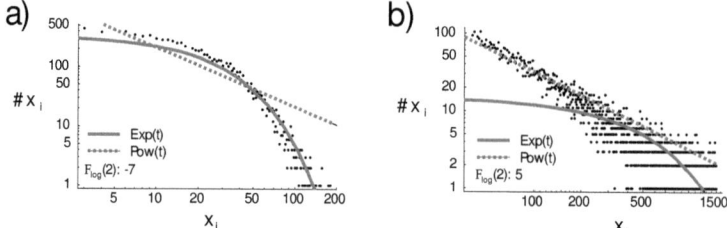

Figure 7.6: Comparing exponential vs. power-law fits of distributions (log-log plots): a) Fitting a exponential distribution. b) Fitting a long-tail distribution.

7. **Linear combination of exponential distributions:** Indicating a change of state in firing of type 1.

8. **Bimodal distributions:** Several kinds are distinguished: 8A indicates burst firing, 8B could result from a change in firing state of type 2, and 8C indicates the presence of two clearly distinguishable length scales possibly due to periodic stimulation.

9. **Multi-modal distributions:** Indicating periodic stimulation. One distinguishes between low-frequency (A) and high frequency (B) stimulation.

10. **L-shaped distribution:** Indicating a cell in different firing states whereas one such state is bursting.

The distinction between exponential and long-tail distributions is of interest, as latter may indicate a power-law that describes the firing of the neuron. The observation of power-law distributions can have far-reaching consequences. For example, for the typically observed decay exponent, the expectation (or mean) does not exist, making it impossible to speak of a mean interspike interval. Power-law decaying probability distributions indicate the presence of long (time-)correlations as exemplified, e.g., by intermittent systems. To quantify, if a certain distribution is closer to a exponential or a long tail type, we use the following procedure: Let N_i be the number of ISI that fall into the i-th bin τ_i ($i = 1, \ldots, n$) of the histogram approximating the probability distribution. To avoid tampering effects of short ISI, the bin number m with $N_m = \max\{N_1, \ldots, N_n\}$ is identified. A cut-off parameter $\kappa \in \mathbb{N}$ determines, up to which bin number κm the left side of the distribution is removed (a reasonable choice is $\kappa = 2$). For the remaining distribution, the best fits (minimizing mean square errors) of an exponential function $Exp(t) = e^{-\alpha t}$ and a power-law function $Pow(t) = t^{-\alpha}$ are calculated (α: fitting parameter). The fit coefficient is defined as:

Definition 7.4 *For the best exponential fit function $Exp(t)$ of the shortened distribution given by $\{(\tau_{\kappa m+1}, N_{\kappa m+1}), \ldots (\tau_n, N_n)\}$ and the best power-law fit function $Pow(t)$, the fit-*

coefficient $\mathcal{F}(\kappa)$ is defined as:

$$\mathcal{F}(\kappa) = \frac{1}{\sqrt{2}(n-\kappa m)} \left(\sum_{i=\kappa m+1}^{n^*} (Exp(i\Delta\tau) - N_i)^2 - \sum_{i=2\kappa+1}^{n^*} (Pow(i\Delta\tau) - N_i)^2 \right)$$

where $\sum_{i=2\kappa+1}^{n^*}$ indicates that the sum only goes over those bins τ_i where $N_i \neq 0$.

The distribution is of long-tail type for $\mathcal{F}(\kappa) > 0$ and of exponential type for $\mathcal{F}(\kappa) < 0$. If $\mathcal{F}(\kappa) \approx 0$, neither function fits the distribution well. As one fit is usually far better than the other, we use the logarithm $\mathcal{F}_{\log}(\kappa) = \text{sig}(\mathcal{F}(\kappa)) \ln(\text{abs}(\mathcal{F}(\kappa))$, for $-1 < \mathcal{F}(\kappa) < 1$: $\mathcal{F}_{\log}(\kappa) = 0$. In Fig. 7.6.a and 7.6.b, we display two distributions generated from model spike trains – one distribution is of exponential type, and one is of long-tail type. The values of $\mathcal{F}_{\log}(\kappa)$ obtained of this example are -7 for the exponential distribution and 5 for the long-tail distribution.

7.3 Correlation Integral based Pattern Discovery

Histogram-based tools for pattern detection require considerable efforts when longer patterns are investigated. Furthermore, the binning introduces an unwanted bias, e.g. when the noise level of the system is investigated. To offer an alternative for interval patterns, we introduced a unbiased statistical approach for pattern detection based on the correlation integral [428, 429]. The method is usually applied to spike trains in the interval representation. If interval patterns with additional spikes are object of the analysis, then the set of all n-th order intervals up to a chosen $n_{\max}$ is investigated.

7.3.1 The Correlation Integral

The correlation integral was originally designed for the determination of the correlation dimension [504]. In this section we explore its potential for the detection of interval patterns in spike trains. Consider a spike train in the interval representation $\{x_1, \ldots x_L\}$ that is embedded in the dimension m. The correlation integral is defined as follows:

Definition 7.5 *For a spike train in the interval representation $\{x_1, \ldots x_L\}$ embedded in the dimension m, the* correlation integral $C_N^{(m)}(\epsilon)$ *is*

$$C_N^{(m)}(\epsilon) = \frac{1}{N(N-1)} \sum_{i \neq j} \theta(\epsilon - \|\xi_i^{(m)} - \xi_j^{(m)}\|),$$

where $\theta(x)$ is the Heaviside function ($\theta(x) = 0$ for $x \leq 0$ and $\theta(x) = 1$ for $x > 0$) and N is the number of embedded points ($N \leq L - m + 1$)

Calculating $C_N^{(m)}(\epsilon)$ scales with $\sim N^2$, making this method computationally inexpensive compared to template-based methods (see section 7.4). Different norms $\|\cdot\|$ can be used to compute $C_N^{(m)}(\epsilon)$. In most cases, the maximum norm is advantageous, as this choice speeds up the computation, and allows an easy comparison of results obtained for different

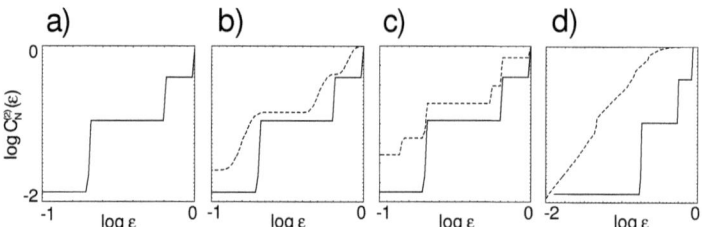

Figure 7.7: Log-log plot steps from different ISI models ($m = 2$, Euclidean norm). a) low-noise case (noise ±1%; solid line in all four plots), b) noise ±10%, c) unstable periodic orbits, d) patterns within noisy background (Figure taken from [429]).

embedding dimensions. Alternatively, the Euclidean norm is used. The connection between $C_N^{(m)}(\epsilon)$ and interval patterns is surprisingly simple: Patterns lead to clusters in the embedding space. For the calculation of $C_N^{(m)}(\epsilon)$, an embedded point $\xi_0^{(m)}$ is chosen at random. Then, the number of points entering its ϵ-neighborhood is evaluated, as ϵ is enlarged. If the chosen point belongs to a cluster, many points will join the ϵ-neighborhood. Once the cluster size is reached, less points are recruited, leading to a slower increase of $C_N^{(m)}(\epsilon)$. When, as required by the correlation integral, an average over different points is taken, pieces of fast increase of $C_N^{(m)}(\epsilon)$ interchange with pieces of slow increase. This leads to a staircase-like graph of the correlation integral. The denser the clustered regions, the more prominent the step-wise structures. Plotting $C_N^{(m)}(\epsilon)$ on a log-log scale not only preserves these structures but enhances the representation of small-scale steps (in the following, we use the logarithm over basis 2). To show the emergence of steps, we constructed a series from a repetition of the sequence {1,2,4}, where the sequence numbers can be interpreted as ISI durations measured in ms. The embedding of this series for $m = 2$ leads to three clusters, represented by the points $P_1 = \{1,2\}$, $P_2 = \{2,4\}$ and $P_3 = \{4,1\}$. Calculating the correlation integral and plotting $\log C_N^{(m)}(\epsilon)$ against $\log \epsilon$ does indeed lead to a clean-cut staircase structure (Fig. 7.7). In practical applications, the steps in the log-log plot generally become less salient due to influences that will be discussed below. In this case, the difference $\Delta \log C_N^{(m)}(\epsilon_i) := \log C_N^{(m)}(\epsilon_{i+1}) - \log C_N^{(m)}(\epsilon_i)$ is a more sensitive indicator of clusters. For small ϵ-neighborhoods, the log-log plot is affected by strong statistical fluctuations. These regions, however, are easily identified and excluded from the analysis.

7.3.2 Smeared log-log Steps

In natural systems, the steps are smeared. We investigated three causes. The first is noise, which is naturally present in measured ISI series. This can be modelled by adding uniform noise to our ISI series {1,2,4,1,2,4,...}. Added noise causes the point clusters in the embedding space to become more dispersed. Consequently the effects of small amounts of

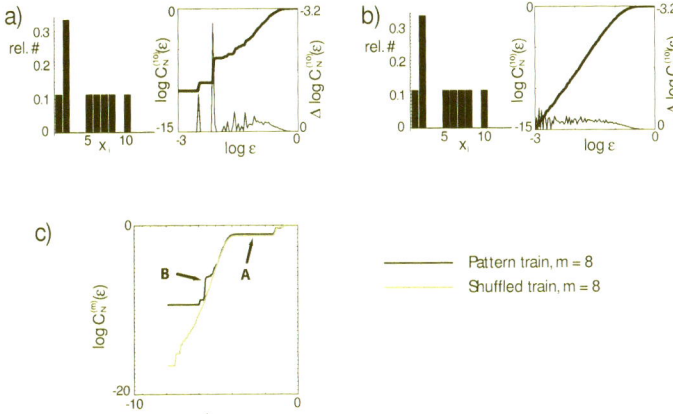

Figure 7.8: Series composed of patterns (a) and series constructed by a random selection of intervals (b), with identical ISI distributions (left). Steps (y-axis: $\log C^{(10)}(\epsilon)$, thick line) only emerge in the presence of patterns. Steps are reflected by peaks in the difference plot (y-axis: $\Delta \log C^{(10)}(\epsilon)$, thin line), respectively ($m = 10$, Euclidean norm. Figure adapted from [429]). c) Comparison between patterned and shuffled trains containing long ISIs ($m = 8$).

noise will only be visible at the step boundaries. As the noise increases, its effects penetrate towards the centers. This is visible in Fig. 7.7.b where the horizontal parts of the steps have become narrower, and the vertical parts less steep. Second, the generator of the ISI series could be chaotic in nature. In this case, a distance from a given unstable periodic orbit grows at $e^{t\lambda}$, where t denotes the time and λ is the (positive) LYAPUNOV exponent of the orbit. This implies that the repetition of any sequence is less likely the larger λ. Moreover, because the decay from the unstable orbit is deterministic, additional (pseudo) orbits will emerge, increasing the number of steps. We can simulate this situation with a simple series composed as follows: With probability $p_1 = 0.5$ we take the whole sequence $\{1,2,4\}$, with $p_2 = 0.31$ the subsequence $\{1,2\}$, and with $p_3 = 0.19$ the subsequence $\{1\}$ (this choice leads to $\frac{p_1}{p_2} \simeq \frac{p_2}{p_3}$). The results (Fig. 7.7.c) show five instead of three steps, indicating that additional orbits have been generated. A third option is that patterns occur within a noisy background. In this case, the pattern only appears intermittently. As a consequence, the fraction of points belonging to clusters in the embedding space is diminished, implying that the steps in the log-log plot become less prominent. To simulate this situation, we took with probability $p = 0.5$ the sequence $\{1,2,4\}$, otherwise, three interspike intervals were randomly drawn from the interval $(0, 4]$. The results (Fig. 7.7.d) show that the number of steps remains unaffected indeed, but the steps themselves have become much less pronounced.

	Embedding dimension m						
	1	2	3	4	5	6	7
Pattern size l 1	0	0	0	0	0	0	0
2	1	1	1	1	1	1	1
3	3	2	1	1	1	1	1
4	6	4	3	2	2	2	2
5	10	8	6	4	2	2	2
6	15	12	9	7	5	3	3

Table 7.2: Maximum number of steps $s(m, n)$ as a function of the embedding dimension m and pattern length l.

In natural systems, more than one pattern may be present. To analyze this situation, we assembled a series by randomly choosing among sequences {2,6,10}, {8,2,1}, {2,7,5}. To contrast this with random firing, we assembled a second series by randomly selecting intervals from the concatenated set {2,6,10,8,2,1,2,7,5}. Thus, both series are based on identical probability distributions. Our analysis (Fig. 7.8 a,b, respectively) shows that steps (peaks in the difference plot) emerge only if patterns are present. However, although the emergence of steps is a necessary condition for the presence of repeated ISI sequences in the data, it is not sufficient: If the series contains ISI of (at least) two clearly distinct length scales and the smaller ISI form the large majority in the data set, two different types of points are present in the embedding space, if the embedding dimension is small (i.e. $m <$ 10): Those whose coordinates consist only of the small ISI, and those that contain at least one large ISI. The distances between the points of the first category are considerably smaller than those between points of the first and second category, because of the large interval. This leads to a pronounced step in the log-log plot. Tho demonstrate this effect, we randomly inserted ISIs of length 100 into the series used for Fig. 7.8.a, such that 5% of all ISI had length 100, and we shuffled the series. In Fig. 7.8.c, the effect of these large ISI is visible: For $m = 8$, the pronounced step A emerges in the patterned series as well as in the shuffled series, whereas the smaller steps B emerging from the sequences are not present in the shuffled train. Remember that the question whether step A should count as an indication of a pattern depends on the chosen background. If not ISI shuffling, but a homogeneous POISSON background is considered as relevant, then the appearance of a particular length scale is also considered a pattern.

7.3.3 Pattern Length Estimation

Once the presence of patterns has been established, an estimate of the pattern length can be given. That this is possible is motivated by the following argument. Using the maximum norm, the distance between two points is defined as the largest coordinate difference. An increase of the embedding dimension yields ever more coordinate pairs, causing the presence of a particularly large difference to dominate. Consequently, the number of steps calculated for pattern length l decreases with increased embedding dimension m. The maximum number of steps $s(l, m)$ can be numerically computed as follows. We start from a series generated by a repetition of a sequence of length l. Additionally, we require that the elements $\{x_1, \ldots, x_l\}$ yield distinct coordinate differences $|x_i - x_j|$. After choosing an embedding dimension m, l distinct embedded points are generated. On this set of points, the maximum norm induces classes of equal inter-point distances. The number of these classes equals $s(l,m)$. The actual calculation of $s(l,m)$ can be done using a computer program (which exhausts ordinary computer capabilities very quickly), or by an unexpectedly involved analytical calculation [620]. The lowest numbers $s(l,m)$ are given in Table 7.2. They clearly confirm the anticipated decrease of the number of steps as a function of m. For the series generated from the sequence $\{5, 24, 37, 44, 59\}$, our correlation integral approach is able to reproduce the predicted decrease of $s(l,m)$ (Fig. 7.9.a): In embedding dimension $m = 1$, all ten possible nonzero differences are visible. As m increases towards 5, the number of steps decreases in accordance with Table 7.2, remaining constant for $m > 5$.

The behavior reported in Table 7.2 only holds if the series are created by repeating a pattern based on distinct inter-coordinate differences. In more general cases, the exact determination of the pattern length is hampered by a basic difficulty: If one single step

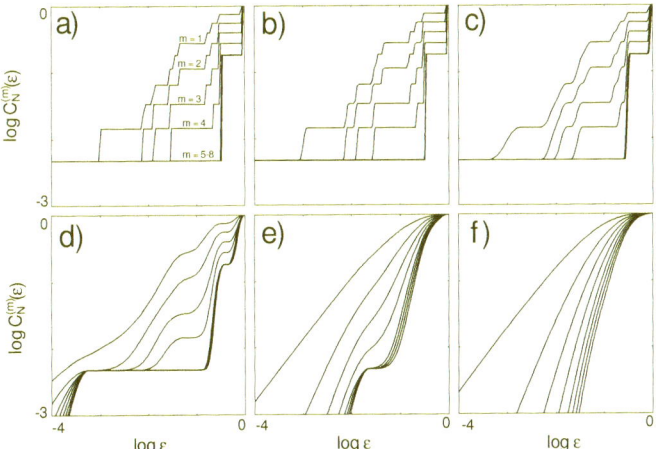

Figure 7.9: Dependence of step appearance from noise level. a): The number of steps decreases as the embedding dimension increases ($m = 1, \ldots, 8$; sequence length: 5). At $m = 1$ there are 10 steps, in agreement with Table 7.2. b-f): Behavior in the presence of additive noise (noise levels 8%, 32%, 128%, 512% and 1024%, see text). The number of steps for $m = 1, \ldots, 4$ decreases for increased noise. The clearest step always emerges for $m = 5$, indicating a sequence of length 5 (Figure taken from [429]).

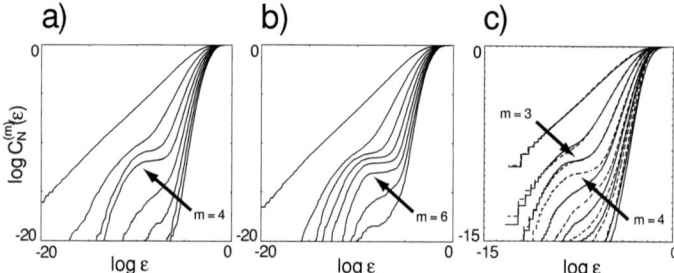

Figure 7.10: Pattern-length indicator $l = m$: Sequences of length n in a homogeneous POISSON background (embedding dimensions $m = 1, \ldots, 8$): a) $l = 4$ – most pronounced step at $m = 4$; b) $l = 6$ – most pronounced step at $m = 6$. c) Sequences of $l = 3$ and $l' = 4$ included with different ratios (3 : 1 solid line, 1 : 3 dashed line) – most pronounced step where m equals the length of the dominating sequence (Figure taken from [429]).

emerges, this can either be due to a interval pattern of length 2 or two interval patterns of length 1. A greater number of steps further complicates this problem. As a consequence, Table 7.2 can only serve as a rough guideline. Fortunately, a helpful indicator for the pattern length exists. A pattern will emerge in the embedded ISI series in its most genuine form (it is neither cut into pieces, nor spoilt by foreign points) if the pattern length equals the chosen embedding dimension ($l = m$). In Fig. 7.9.a, the most pronounced steps appear at $m = 5$, correctly indicating a pattern of length $l = 5$. To investigate the reliability of the criterion in natural settings, noise was added to the series generated from the sequence {5, 24, 37, 44, 59}. For the following, we define the noise strength as the ratio of the noise sampling interval over the shortest sequence interval. The results (Fig. 7.9 b-f) demonstrate that the pattern length can be reliably estimated up to a noise level of 512% (Fig. 7.9.e), where the most pronounced step still appears at $m = 5$. The number of steps for $m < 5$ is affected by the noise: For $m = 1$, for example, 9 steps are present at 8% noise (Fig. 7.9.b), 7 steps at 32% (Fig. 7.9.c) and 3 steps at 128% (Fig. 7.9.d). The step-structure disappears, if the noise level reaches the size of the largest sequence element (Fig. 7.9.f). Thus, the observation that the most pronounced step appears at $l = m$, yields a valuable criterion for estimating the pattern length.

We found that this criterion also extends to less ideal settings. To illustrate, we injected the sequences {5, 25, 10, 2} and {5, 25, 10, 2, 17, 33}, each with probability $p = 0.06$, into a noisy background generated by a homogeneous POISSON process with refractory period. The POISSON distribution was tuned to produce a mean identical with that of the patterns. The clearest steps emerge at the embedding dimensions 4 and 6 (Fig. 7.10.a,b) showing that also in this case the pattern length can be estimated. We refined our investigation by varying the injection probabilities. Using the sequences {4, 17, 12} and {5, 25, 10, 2}, the first sequence

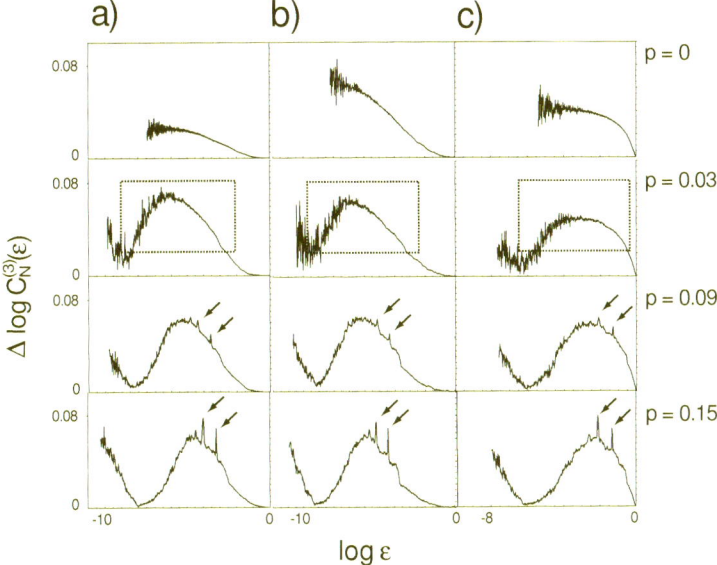

Figure 7.11: Detection of sequences injected (with $p = 0, 0.03, 0.09, 0.15$) into random backgrounds based on the difference quotient: a) Homogeneous POISSON background, b) inhomogeneous POISSON background, c) uniform background. Already at a low injection probability of $p = 0.03$, a hump emerges (dashed boxes). At $p = 0.09$, smaller peaks indicate the statistically significant accumulation of pattern-induced distances ($m = 3$) (Figure taken from [429]).

was chosen with $p = 0.12$ and the second with $p = 0.04$. We compared this series with a series based on interchanged probabilities. The outcome shows that the clearest steps emerge for $m = n$ where the pattern with the higher probability dominates (Fig. 7.10.c). If the two probabilities are similar, the estimation may be hampered by effects of interference. A means of quantifying the 'clarity' of a step, is to calculate the ratio between the slopes of the flat and of the steep part of the steps. Consistently, the embedding dimension for which the slope-ratio reaches a minimum coincides with the pattern length.

Currently, alternative models of noisy backgrounds exist. To show that our results hold regardless of the model applied, we injected the sequence $\{33, 14, 22\}$ into backgrounds generated by a) a homogeneous POISSON process with refractory period, b) an inhomogeneous, sinusoidally modulated POISSON process with refractory period, and c) a uniform random process on the interval $(0, 46]$, using injection probabilities $p \in \{0, 0.03, 0.09, 0.15\}$. The re-

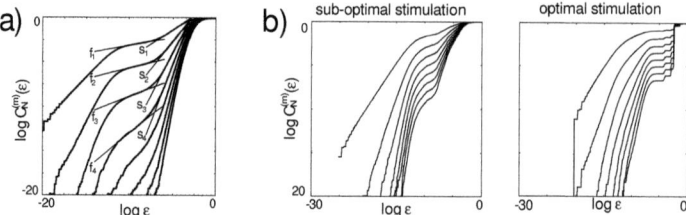

Figure 7.12: Illustration for example 7.2: Log-log plot of cat V1 data. a) Pattern-length estimation: The most pronounced steps appear at $m = 2$ and $m = 3$, indicating patterns of length 2 and 3. b) Optimal stimulation leads to a pattern sharpening effect, clearly visible in the log-log plot (Figure adapted from [429]).

sults show that the nature of the noisy background has a negligible influence on the pattern detectability. Instead, the injection probability is decisive (Fig. 7.11). Whereas two sharp peaks are obtained for $p = 0.15$ and $p = 0.09$ (arrows), only one large hump emerges for $p = 0.03$. A single broad peak indicates a reduced frequency of short intervals, which is the first indicator of patterns at lowest injection probability. Two narrow peaks indicate the pattern-generated statistically significant accumulation of particular distances.

Example 7.2: We demonstrate the use of correlation integral based pattern discovery using data obtained from striate cortex of cats (for details see section 8.1). First, we show the determination of the length of interval patterns: In Fig. 7.12, we display the log-log plot of the correlation integral for the embedding dimensions $m = 1, \ldots, 8$ for one spike train. It can be seen that the steps are expressed differently in the different embedding dimension. To quantify these differences, the ratio of the slope of the flat part on the left side of the step f_m and the slope of the step s_m is calculated for different embedding dimensions m (Table 7.3.3). This analysis reveals ratio minima at $m = 2$ and $m = 3$, indicating that patterns of length 2 and 3 are present. As a second example, we display the log-log plots for spike trains from a neuron under different stimulus conditions. Here, a pattern-sharpening effect can be observed in the log-log plot [704]. These examples demonstrates how information about pattern length and noise level can be obtained by comparing the plots for different embedding dimensions.

m	1	2	3	4
$\frac{f_m}{s_m}$	0.27	0.19	0.20	0.30

Table 7.3: Ratio of the flat slope (f_m) over the steep slope (s_m) of a step, as a function of the embedding dimension m, of the data shown in Fig. 7.12.

7.4 Template-based Pattern Recognition

A template-based approach for pattern detection predefines a template and then counts the occurrence of matchings of the template within the data. Therefore, template-based approaches fall into the category of pattern recognition procedures – unless all *possible* templates are tested. In this section, we introduce template-based methods for pattern recognition in spike trains and discuss details of the shortcomings of template-based methods. We only discuss single train interval patterns of single spikes. For any other pattern type, the procedure is analogous [423, 490].

7.4.1 The Template Approach

Assume a spike train given in the interval representation $\{x_1 \ldots, x_L\}$ and a template determined as an ISI sequence of the form $\bar{x} = \{\bar{x}_1, \ldots \bar{x}_l\}$. The template approach requires the comparison of the template with the data. Due to the noise in the system, perfect matching between a sequence $\{x_i, \ldots x_{i+l-1}\}$ and the template sequence $\{\bar{x}_1, \ldots \bar{x}_l\}$ so that $\bar{x}_{j-i+1} = x_j$, $\forall j = i, \ldots i + l - 1$ cannot be expected. Therefore, one must find a way to *quantify* the degree of matching between the template and the data. This is done by defining a kernel function over a matching interval, and a matching threshold [444].[3] The counting procedure (see below) is performed on the spike train in the timing representation.

Definition 7.6 *A template kernel $K_x(t)$ is a function of the form*

$$K_x(t) = \begin{cases} K(t) & if \quad -b \leq t \leq b \\ 0 & otherwise \end{cases}$$

The interval $[-b, b]$ is called matching interval. *The function $K(t)$ can have different forms (e.g. rectangular, triangular or gaussian).*

Definition 7.7 *For a interval sequence $\{\bar{x}_1, \ldots \bar{x}_l\}$ and a template kernel $K_x(t)$, a template $\mathcal{T}_l(t)$ of length l is*

$$\mathcal{T}_l(t) = K_x(t) + \sum_{k=1}^{l} K_x\left(t - \sum_{i=1}^{k} \bar{x}_i\right)$$

The interval sequence $\{\bar{x}_1, \ldots \bar{x}_l\}$ used to define the template is called template sequence.

Definition 7.8 *For a spike train $\{t_1, \ldots, t_L\}$ and a template $\mathcal{T}_l(t)$, the matching function $M(t^*)$ of the template set at position t^* of the spike train is*

$$M(t^*) = \sum_{\forall t_i \geq t^* - b} \mathcal{T}_l(t_i - t^*)$$

Definition 7.9 *For a matching function $M(t^*)$ and a matching threshold M_{tresh}, the matching $\mathcal{M}(t^*)$ of the template put at position t^* of the spike train is*

$$\mathcal{M}(t^*) = \begin{cases} 1 & if \quad M(t^*) - M_{tresh} \geq 0 \\ 0 & otherwise \end{cases}$$

---

[3]In the original proposal [444], the template kernel has not been applied to the first spike of the template. This underestimates the number of template matches.

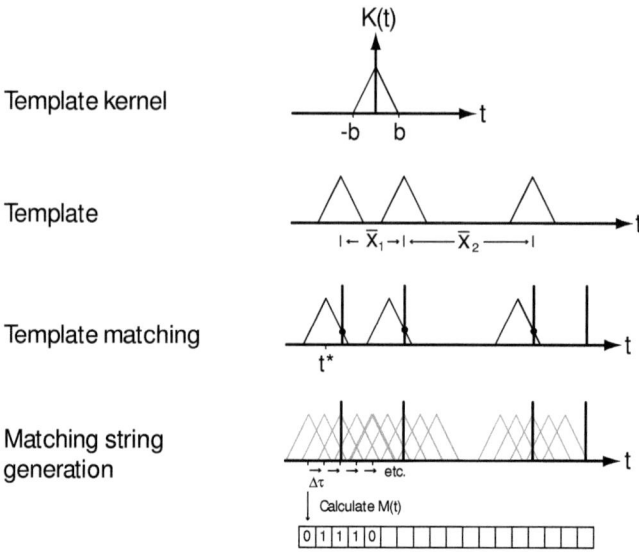

Figure 7.13: The basic template matching algorithm (from top to bottom): First, a template kernel $K(t)$ with matching interval $[-b, b]$ is defined. Then, by choosing a template sequence $\{\bar{x}_1, \bar{x}_2\}$, the template is generated. Then, after defining a matching threshold, the matching function for a certain t^* determined if the template matches a spike train ($\mathcal{M}(t^*) = 1$ in the example). By applying this procedure from the beginning of the spike train and for steps $\Delta \tau$, one obtains the the matching string $\bar{\mathcal{M}}$. The number of switches $0 \to 1$ defines the number of template matches.

Usually, the matching is calculated for succeeding values of $t^* = t_1^*, \ldots, t_n^*$ and the obtained values 0 or 1 are then written in a matching string $\bar{\mathcal{M}} = \{\mathcal{M}(t_1^*) \ldots \mathcal{M}(t_n^*)\}$.

The procedure to calculate the number of template matches is described as follows (see Fig 7.4.1): For a given spike train and a template, a bin-size $\Delta \tau$ (e.g. 1 ms) and a matching threshold M_{tresh} are chosen. Then, the matching string is calculated for $t^* = i \cdot \Delta \tau$ where $i = 0, \ldots, n$ (n is the number of bins that cover the measurement interval of the spike train). The number of switches $0 \to 1$ in the matching string defines the number of template matches. If the matching string already starts with a '1', then 1 is added to the number of switches. Note, that this general procedure allows the detection of patterns where additional spikes are allowed. It is certainly possible to restrict the template approach to detect interval patterns where no such spikes are allowed.

7.4.2 Problems with Templates

Template based methods suffer from two fundamental difficulties. First, the detection relies on the set of pre-chosen templates. As the patterns are *a priori* unknown, large template sets are required that include all potential patterns. Second, a bias is introduced by choosing the template kernel (choice of $K(t)$ and $[-b, b]$). Especially the choice for a tolerance for template matching is problematic. Adopted matching intervals range from fractions of one [714] to a few milliseconds [720], demonstrating the difficulty in determining the required accuracy. We illustrate both aspects separately.

Combinatorial explosion: Unbiased template-based pattern detection would require testing all possible templates (the *ad hoc method* [353]). The number of operations required for unbiased testing can be estimated as follows: Let $\{x_1, \ldots, x_L\}$ be a spike train in the interval representation with $x_{\min}$ denoting the smallest and $x_{\max}$ the largest element. Let $\{\bar{x}_1, \ldots, \bar{x}_l\}$ be the template sequence and $[-b, b]$ be the matching interval. An optimal bin size of $\Delta\tau = 2b$ yields $N = \lceil \frac{x_{\max} - x_{\min}}{\Delta\tau} \rceil$ values to test, and N^l templates to match. Moreover, an unbiased template analysis requires choosing a set B of distinct matching intervals. This leads to $\sim lB(L-l+1)N^l$ templates that need to be compared. The problem is aggravated, when interval patterns in multiple spike trains are target of the analysis, as the templates have to be applied to all possible combinations of spike trains. It is evident that an efficient template analysis can only be performed for small l. This problem can be reduced using the *bootstrap method* [353]: The basic idea is to perform the template analysis only for small l. Longer templates are then created by using only those intervals that have been found to form short interval patterns.

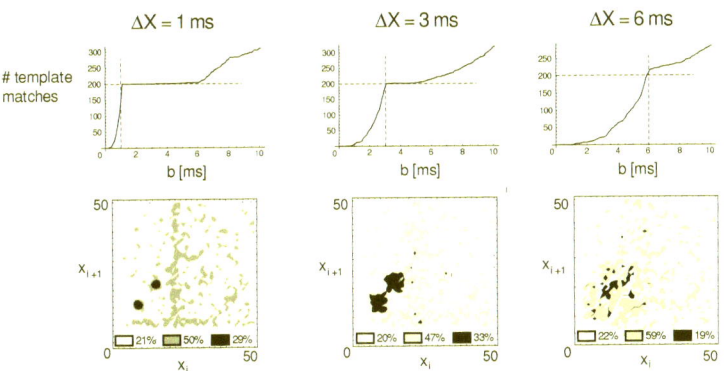

Figure 7.14: The influence of the matching interval $[-b, b]$ of a template on the number of template matches. First row: Pattern with variance $\Delta I = 1$ ms. Second row: Pattern with variance $\Delta I = 3$ ms. Third row: Pattern with variance $\Delta I = 6$ ms. Below the 2D histogram.

Matching interval: A way to minimize the number of templates to be tested is to get a valuable estimation of the size of the matching interval: This can be obtained by applying the 2D histogram or the correlation integral method. If the noise that affects the pattern is small the clusters in the embedding space are dense, which shows up accordingly in the 2D histogram or in the slope of the steps in the correlation integral. On the other hand, by systematically varying the matching interval for a given template, one is also able to obtain an estimation for the noise level, as the number of matchings of templates reaches a plateau for the matching interval that corresponds to the noise level [564].

Example 7.3: We show the relation between matching interval and noise level using the following example: We generated a spike train from a homogeneous POISSON process with refractory period and inserted the sequence $\{(9 \pm \Delta X), (15 \pm \Delta X), (22 \pm \Delta X)\}$ such that 200 sequences where in each test train. For generating the first train, the interval variance was set $\Delta X = 1$ ms, for the second train, $\Delta X = 3$ ms and for the third train, $\Delta X = 6$ ms. We then applied a template algorithm (not allowing inserted spikes) and varied b from 0 to 10 ms in steps of 0.1 ms. Figure 7.14 displays the result: For a low variance ($\Delta X = 1$), the number of template matches in dependence from b shows a long plateau phase that gets shorter for increased variance. The plateau indicates the presence of a pattern and its length indicates the noise level.

7.5 Clustering-based Pattern Quantification

7.5.1 The Clustering Algorithm

Using template methods for pattern quantification requires prior knowledge on pattern length and noise level. In this section, we introduce an alternative approach to quantify the occurrence of patterns. As in the previous section, we restrict ourselves to interval patterns of single spikes in a single spike train. The method requires the embedding of a spike train in the interval representation and the application of the sequential superparamagnetic clustering algorithm. The latter is unbiased in the sense that neither the size nor the number of clusters has to be predefined. It delivers excellent results for various problem types [630].

Sequential Superparamagnetic Clustering: The conceptual idea of superparamagnetic clustering can be outlined as follows: Data points are interpreted as particles to which a POTTS-spin is assigned. Each particle can interact via its spin with the particles of a defined neighborhood (usually k-nearest neighbors) . The particles tend to align the direction of their spins depending on the interaction strength, whereas thermal fluctuation opposes this tendency. The interaction strength is a decreasing function of the distance between the points. Groups of particles with aligned spins form clusters, whose size diminish with increasing temperature T. Groups of particles with strong interaction are able to resist this tendency of disintegration. Thus, the size of this cluster is stable over a broad range of T. Clustering is applied for a certain range of $T = 0, \ldots, T_{end}$ in steps of ΔT. Usually at $T = 0$, one cluster is present, which breaks up into smaller clusters for increasing T. The sequential approach allows to take inhomogeneities in the data space into account: The data points of the densest cluster are removed and the clustering algorithm is reapplied to the remaining data set as well as to the removed cluster. The application of superparamagnetic clustering algorithm requires the determination of several parameters. The most important ones are minsize (the minimal size of clusters) and s_θ (the minimal required cluster stability), which define the resolution of the clustering procedure. The other parameters are only of minor interest and basically influence the efficiency of the algorithm. In this way, the clustering algorithm comes equipped with an intrinsic measure for cluster stability s, with $0 \leq s \leq 1$. It sequentially reveals clusters according to their stability, i.e. the most stable cluster is detected first and the remaining cluster is the least stable one. The result of clustering is displayed in a dendrogram that indicates how larger clusters break apart into smaller clusters. Furthermore, the size of the cluster N, T_{max} (the temperature, where all clusters have disintegrated), the cluster stability s (the temperature range over which the cluster remains stable relative to T_{max}), T_{cl} (the temperature range over which the cluster remains stable) and T_{ferro} (the temperature, where the cluster is still in the ferromagnetic phase, i.e. all spins are aligned) are displayed. For a formal description of the algorithm, we refer to Ref. [629].

The superparamagnetic clustering algorithm delivers the size of a cluster and its center of gravity, the standard deviation for each coordinate of the center of gravity and the index numbers of the points which form the cluster. It provides the following information:

- When stable clusters emerge, they indicate the presence of patterns (see below).

- The coordinates of the centers of gravity of the clusters provide information about the interval pattern.

- The standard deviations of the coordinates of the centers of gravity provide information about the cluster density and thus about the noise level.

- By using the index numbers of the points one gains information about the exact occurrence of the patterns within the spike train.

7.5.2 Relating Patterns and Clusters

We have already noticed (section 7.3), that interval patterns in spike trains show up as clusters in the embedding space. In this paragraph, we outline this connection in more detail. We assume that the interval pattern $\{(X_1, \Delta X_1), \ldots, (X_l, \Delta X_l)\}$ is present in the spike train $\{x_1, \ldots x_L\}$. Embedding the spike train leads to points $\xi_k^{(m)} = (x_k, \ldots, x_{k+(m-1)})$, $k = 1, \ldots, N$, where k is the index number of the point, m is the embedding dimension and $N = L - m + 1$ is the number of generated points. The clusters that emerge due to the presence of the pattern consist of points of the form $\xi^{(m)} = \{(X_i \pm \Delta X_i), \ldots (X_j \pm \Delta X_j)\}$ where $j - i = m - 1$. The number of points forming the cluster equals the number of repetitions of the pattern in the spike train. In other words, if the embedding of a spike train leads to clusters in the embedding space, the centers of gravity of the clusters indicate what patterns may be present. The density of the clusters indicate the noise level.

We outline in the following how one infers a pattern from clusters obtained in different embedding dimensions (see also example 7.5). We assume that we have embedded a spike train in the interval representation that contains only one unknown pattern of length l in different embedding dimensions $m = 2, \ldots, M$ (the case $m = 1$ is in most cases uninteresting). For $m = 2$, $l - 1$ clusters will emerge, whose centers of gravity can be described as $(c_1^1, c_2^1), \ldots, (c_1^{l-1}, c_2^{l-1})$. In the symbol c_j^i, i denotes the number of the cluster and j denotes the coordinate number within each center of gravity. When (as we assume) the clusters emerge from only one pattern, they can be re-numbered such that $c_2^1 \approx c_1^2$, $c_2^2 \approx c_1^3, \ldots,$ $c_2^{l-2} \approx c_1^{l-1}$. For $m = 3$, $l - 2$ clusters will appear where a similar relation can be obtained for the coordinates c_j^i of the centers of gravity. For $m = l$, only one cluster will emerge and for $m > l$, no more cluster will appear. The systematic comparison of the centers of gravity of clusters obtained in different embedding dimensions will thus result in the pattern present in the data (see also example 7.5 below). This approach may become complicated if many different patterns are present in the data as several combinations have to be tested. This is, however, a problem for all pattern detection methods.

Cluster stability: In practical applications the parameter stabmin provides additional information on the noise level of the system: if clusters remain stable for large values of stabmin, the noise level is low. It is, however, important to notice, that the embedding of patterned data will automatically lead to stable *and* unstable clusters. To discuss this point in more detail, we again assume that repeating sequences of the form $(X_1 \pm \Delta X_1, \ldots, X_l \pm \Delta X_l)$ are present in a spike train. Thus, points $(X_{l-j+1}, \ldots, X_l, x_{i+j}, \ldots, x_{i+(m-1)})$ and $(x_i, \ldots, x_{i+(m-1)-j}, X_1, \ldots, X_j)$, $0 < j < m$ and $m \leq l + 1$ will emerge in the embedding space of dimension m, where only the first or last $j < m$ coordinates derive from the pattern (see Fig. 7.15 as an example). These points form unstable clusters, as $m - j$ coordinates emerge from the random background and not from the pattern. Unstable means, that the clusters decrease rapidly in size for increasing temperature T. This fact is expressed in a much higher standard deviation of *some* coordinates of the centers of gravity of unstable clusters – namely those coordinates which are not related to the pattern. Clusters related to patterns, on the other hand, show a substantial decrease of the standard deviation for *all* coordinates of the center of gravity of the cluster at some critical temperature T' (due to the split of a larger cluster) and an almost constant size for a range [T', T''].

Example 7.4: Starting point is a ISI series provided by a uniform random process on the interval $[1,20]$, into which the sequence $(2,15,8)$ and all its subsequences $\{(2,15), (15,8), (2), (15), (8)\}$ have been injected. Furthermore, uniform noise has been added to the series. Its embedding in $m = 1, 2, 3$ (see Fig. 7.15.a) and the application of the clustering algorithm resulted in several clusters. The centers of gravity of the stable clusters are (as expected) $(2.0), (15.0), (8.0)$ for $m = 1$, $(2.0, 15.1), (15.1, 8.0)$ for $m = 2$ and $(2.0, 15.0, 8.0)$ for $m = 3$ (encircled clusters in Fig. 7.15.a). The pattern group is thus reconstructed as $\mathcal{P}_3 = \{(2), (15), (8), (2,15), (15,8), (2,15,8)\}$. Note, that one cannot expect perfect matches when comparing according coordinates of the centers of gravity in different embedding dimensions ($15.0 \simeq 15.1$ in the example).

To demonstrate the criterion of cluster-stability, we consider the emergence of the stable cluster indicating the sequence $(2,15,8)$ in $m = 3$ (Fig. 7.15.b). In this cluster, all coordinates of the center of gravity have settled down to the values 2, 15, and 8, their standard deviations remain small and the cluster size remains largely unchanged for $0.05 \leq \mathsf{T} \leq 0.12$. In clusters that emerge from points, where only the first two coordinates emerge from the pattern, the situation is different: In the example, the third coordinate of the center of gravity changes its value considerably for increasing T, its standard deviation stays large compared to those of the other coordinates and the cluster size also diminishes considerably for increasing T. If stabmin is chosen appropriately, these unstable clusters will not emerge as clusters in the algorithm.

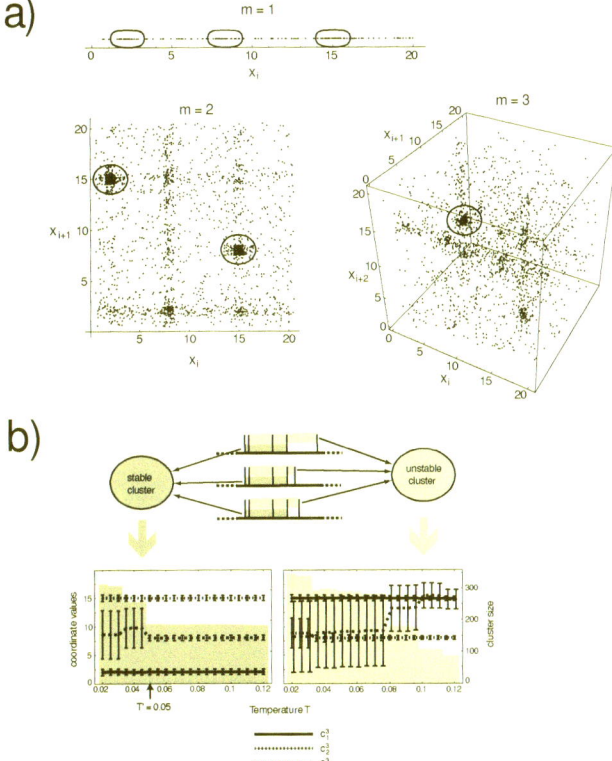

Figure 7.15: Relating patterns and clusters. a) Embedding of a spike train containing one pattern (example 7.5) in $m = 1, \ldots, 3$. Encircled are clusters, whose centers of gravity correspond to the sequences forming the pattern. c) Identifying clusters related to patterns using the criterion of cluster-stability for $m = 3$ (see text).

As the clustering approach is able to quantify a whole pattern group, the method is suited to determine the stability of a pattern (see section 6.1.3). If, however, the number of patterns present in the data as well as the noise level are high, the clustering based method may only be used as a tool to find an appropriate template.

7.6 The LZ-Distance of Spike Trains

A thorough classification of spike trains requires two ingredients: The choice of an appropriate distance measure, and an efficient clustering algorithm [477]. In this section, we focus on the first step. For the second step, we use the sequential superparamagnetic clustering algorithm (see previous section). We introduce a novel distance measure based on the Lempel-Ziv-complexity (LZ-complexity) [143] that does not require tuning parameters, is easy to implement and computationally cheap. The measure is applied to spike trains transformed into bitstrings and considers spike trains with similar compression properties as close. Used in spike train clustering, the measure groups together spike trains with similar but possibly delayed firing patterns (Fig. 7.16). This is important as neurons that receive similar input or/and perform similar computations might have a tendency to respond by similar patterns. Due to the complex neuronal connectivity of the cortex, similar firing patterns may occur delayed in different neurons and distance measures focussing on synchronized firing would not be able to classify such cells as similar. Furthermore, when firing reliability of single neurons under *in vivo* conditions is considered, the measure again allows to deal with different delays of firing patterns that may appear in the different trials of the same experiment. Such differences may result from influences to the neuron under investigation that are, in the *in vivo* condition, beyond the experimenter's control. This is of special importance when the firing reliability of neurons in higher cortical areas is considered as it

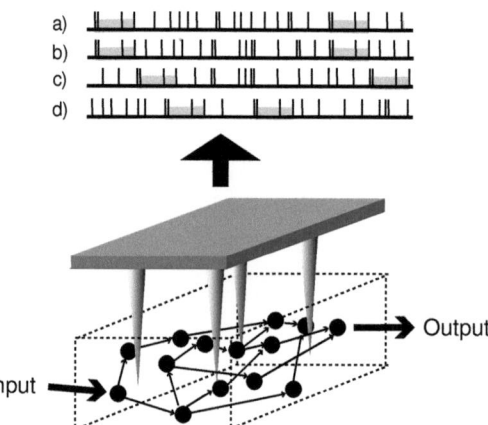

Figure 7.16: Outline of the spike train clustering problem: Multi electrode array recordings probe the spatio-temporal activity within a neural net. Correlation-based distance measures allow to group together spike trains with synchronous spike patterns (a,b; pattern marked with grey bar), but fail to group trains with delayed patterns (c,d).

is known that timing reliability, measured by distances focussing on synchronized firing, will be lost. Thus, the LZ-distance measures an additional aspect of neuronal firing reliability.

7.6.1 The LZ-Distance Measure

The LZ-distance measure is applied to spike trains in the bitstring representation (see section 5.1.1). This string can be viewed as being generated by a more general information source. For this source, we want to find the optimal coding [437]. This coding is based on *parsing*, a procedure to partition the string into non-overlapping substrings, according to some procedure. Based on the concept of LZ-complexity, two distinct parsing procedures have been introduced [143, 291]. Both of them follow the same basic idea: strings are sequentially parsed into sequences that have not occurred. To explain the differences among the two procedures, we introduce the following definitions:

Definition 7.10 *For a bitstring $X_n = (x_1 \ldots x_n)$ of length n ($x_i \in \{0,1\}$), a procedure that partitions X_n in non-overlapping substrings is called a* parsing. *If all substrings of a certain string X_n generated by a parsing are distinct, then the parsing is called a* distinct parsing.

Definition 7.11 *For a bitstring X_n of length n, a substring starting at position i and ending at position j of the string which is the result of a parsing procedure is called a phrase $X_n(i,j)$. The set of phrases generated by a parsing of X_n is denoted with P_{X_n} and the number of phrases $|P_{X_n}|$ is denoted by $c(X_n)$*

Definition 7.12 *If X_n is a bitstring of length n, the set of all substrings of X_n is called* vocabulary, *denoted V_{X_n}.*

Assume a bitstring X_n that has been parsed up to position i, so that $P_{X_n(1,i)}$ is the set of phrases generated so far and $V_{X_n(1,i)}$ is the vocabulary of the parsed substring $X_n(1,i)$. The question is: What will be the next phrase $X_n(i+1,j)$? According to the originally proposed parsing procedure [143], it will be the first substring which is not yet an element of $V_{X_n(1,i)}$ (LZ-76). According to the second proposed parsing procedure [291], it will be the first substring which is not an element of $P_{X_n(1,i)}$ (LZ-78). As an illustration, take the string 0011001010100111. Using the LZ-76 procedure, it will be parsed as 0|01|10|010|101|00111, whereas it will be parsed as 0|01|1|00|10|101|001|11 using the LZ-78 procedure.

Definition 7.13 *If X_n is a bitstring, the* Lempel-Ziv-complexity $K(X_n)$ *(LZ-complexity) of X_n is defined as*

$$K(X_n) = \frac{c(X_n)\log c(X_n)}{n}$$

If a bitstring X_n is the result of a stationary, ergodic process with entropy rate H, then the LZ-complexity is asymptotic optimal, i.e. $\limsup_{n\to\infty} K(X_n) \leq H$ with probability 1 [437]. Stationarity, however, which limits the use of the LZ-complexity for calculating the entropy rate of a spike train [364], is not critical for calculating the LZ-distance. Non-stationarity in firing would just increase the mean distance of spike trains if, for example, the reliability of neuronal firing is calculated. The basic idea of the LZ-distance is as follows: Consider two strings X_n, Y_n of equal length n. From the perspective of LZ-complexity, the amount of information Y_n provides about X_n is given as $K(X_n) - K(X_n|Y_n)$, where $c(X_n|Y_n)$ is the size of the difference set $P_{X_n} \setminus P_{Y_n}$. If Y_n provides no information about X_n, then the sets P_{X_n} and P_{Y_n} are disjoint, and $K(X_n) - K(X_n|Y_n) = 0$. If Y_n provides complete information about X_n, then $P_{X_n} \setminus P_{Y_n} = \emptyset$ and $K(X_n) - K(X_n|Y_n) = K(X_n)$. For our definition of the LZ-distance, the numerator $K(X_n) - K(X_n|Y_n)$ is the information Y_n provides about X_n and the denominator X_n serves as a normalization factor such that the distance $d(X_n, Y_n)$ ranges between 0 and 1. The LZ-complexity approximates the Kolmogorov complexity $K_K(X_n)$ of a bitstring and a theorem in the theory of Kolmogorov complexity states that $K_K(X_n) - K_K(X_n|Y_n) \approx K_K(Y_n) - K_K(Y_n|X_n)$ [571]. In practical applications, however, this equality does not hold and we have to calculate $K(X_n) - K(X_n|Y_n)$ as well as $K(Y_n) - K(Y_n|X_n)$ and we take the smaller value in order to ensure $d(X_n, X_m) > 0$ for $n \neq m$. This leads to the following definition of the *LZ-distance*:

Definition 7.14 *For two bitstrings X_n and Y_n of equal length, the Lempel-Ziv-distance $d(X_n, Y_n)$ is:*

$$d(X_n, Y_n) = 1 - \min\left\{\frac{K(X_n) - K(X_n|Y_n)}{K(X_n)}, \frac{K(Y_n) - K(Y_n|X_n)}{K(Y_n)}\right\}$$

The definition satisfies the axioms of a metric [427]. The LZ-distance compares the set of phrases generated by a LZ parsing procedure of two bitstrings originating from corresponding spike trains. A large number of similar patterns appearing in both spike trains should lead to a large overlap of the sets of phrases. We predict that distances between spike trains with similar patterns are small, whereas distances between trains with different patterns are large. Thus, the LZ-distance should allow a classification of spike trains according to temporal similarities unrestricted with regards to temporal synchrony.

7.6.2 Assessment of the LZ-Distance

To verify our predictions, we shall evaluate three test cases. We first compare the parsing procedures LZ-76 and LZ-78 from the practical aspects point of view. We then analyze whether the LZ-distance classifies spike trains into physiologically meaningful classes. We finally compare the LZ-distance with correlation-based distance measures, showing its superiority in the presence of delayed patterns.

Choosing the parsing procedure: We calculated the LZ-distance for five pairs of model spike trains (see Fig. 7.17.a) using both parsing procedures. I: two periodicity-1 spike trains (interspike interval, ISI, = 50 ms) with equal period length and phase shift (25 ms). II: two periodicity-3 spike trains (ISI-pattern (10,5,35)) with no phase shift but spike jitter (± 1 ms). III: two spike trains obtained from an uniform random process on the interval [1, 50]. IV: a periodicity-1 (ISI = 50 ms) and a periodicity-3 (ISI-pattern (10,5,35)) spike train with coincident spikes. V: two periodicity-1 spike trains with different period lengths (50 ms,15 ms). In order to analyze the convergence behavior, for LZ-76 parsing, the length of the trains have been increased from 100 ms up to 4000 ms in steps of 100 ms. For LZ-78 parsing, the lengths of the trains have been increased from 1 sec up to 25 sec in steps of 1 sec. The results (Fig. 7.17.b) show that the distance based on LZ-76 parsing converges

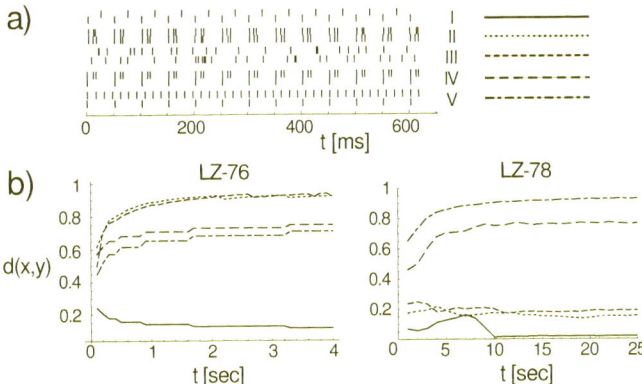

Figure 7.17: Distances of pairs of spike trains. a) Raster plot of pairs I-V. b) Distances of pairs I-V in dependence from the length of the spike train for LZ-76 and LZ-78 parsing.

faster than LZ-78 parsing. Furthermore, same pairs lead to different distances: Using LZ-76 parsing, the trains I are close, the trains IV and V are rather distant and the trains II and III are most distant. Using LZ-78 parsing, the trains I are close as well, but the trains II and III are less close, and the trains IV and V are most distant. This demonstrates, that the LZ-distance based on LZ-78 parsing is more noise-robust than LZ-76 parsing. The latter considers similar, but noisy trains (II) as most distant, whereas the first measure considers spike trains with distinct firing behaviors (IV, V) as most distant. The evaluation of the parsing procedure explains this difference: In LZ-76 parsing, the lengths of the phrases increase much faster during the procedure than in the LZ-78 parsing, because $|V_{X_n(1,i)}| \gg |P_{X_n(1,i)}|$. Therefore, the probability that strings obtained from two similar but noisy spike trains contain many different phrases is higher for the LZ-76 if compared to the LZ-78 parsing. As noise robustness is important when dealing with neuronal data, LZ-78 is better suited for practical purposes, also because it is computationally cheaper.

Spike train classification: To test whether the LZ-distance sorts spike trains in physiologically meaningful categories, we generate a set of spike trains with *in vivo* and model data of comparable firing rate (80-90 spikes/second), as it is obvious that groups of spike trains that differ in firing rate are easily recognized using the LZ-distance. Our multi train data set contains five classes with nine spike trains of 2400 ms length each. Class A: Spike trains of a complex cell (macaque monkey visual cortex data, for further explanation see next section) driven by drifting gratings of 6.25 Hz. Class B: Spike trains of a simple cell driven by drifting gratings of 12.5 Hz. Class C: Spike trains of a homogeneous Poisson process with refractory period that models the firing of the recorded complex cells. Class

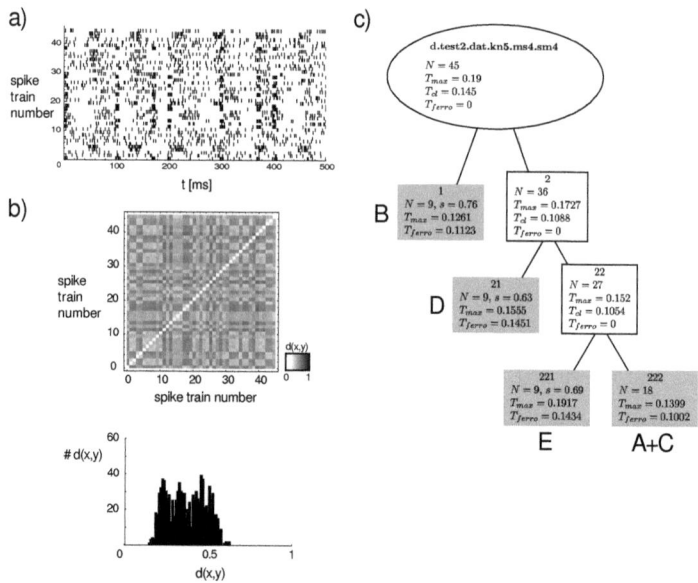

Figure 7.18: Clustering of simulated multi-train data: a) Raster plot of initial spike set. b) distance matrix and histogram of distances obtained after calculating the pairwise LZ-distance. c) Dendrogram outlining the result of clustering.

D: Spike trains of an inhomogeneous, step function driven (12.5 Hz) Poisson process with refractory period that models the firing of the recorded simple cells. Class E) Poisson spike trains with noisy, synchronized bursts. We randomized the order of the spike trains in order to obtain a multi train data set (Fig. 7.18.a). After calculating the LZ-distance between all trains (for the resulting distance matrix and the distribution of distances see Fig. 7.18.b), clustering led to the following result: The classes B, D and E have been separated, whereas the classes A and C fell in the same cluster (Fig. 7.18.c). If the algorithm is applied to the remaining cluster for a decreased s_θ, only an incomplete separation between spike trains of the classes A and C is possible, as two smaller clusters of 5 (spike trains of class C) and 13 elements (spike trains of classes A and C) emerge (not shown). Two main conclusions can be drawn from this result: First, we are able – up to a certain degree set by s_θ – to correctly classify spike trains with comparable firing rate, but differing temporal structures. Second, we are able to interpret classification failures as in the incomplete separation between spike trains of the classes A and C. Here, the spike trains C derive from a model of the firing of a

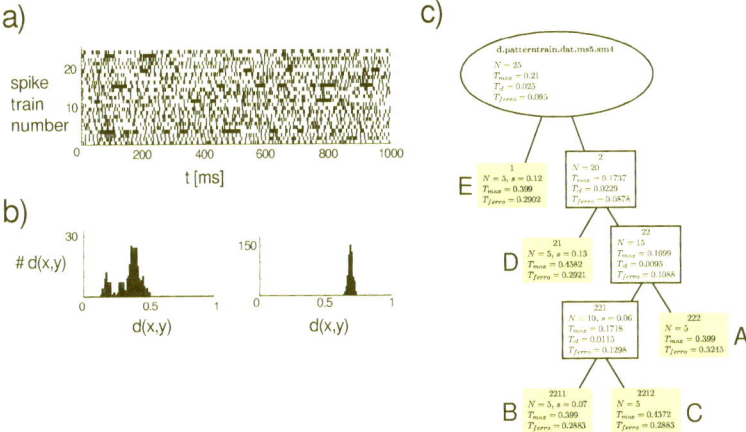

Figure 7.19: Comparison of distance measures. a) Test trains b) Distributions of LZ-distances (left) and C-distance. c) Dendrogram of clustering the spike train set using the LZ-distance: All five classes are clearly recognized.

complex cell and the incomplete separation indicates that the firing behavior of the complex cell appears to be properly modelled by a Poisson process with refractory period. The firing of the simple cell, on the other hand, is not properly modelled by the inhomogeneous Poisson process, as the two classes B and D are separated by the clustering process.

Comparison with C-distance: In a second step, we cluster model spike trains of similar rate that contain different types of repeating sequences of interspike intervals (interval patterns). The repeating sequences within spike trains of a single class characterized by one type of interval pattern are not synchronized, but randomly distributed within each train. This challenging task for spike train clustering is performed using the LZ-distance and the C-distance ([663], see 7.1.3), the latter is a common distance measure for spike train clustering problems [477]. We generated five classes of spike trains, characterized by the following interval patterns. Class A: (4,4), class B: (13,13,13), class C: (5,20,3), class D: (3,16,3,16), and class E: (1,4,7,2,6,11). Each spike train (five per class) was generated such that 50% of the interspike intervals of the train originate from the sequence and 50% from a homogeneous Poisson process. The rate of the process has been adapted for each class in order to generate almost identical mean firing rates for all spike trains (92-94 spikes per second). The order of the spike trains was again randomized to generate our multi train data set (Fig. 7.19.a). To this data set, sequential superparamagnetic clustering has been applied, using both the LZ- and the C-distance. The result shows a clear difference between the two distance measures: Whereas the LZ-distance allowed a clear-cut separation of all five classes (Fig. 7.19.c), the use of the C-distance did not lead to any classification (data not shown). This noticeable difference in performance becomes transparent, if the distribution of the distances is compared (Fig. 7.19.b). For the LZ-distance, their range is ∼0.4 with a multi-modal distribution (indicating the structure within the data set), whereas for the C-distance their range is ∼0.1 with a unimodal distribution. The latter observation implies that a re-scaling would not allow a performance increase. This demonstrates that the LZ-distance is a more general measure allowing to classify spike trains with delayed patterns. The usual correlation measures that focus on synchronized firing fail in such situations.

7.7 Pattern Detection: Summary

We conclude this chapter by providing an integrative overview of the presented pattern detection methods. We distinguish two basic problem types: The reliability of firing discussion and the pattern detection problem. The problem of firing reliability subdivides in two cases: First, the question to what extent spike trains of several trials of the same experiment differ (see section 6.3.3). Second, the question to what extent a single neuron fires reliably during a certain period. The adequate way to solve the former problem is to define an appropriate distance measure and to calculate the distances of the different trials. The higher the variance of the distances, the less reliable is the firing. The latter problem is solved by calculating the pattern stability (see section 6.1.3).

The pattern detection problem can also serve two purposes: First, classifying sets of spike trains according to similarities in firing. Second, detecting patterns and putting them in a coding/computation framework. The first sub-case, the spike train clustering problem, relies on the choice of a distance measure and a clustering algorithm (see section 7.6). The second sub-case requires a more detailed analysis: One first has to determine the pattern type of interest, which leads to an appropriate transformation of the original data (section 6.1.1). If for example burst patterns are of interest, then one transforms a spike train in a series that indicates only the occurrence times of bursts, neglecting the occurrence times of single spikes. Second, one has to choose an appropriate way of randomization (section 6.2.3). In order to evaluate the degree of randomness introduced in the data, the distance between the original train and the shuffled train can be calculated, using an appropriate distance measure. Third, one has to choose the appropriate method by taking care of the bias they introduce: The correlation integral based pattern discovery is less biased than the histogram-based approaches (the latter needs the definition of a bin-size), although it provides a more detailed analysis about the elements of the pattern. For pattern quantification, the clustering approach is less biased than the template-based approach, unless all possible templates and matching intervals are tested. The template approach, however, may allow better comparison of different data sets. The statistical significance of the results obtained is tested by applying the same methods to the randomized trains. The significance in context of a code relation, finally, has to be evaluated in specific experiments. Figure 7.20 provides an overview of the pattern detection problem.

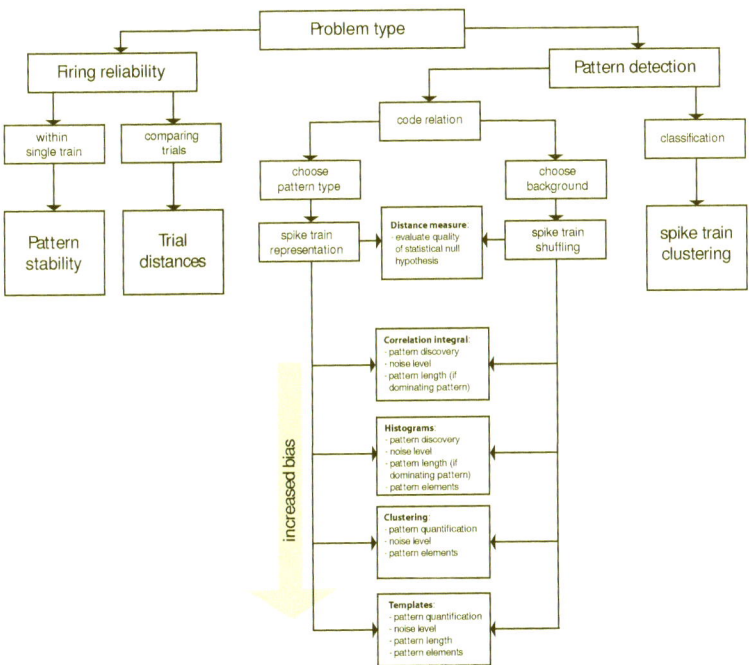

Figure 7.20: The pattern detection problem: Overview of problem types and methods.

Chapter 8

Investigating Patterns

In this chapter we test the methods of chapter 7 and evaluate the predictions that emerge from the STOOP-hypothesis presented in section 6.6.5: We investigate the firing statistics in the visual system of cat and macaque monkey, and the olfactory system of rat, we analyze the firing reliability in the visual system of the monkey, we perform clustering of cell groups in the olfactory system, we test for patterns in the visual system of cat and monkey, and probe the pattern stability for selected cells in the visual system of macaque monkey.

8.1 Description of the Data

8.1.1 Rat Data: Olfactory System (mitral cells)

The rat data derive from experiment performed by ALISTER NICOL of the Cognitive & Behavioral Neuroscience Lab, *Babraham Institute*, Cambridge (UK). Neuronal activity of anaesthetized rats was sampled by means of a micro-electrode array positioned in the olfactory bulb (25% urethane, intraperitoneal, 1.5g per kg body weight). Micro-electrode arrays comprised 30 electrodes (6 × 5, 350 μm separation). Odors, carried as saturated vapors in nitrogen gas (odorless), were mixed in various concentrations with dry air and were delivered for 10 seconds to the rat via the mask. Breathing was monitored and onset of odor delivery was timed to mid-expiration. Neuronal activity was sampled in the 10 second period before odor onset (the pre-stimulus period) and the 10s period of odor presentation (the during-stimulus period). Spikes were detected when a triggering threshold was crossed by the recorded signal from each electrode. This threshold was set at $\geq 2 \times$ the background noise level. The activities of single or multiple neurons were detected in the activity sampled by each electrode. The activities of individual neurons were discriminated from multiple neuron activity using a KOHONEN network to cluster principle components derived from the action potential waveforms, allowing discrimination of activity from 1-6 neurons at each active site. In this way, simultaneous recordings from 54 neurons were made. Odor presentations were 10s in duration. Four different odors (amyl acetate, M-Butanol, DL-camphor, cineole) were

Area	Animal	# cells	# stimuli	# files analyzed
LGN	cat	6	3	18
	monkey	9	6-7	27
V1	cat	4	3-6	17
	monkey	28	5	140
MT	monkey	7	16	35
olfactory bulb	rat	54	16	∼5100

Table 8.1: Overview data set. All measurements have been performed in the *in vivo* condition

presented in four different concentrations (5.42 $\times 10^{-8}$ mol, 2.71 $\times 10^{-7}$ mol, 1.36 $\times 10^{-6}$ mol, 5.42 $\times 10^{-6}$ mol). Thus, pre- and during-stimulus recordings were made under a total of 16 different stimulus conditions. For each condition, there were three trials, resulting in a total of 96 spike trains per neuron. The total number of files analyzed by us was ∼5100, as not all cells have been recorded in all stimulus conditions. All procedures used for data collection conformed to the Animals (Scientific Procedures) Act, 1986 (U.K. Home Office).

8.1.2 Cat Data: Visual System (LGN, striate cortex)

The cat data used in the analysis derive from two sources: The LGN data derive from *in vivo* measurement of an anesthetized cat performed by VALERIO MANTE and MATTEO CARANDINI from the *Smith-Kettlewell Eye Research Institute*, San Francisco (USA). The detailed experimental procedures are outlined in Ref. [485]. We analyzed the responses of six cells to three classes of stimuli: sinusoidal gratings with randomly changing spatial frequency and direction, a cartoon movie, and a natural stimulus video. Each stimulus class consisted of 2-4 different stimuli of 10-30 seconds duration. The stimuli were shown repeatedly (15-25 times in randomized order). To avoid stimulus onset artifacts, the first 50 ms of each trial were deleted. By concatenating these recordings we obtained 18 cell, and stimulus, specific ISI series. The V1 data derive from a series of *in vivo* experiments dedicated to the study of contrast dependence of neural activity in cat visual cortex performed by KEVAN MARTIN of the *Institute of Neuroinformatics*, University and ETH Zurich (Switzerland). The detailed experimental procedures are outlined in Ref. [599], with the modifications according to Ref. [360]. An analysis of the data from our side has been published in Refs. [701, 704]. We used 17 spike trains from 4 unspecified neurons from unspecified layers (12 trains display evoked activity, 5 spontaneous activity) for our analysis.

8.1.3 Monkey Data: Visual System (LGN, V1, MT)

The monkey data derive from measurements in the LGN, V1 and area MT of macaque monkey, performed by ADAM KOHN of the *Center of Neural Science*, New York University (USA). The detailed experimental procedures are outlined in [547]. Measurements of LGN cells have been performed in sufentanil citrate anesthetized macaque monkey to 2-3 trials of drifting sine-wave gratings of increasing temporal frequency (6 to 7 different frequencies). The duration of each measurement was 5 seconds, 9 cells have been measured. For most tests, we only considered the three most responsive stimulus conditions and generated our test files by concatenating the trains of the two or three trials. The measurements of V1 cells have been performed in two different experimental conditions: A total of 10 complex and 6 simple cells has been recorded in sufentanil citrate anesthetized macaque responding 60-125 trials of drifting sine-wave gratings of five different orientations (usually in 22.5° steps). The stimuli typically last for 2.5 sec. and are separated by intervals of 3 sec (gray, isoluminant screen). The five different orientations have been shown in block randomized fashion. The second set of recordings derives from six pairwise simultaneously measurements of, leading to a total of 12 complex cells. The measurements of MT cells (7 in total) have been performed in sufentanil citrate anesthetized macaque with drifting sine-wave gratings. 16 orientations of the stimulus and one presentation of a blank screen to measure the spontaneous rate have been performed, the stimuli lasted 1.28 seconds. We took the 5 most responsive measurements and concatenated the single trials to in total 35 files. Table 8.1 gives an overview of the complete data set.

8.2 First Prediction: Firing statistics

To test the first prediction, whereas only a minority of neurons display a firing statistics that is consistent with the POISSON hypothesis (see section 6.2.2), we investigate all spike trains of our data set by using the fit-coefficient $\mathcal{F}(\kappa)$ according to Def. 7.4.

> **Methods:** For the neurons from the visual system, we only took the files deriving from the three (in one case: two) most responsive stimulus conditions in order to have enough ISI for a proper statistics (for the cat V1 data, we took all files). This led to a set of in total 166 files. Inappropriate files were excluded using the following criteria: All files that display a *bimodal distribution* of the type 8C (TUCKWELL classification) have been excluded (32 files). In order to check for *stationarity*, the trains have been partitioned into 2 or 4 parts (depending on the number of ISI per file). If the variance of the firing rate within the parts of one train was high ($c_v > 0.1$), we calculated $\mathcal{F}_{\log}(\kappa)$ for each part separately and excluded files where the differences of $\mathcal{F}_{\log}(\kappa)$ were larger than 2 (5 in total). We finally checked for a *cut-dependence* of the classification (see below). We calculated $\mathcal{F}(\kappa)$ for $\kappa = 2$ and 3. If $\mathcal{F}_{\log}(3)$ led to a different classification than $\mathcal{F}_{\log}(2)$ (see below), the file was excluded (4 in total). For the remaining 125 files, $\mathcal{F}_{\log}(2)$ has been calculated and the files were classified as follows:
>
> - Exponential (exp): $\mathcal{F}_{\log}(2) < -2$
> - Not attributable (na): $-2 \leq \mathcal{F}_{\log}(2) \leq 2$
> - Power-law (pow): $\mathcal{F}_{\log}(2) > 2$
>
> Neurons, where the majority of the assigned files was in either class, were classified accordingly. Neurons, were some of the assigned files were exponential and others power-law, were classified as 'switcher'. For the olfactory system, we only used files with a firing rate of at least 20 spikes/second that emerged from cells that have been measured in at least 90% of all 96 trials. The total set consisted of 25 cells, none of which showed a ISI distribution of type 8C. As most of the files were still rather short, we could not test for stationarity and cut-dependence.

The result of our analysis is plotted in Fig. 8.1 (visual system) and Fig. 8.2 (olfactory system). We find that the majority of the cells display long-tail ISI distributions. This is in contrast to the POISSON hypothesis. More than half of the cells fall into the category

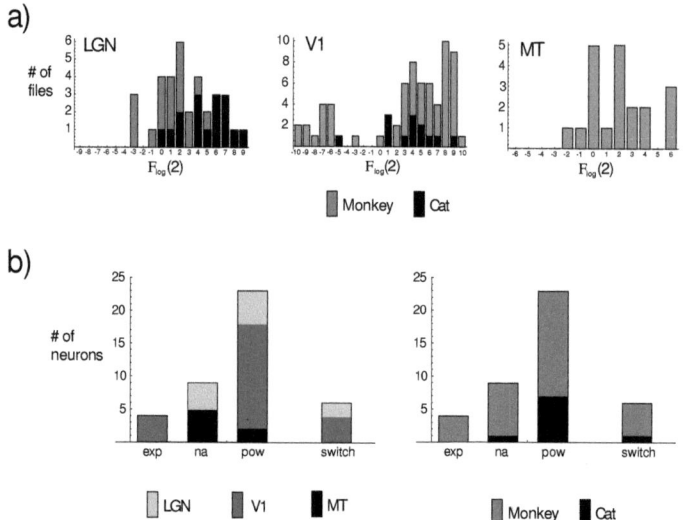

Figure 8.1: Calculating the fit-coefficient for neurons in the visual system: a) $\mathcal{F}_{\log}(2)$ for spike trains from neurons of the cat and monkey visual system, classified according to recording site along the visual pathway. b) Histograms of neurons classified as exponential (exp), not attributable (na) or power-law (pow; right: classified according recording sites; left: classified according animals).

'power-law' and only around 10% are clearly identified as 'exponential'. For latter cells, the coefficient of variation was around 1, and the 2D histogram (not displayed) as well as correlation-integral pattern discovery did not indicate any preferred serial order of the ISI. Thus, only these cells fulfill the criteria of the POISSON hypothesis.

Furthermore the distribution of the values of $\mathcal{F}_{\log}(2)$ show a remarkable dependence of the recording site along the visual pathway: In V1, two clearly distinguishable populations of neurons are visible: a majority of long-tail firing type and a minority of exponential firing type. This distinction is far less pronounced in LGN and MT. In LGN and V1 also 'switchers' are present – neurons that change their firing behavior from a long-tail type to an exponential type for changing stimuli. The comparison between neurons of the two species show that, to a first approximation, the ratios of neurons of either firing type is comparable, although the low number of neurons does not allow to make any concise statement.

In the olfactory system, we found a high variability in firing rates of the individual trials (see section 8.4), as well as in the $\mathcal{F}_{\log}(2)$-values. We neglected the different types and concentrations of stimuli and classified the cells according to the mean of the fit-coefficient in

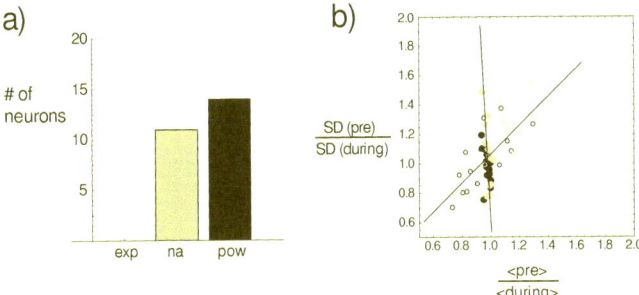

Figure 8.2: Calculating the fit-coefficient for neurons in the olfactory system. a) Histogram of neurons classified as exponential (exp), not attributable (na) or power-law (pow). b) Comparing the ratio of means and standard deviations before and during stimulus presentation of each neuron for the firing rate (filled dots) and the fits (circles). Black: 'pow'-neurons. Grey: 'na'-neurons.

the pre and the during stimulus condition. We found no neuron that is classified as 'exponential', whereas more than half is classified as 'power-law' (Fig. 8.2.a). As the variability measured by the standard deviation of the fits of the ISI distributions and the firing rates for each neuron was high, we were interested if differences between the variability were associated to whether a stimulus was present or not. We plotted for each neuron the fraction of the means of the firing rates before the stimulus and during the stimulus against the fraction of the standard deviations before and during stimulus (Fig. 8.2.b, filled dots). The same was done for the means and standard deviations of the fits (Fig. 8.2.b, circles). For the firing rate, we found for all neurons a ratio of the means of almost one. The trial variability of the firing rates when comparing the pre- and during-stimulus condition is only expressed in the standard deviations. For the fits of the ISI distributions, however, the variability is expressed in variable standard deviations *and* variable means. This indicates, that neurons display changes in their firing statistics when a stimulus is applied, which are not reflected in changes in firing rate – i.e. some neurons display a better mean fit to a power law distribution, whereas other neurons display a worse fit. The results confirm older studies (see section 6.5 claiming that the firing rate is not the relevant parameter for understanding olfactory coding (see also section 8.4).

8.3 Second Prediction: Firing Reliability

According to the second prediction, we expect different degrees of firing reliability in different stages of neuronal information flow (see also section 6.3.3). This aspect is investigated using the data obtained along the visual pathway of neurons in the macaque monkey (LGN, area V1 and area MT). We use the LZ-distance and the C-distance (see section 7.6) in order to investigate firing reliability in terms of the mean distance between a set of spike trains of one neuron obtained under equal stimulus conditions. The LZ-distance measures reliability in terms of similar firing patterns, whereas the C-distance indicates the similarity of firing in terms of spike timing.

Methods: We use recordings of 9 LGN neurons, 10 complex neurons in V1, 6 simple neurons in V1 and 7 neurons in area MT as described in section 8.1.3. For the LGN-cells, 6-7 stimulus conditions have been evaluated. However, only 2-3 trials per stimulus have been performed. This limits the significance of the results obtained for this class of cells. For the V1 cells, 10 trials have been taken per cell and per stimulus (5 orientations). For the MT cells, 6 trials have been taken (16 orientations). For each set of spike trains emerging from trials performed on a specific cell and using a specific stimulus, the mean C- and LZ-distance between the trains have been calculated (the standard deviation of the Gaussian kernel of the C-distance was 1 ms).

We expect an interrelation between the mean rate and the mean distance for each set of spike trains, because a higher firing rate increases the chance of coincident spikes and leads to smaller mean C-distances just because the rate is higher. We use the following approach to compare the reliability of neuronal firing independently of the firing rate. It is well-known that the most random distribution of events in time is provided by a POISSON process. Mean distances obtained by analyzing real data of a specific neuron can then be compared with mean distances of a set of POISSON spike trains with similar rate. The larger the deviation, the more reliable (in terms of the distance used) fires this neuron. To investigate this aspect in more detail, we generated by means of a POISSON process in total 72 classes of 10 spike trains each (duration: 10 seconds per train) with mean firing rates ranging from 0 spikes/second to 1000 spikes/second sampled with a resolution of 1 ms. The majority of the trains had (physiologically meaningful) firing rates of 1-100 spikes/second. To analyze the limit behavior, also higher firing rates were used. For each class, the mean C- and LZ-distance have been calculated (Fig. 8.3). For the C-distance we see that it is basically linearly related to the firing rate for small rates and that it asymptotically approaches $d(t, t^*) \simeq 0$ for large firing rates. This is plausible, as the probability that the convolved trains overlap scales linearly with the increase in number of (randomly distributed) spikes for small firing rates and saturates asymptotically for large firing rates. By testing several fit-functions (polynomials, exponential and power-law functions) we found that a fourth order polynomial of the type $f(x) = ax^4 + bx^3 + cx^2 + dx + e$ provided the best fit in terms of minimized mean-square errors (for a firing rate of 0, the function has a point of discontinuity, as the mean distance would be zero). This function is taken as the Poisson reference for the C-distance. The relation between the LZ-distance and the firing rate is more complex. For small firing rates (up to ∼10 spikes/second) we find a steady increase of the mean distances up to ∼0.2, which is followed by a very slow increase for larger firing rates until ∼500 spikes/second. For larger firing rates, we find a symmetry along the vertical axis $x = 500$, which is plausible, because in the bit-coded spike trains basically a switch between the letters '0' and '1' is observed as the spike train predominately consists of spikes. Again, several fit-functions have been tested. We obtained the best fit by using a fourth-order polynomial for the interval [0,20] spikes/second and a quadratic function for the interval [20,980] spikes/second. We use this composed function as the Poisson reference for the LZ-distance. The degree of discrepancy of the neuronal data to these two Poisson reference functions is

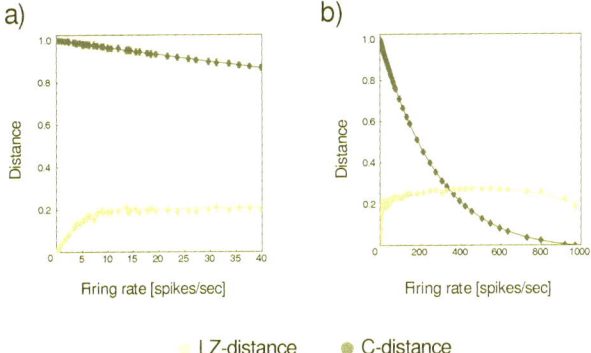

Figure 8.3: Relation between the C/LZ-distances and the firing rate for Poisson spike trains: a) Small firing rates. b) Asymptotic behavior for large firing rates.

our measure for the reliability of firing in terms of the C/LZ-distances. The discrepancy is measured as the average mean distance of all data points from these functions.

The results of the analysis of the four cell types investigated are displayed in Fig. 8.4. Each point in the graph reflects the mean distance of all trials of a single stimulus parameter (orientation) for a single neuron calculated using the LZ-distance (light grey) or the C-distance (dark grey). The curves shown are the Poisson reference functions. We then determined the deviations of the mean distances of the trials from the Poisson reference for each distance measure and for each class of neurons in the different brain areas. As the standard deviations of the data sets obtained in this way (LGN, V1C, V1S, MT) were high, we pairwise tested the null-hypothesis that the two sets derive from the same distribution using the non-parametric Wilcoxon-Mann-Whitney u-test with significance levels $p < 0.05$ and $p < 0.001$ (two-sided). The results are displayed in Fig. 6. For the C-distance, the data sets obtained from the MT-cells and the V1 complex cells are not distinguished for $p < 0.05$ and $p < 0.001$, whereas the other sets are indeed recognized as distinct distributions. For the LZ-distance and a significance level of $p < 0.05$, the data sets V1C and V1S, as well as V1S and LGN are not distinguished, but MT was classified as distinct. For the higher significance level $p < 0.001$, the data sets V1C, V1S, and LGN, as well as LGN and MT were not distinct. Taken together, the results suggest the following interpretation: In terms of the 'timing deviation' of the data from the Poisson reference, the V1 simple cells have the highest value, i.e. have the highest timing reliability, followed by the LGN cells and the MT/V1 complex cells. In terms of the 'pattern deviation' of the data from the Poisson

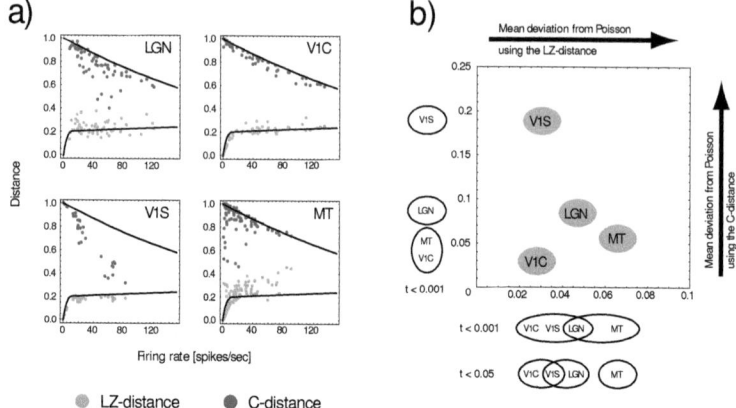

Figure 8.4: Firing reliability determination of visual neurons: a) Relation between the C/LZ-distances and the firing rate for spike trains obtained from neurons of areas LGN, V1 (comples: V1C, and simple: V1S), and MT in macaque monkey. b) Timing and Pattern deviation from the Poisson reference of neurons measured in areas LGN, V1 (complex/simple cells) and MT.

reference, the MT neurons have the highest value, i.e. have the highest 'pattern reliability', followed by the LGN, V1 simple and complex cells. Thus, the complex cells are closest to the Poisson model in both respects, whereas the MT cells have a considerable 'pattern reliability' and the V1 simple cells a considerable 'timing reliability'. The LGN cells display both characteristics up to some degree, although we have to remind that the number of trials per stimulus conditions in this case is low and requires a careful interpretation.

8.4 Third Prediction: Neuronal Clustering

We investigate the third prediction by analyzing the functional clustering within a neuronal network of an olfactory sensor when confronted by a specific odor. We address this question using the LZ-distance and the C-distance as population criteria and the sequential superparamagnetic clustering paradigm.

Methods: We use the LZ- and C-distance measure and the sequential superparamagnetic clustering algorithm as explained in sections 7.6 and 7.5 on the complete olfactory data set described in section 8.1.1. The clustering parameters have been chosen as follows: `minsize` = 2 and $s_\theta = 0.04$. For the C-distance, spike trains that share a large number of synchronous spikes are considered to be close. Therefore, the measure reflects the criterion of synchronization in neuronal group formation. As the intertrial-variability in the data set was high, and the number of trials per stimulus was low, we focussed our analysis on differences between the pre-stimulus and the during-stimulus condition of the neuronal network, disregarding the identity and concentration of the applied odors. The analysis has been performed in a three-step procedure: First, we determined the number of clusters of neurons in each of the 96 periods of recording, using both the LZ-distance and the C-distance. Second, based on this result, the mean and the standard deviation of the number of clusters across trials for both conditions was calculated. For each neuron we then determined the mean size of the clusters to which it belonged in the activity in the pre-stimulus period, and in the during-stimulus period. Periods of recording were excluded if no clusters formed. Such periods occurred in all four odors as well as in the pre- and during-stimulus period using both distance measures (LZ-distance: 11 times / C-distance: 18 times). Third, to quantify the interactions of each neuron with each other neuron in the during-stimulus period compared to the pre-stimulus period, we assigned to each neuron a vector, whose components indicate the number of times the specified neuron finds itself in a cluster with another neuron. For example, for the i-th neuron, the vector has the form $\vec{n}_i = (x_1, \ldots, x_{54})$, where x_j indicates the number of times the i-th neuron is in the same group as the j-th neuron. The distance between two such vectors $\vec{n}_i$ and $\vec{n}_j$ is simply

$$d(\vec{n}_i, \vec{n}_j) = 1 - \frac{\vec{n}_i \cdot \vec{n}_j}{\|\vec{n}_i\| \|\vec{n}_j\|}$$

Using this measure, the proximity of two neurons indicates that they often participate in the same cluster. Clustering with this distance measure works out the degree of interrelation of neurons within the network, averaged over all trials ('clusters among partners'). We determined the dendrogram for $s_\theta = 0.02$ such that also clusters of low stability are detected. Based on the dendrogram showing the breakdown in clusters for decreasing cluster stability s, we identified the values of s, where new clusters emerged. The result (the dependence of the number of clusters on a required stability s) is displayed in diagrams for the pre- and during-stimulus condition. The diagrams indicate the 'stability behavior' of the network in terms of the mean stability of interrelationships between neurons expressed via the LZ-distance or the C-distance. If there are many clusters for high values of s, then the interrelatedness of the neurons in the network is stable. Note, that in this analysis two different kinds of clusters are investigated. In the first step, the resulting clusters reflect the degree of relatedness in a single period of recording. In the third step, the resulting clusters reflect the mean degree of relatedness between the neurons of the network over all stimulus conditions.

We first estimate the intertrial-variability of single cell firing by using a moderate criterion, the change in firing rate over a coarse time scale. We calculated the firing rate vector of each spike train by partitioning the spike train in 10 ms segments and calculating the firing rate (spikes/sec) in each segment. The vectors obtained in this way reflect the change in firing rate of each spike train. For each neuron, the distances between all 96 firing rate vectors have been obtained by calculating the pairwise dot product. The resulting distance matrix has been used as input for the clustering algorithm. In neither case did clusters emerge, demonstrating that even when the same odor was applied in the same concentration, the intertrial-variability in firing rate was too high and prevented the formation of clusters in the three periods of activity recorded under matched stimulus conditions. For some neurons, we directly calculated the coefficient of variation in the average firing rate for spike trains obtained unter matched stimulus conditions. In the large majority, the coeffi-

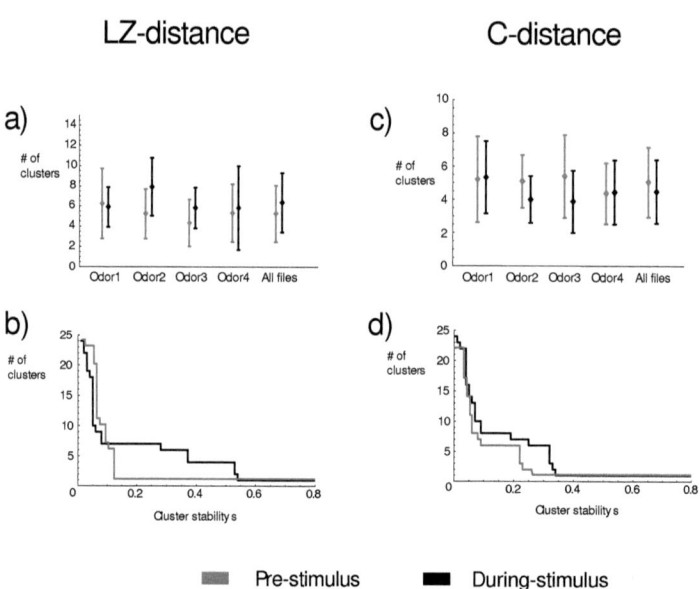

Figure 8.5: Olfactory neural network clustering, measured by the LZ-distance (left) and the C-distance: a/c) Number of clusters in the pre- and during-stimulus condition. b/d) Stability of 'clusters among partners' (see text).

cient lay between 0.5-1.5, further indicating high intertrial-variability. This high variability in the network requires an approach that compares the mean behavior of the network in the presence or absence of an odor. It also confirms that, in this data, the firing rate of the individual neuron is not an adequate means for understanding olfactory coding.

We performed the procedure described in the methods section using the LZ-distance as a similarity measure for clustering. The result of our analysis is shown in Fig. 8.5.a/b: We find that the number of clusters emerging on stimulus presentation does not per se indicate whether the network is in a pre-stimulus or a during-stimulus condition. Also odor identity cannot be determined in this way. The large standard deviations reflect the intertrial-variability of the network behavior (Fig. 8.5.a/c). The analysis of the stability-behavior of the network (Fig. 8.5.b) provides new insight: the number of stable clusters (large values of **s**) is lower in the pre-stimulus condition than in the during-stimulus condition; the number of unstable clusters (small **s**) tends to be larger for a given **s** in the pre-stimulus condition when compared with the during-stimulus condition. For **s** = 0.54, four clusters emerge in the during-stimulus condition. This number increases to seven until an abrupt increase in the number of clusters occurs at $s = 0.08$. In the pre-stimulus condition we observe a single cluster. Here, an even more abrupt increase in number sets in at $s = 0.12$. We performed the same analysis using the C-distance for clustering. In this way, we focus on groups of synchronously firing neurons. As in the previous case, the number of pre- or during-stimulus clusters provide no information per se about stimulus presence or absence (Fig. 8.5.c). This result is rather surprising, as the emergence of synchronization is considered as a possible indicator for the presence of an odor. In this data, however, this seems not to be the case. The analysis of the stability behavior of the network shows that there is greater similarity between pre- and during-stimulus activity when considered in terms of synchronization of the network than when considered in terms of the LZ-distance (Fig. 8.5.d). Although a slightly higher degree of cluster stability is visible in the during-stimulus condition than for the pre-stimulus condition (6 clusters emerge for $s = 0.32$ compared to $s = 0.22$ for the pre-stimulus condition), the effect is much less pronounced compared to that in the case of the LZ-distance. This again indicates that synchronization effects in network behavior do not necessarily reflect stimulus onset. In summary, our electrode array recordings from the olfactory bulb (mitral cell layer) of the rat reveal that stable inter-neuron relationships, expressed by the LZ-distance, emerge during odor presentation, and that these relationships are more stable than the synchronization of the neurons. These clusters reflect stimulus-evoked stabilization of the network-dynamics as expected by our hypothesis.

8.5 Fourth Prediction: Classes of Firing

Earlier investigations (see section 6.6.4) suggest the existence of three classes of neuronal firing [429, 705]. According to our fourth prediction (section 6.6.5), we expect to reproduce this finding when analyzing the extended data set.

> **Methods:** We apply the correlation integral based pattern discovery method as described in section 7.3 to our data set. We restrict ourselves to the data obtained in the visual system, because the spike trains are generally much longer than those obtained from the olfactory system and thus lead to statistically more reliable results (45 files of LGN, 157 files of area V1 and 35 files of area MT). Eight files with a very small number of ISI have been excluded from the analysis. The spike trains were embedded in dimensions $m = 1, \ldots, 10$. To distinguish steps in the log-log plot that emerged as a result of different length scales (which disappear for POISSON randomization) from steps that emerge from repeating sequences, we randomly shuffled the ISI of each train and re-applied the correlation integral algorithm to the shuffled data. Each file has been classified according to the classes described in section 6.6.4. The most often represented class in all five trials of a neuron determined, to which class the neuron belongs. In eight neurons we found files, where different stimulus conditions led to a classification into groups I or III respectively. These cells were excluded from the analysis.

We found in all areas of the visual system of cat and monkey from which data was available files and neurons of all three classes (fig. 8.6) The first and the third class are more or less equal in size, whereas the second class is usually smaller. The analysis of the shuffled files (not displayed) showed that the majority of files that are categorized into classes I, II or III remain in the same class after shuffling, indicating that the steps result

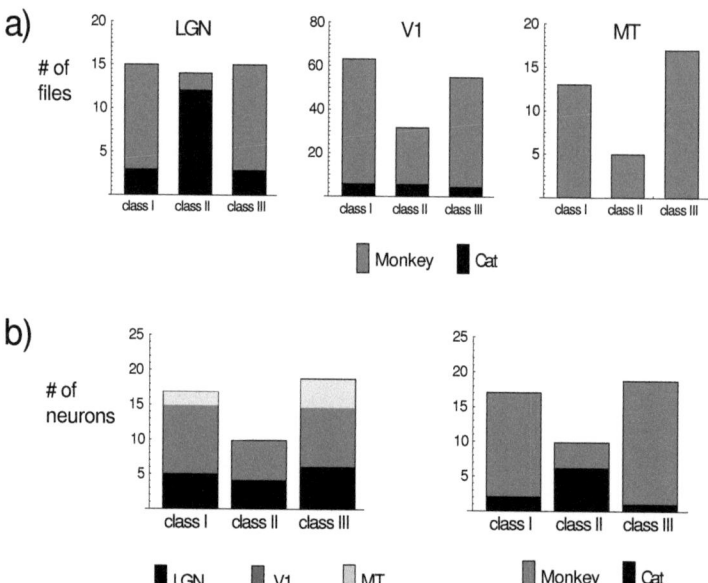

Figure 8.6: Pattern discovery in visual neuron spike trains: a) Number of files classified as I, II or III in each visual area. b) number of neurons in each class, subdivided according to visual area and species.

from pronounced differences in lengths scales and not from repeating sequences. However, differences between the areas are visible: In LGN, none of the 45 files was classified in another class after shuffling. In V1, 20 of 149 files that were classified in classes III or II fell after shuffling in the classes II or I. In area MT, 4 of 35 files were re-classified after shuffling. This indicates, that interval patterns of $l > 1$ are present in at least some files in higher cortical areas. As the number of neurons that we investigated is still rather limited, we can not make any decisive statement about the relative sizes of each class in the several areas of the brain investigated. Furthermore, the stimulus used was in most cases not optimal for our analysis: neurons (V simple cells and some LGN neurons) that reproduce in their firing the temporal structure of the stimulus (drifting gratings) certainly show two different length scales in the resulting spike trains – which again shows up as steps in the log-log plot. To avoid this problem, more and different types of stimuli (as in the cat data) should have been used. Unfortunately, we did not have access to this type of data.

8.6 Fifth Prediction: Pattern Stability

We apply our definition of pattern stability on data obtained from two complex V1 cells of macaque monkey. Both cells show a sharp peak at ~100 Hz in the power spectrum (not displayed) that indicates oscillatory firing. The shift predictor indicates, that the oscillation of one neuron (neuron A) is driven by an external stimulus (possibly the frame rate of 100 Hz of the monitor on which the stimulus has been presented), whereas the second neuron (neuron B) is an intrinsic oscillator. We are interested, whether we find differences in the stability of the patterns associated with oscillatory firing.

Methods: We apply the definition of pattern stability of section 6.1.3. For pattern quantification, we use a template-based approach as described in section 7.4 (no additional spikes were allowed within a sequence). To identify the template sequence and the matching intervals adjusted to the individual noise levels of the files, we use sequential superparamagnetic clustering as described in section 7.5. For the clustering algorithm, the parameters were set as: `minsize`: 25 and s_n: 5. We applied the clustering algorithm to two files per neuron, where the most stablest clusters have been found. All files had large, stable clusters of points of the type (a, a), and smaller clusters of the type (a, b) and (b, a) (a, b indicate ISI of different length scales). This indicates, that the firing of both neurons is characterized by a firing of two periodicities: a periodicity-1 firing – indicated by a pattern of the type $\{a, a, a, a, \ldots\}$ – and a periodicity-2 firing – indicated by a pattern of the type $\{a, b, a, b, \ldots\}$. The noise affecting the firing of neurons A and B, respectively, was different. In neuron A, the standard deviation of the coordinates of the stable clusters was in the order of 3 ms, whereas in neuron B, the standard deviation was in the order of 1 ms for the smaller coordinate and 2 ms for the larger coordinate. This shows that the firing of neuron A is noisier as the firing of neuron B, which is also clearly visible in the 2D histogram (Fig. 8.7.a/b). Therefore, for the template algorithm, the following sequences were investigated. For periodicity-1 firing of cell A, the template had the form $\{9, 9, 9, \ldots\}$ with matching interval $[-3, 3]$ and for cell B, the template sequence was the same but the matching interval was $[-2, 2]$. To test the influence of the matching interval on the result, we also applied the algorithm to the files of A with matching interval $[-2, 2]$ and to the ones of B with $[-3, 3]$. The result (not shown) shows that the matching interval suggested by clustering lead to the higher values in either case. This supports our claim, that the clustering algorithm provides optimal value for constructing the template sequence and the matching interval.

After calculating the stability of the patterns associated with firing of periodicity-1 and periodicity-2 for both files of each neuron, we find no difference of pattern stability for periodicity-1 firing between the driven and the intrinsic neuronal oscillator (Fig. 8.7.c right). For neuron B (intrinsic oscillator) the stability values obtained for both files are comparable (around 0.1), whereas for the driven oscillator, the stability values of the two files are rather different (005 to 0.15). For periodicity-2 firing, a different picture emerged. Here, the template algorithm was applied to the two files of cell A using template sequences of the form $\{9, 19, 9, 19, 9, \ldots\}$ with matching intervals $[-3, 3]$ for both length scales as suggested by the result of clustering. For the files of cell B, the template sequence had the form $\{9, 2, 9, 2, 9, \ldots\}$ with matching intervals $[-2, 2]$ for the larger, and $[-1, 1]$ for the smaller length scale (as suggested by clustering). Here (Fig. 8.7.c left), the intrinsic oscillator shows a three times more stable firing pattern of periodicity 2 than the driven oscillator. The

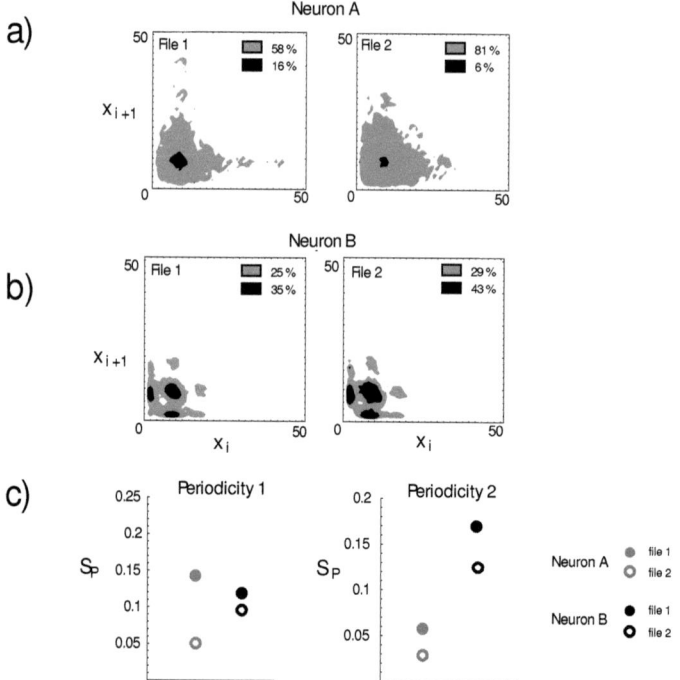

Figure 8.7: Pattern stability: a) 2D histogram of driven neuronal oscillator (neuron A). b) 2D histogram of intrinsic neuronal oscillator (neuron B). c) Stability of all cells of the periodicity-1 oscillatory pattern for different matching intervals $[-b, b]$ (right) and the stability of the periodicity-2 oscillatory pattern for optimal matching intervals (left).

periodicity-2 firing of the intrinsic oscillator (cell B) is therefore a more stable phenomenon, as one would expect it in the framework of the STOOP-hypothesis, where higher periodicities may emerge as a result of a computation.

Chapter 9

Conclusions

In this chapter, we summarize the main achievements of the scientific part of the thesis: We show to what extend the results of the previous chapter support the STOOP-hypothesis, we put this result in the general context of neural coding and computation, and include the more specific points on spike patterns, noise and their functional role. We finalize this chapter by outlining possible experiments in order to approach the open questions.

9.1 Support for the Stoop-Hypothesis

We have shown in the previous chapter that our predictions formulated in section 6.6.5 have been confirmed, although the available data were not optimal in all cases:

- The analysis of the ISI distributions demonstrated that the POISSON hypothesis does not cope with the majority of the data we investigated. The number of 'POISSON neurons' – i.e. neurons that are compatible with all conditions for claiming the validity of the POISSON hypothesis – was indeed surprisingly small. This finding is, however, not necessarily contradicting the prediction of the STOOP-hypothesis that a considerable number of 'noisy neurons' (class I neurons) should be present, as class I neurons must not necessarily display a distinct POISSON firing statistics. When comparing the result of the ISI distribution analysis with the result of correlation integral pattern discovery, we find that neurons classified as 'not attributable' often also fall into class I. Furthermore, the analysis of the firing statistics revealed a certain dependence on the firing statistics from the stimulus. In the visual system, we found neurons classified as 'switcher' – neurons that display exponential or long-tail firing statistics for different stimulus conditions. In the olfactory system, we found a stimulus-related change of the firing statistic variability. Both findings may be further analyzed when better data are available. We have to remind that, in the olfactory system, the spike trains available for calculating the distribution were rather short. In the visual system, on the other hand, almost all files analyzed emanate from evoked and not from spontaneously active neurons. Although non of the few files of spontaneous activity that were available to us were classified as exponential, it would be advantageous to include spontaneous spike trains into the analysis, since many older studies on this matter based their findings on both spontaneous and evoked firing. Finally, the data that were available for us were not sampled in order to obtain a representative set of neurons in respect to firing statistics. Our result therefore do not allow a quantitative estimation of the sizes of the populations of exponential-firing or long-tail-firing neurons. It would be advantageous to have data that emerge from experiments with a broader class of stimuli.

- The analysis of firing reliability in the different visual areas also confirmed our expectation that the reliability in respect of timing and patterns depends on whether the neuron acts more as a 'information transmitter' or as a 'information processor'. Some aspects need, however, a careful interpretation. Our measure of reliability is based on the deviation from a POISSON reference. Using the LZ-distance, the mean distance of some spike trains obtained in multiple presentations of a single stimulus condition is higher than the POISSON reference – especially in area MT. Thus, the neurons may deviate from the reference function not in the sense that certain specific firing patterns are replicated in the different trials, but that *different* firing patterns are present in different trials of a single stimulus condition. This interpretation may be explained within the framework of the STOOP-hypothesis, as we expect that higher cortical areas, where different types of computation may take place – as area MT –, display different types of patterns, whose actual presence and weighting in different trials of a single stimulus condition may change in the *in vivo* condition. This type of firing would, however, be well distinguished from POISSON firing.

- The result obtained by clustering of olfactory spike trains offers a promising approach for analyzing the behavior of large neuronal networks, as it introduces an alternative approach for studying neuronal population coding in the olfactory system. We found that stabilization of neural firing patterns, rather than synchronization, may faithfully represent an odor. This stabilization of the network can be interpreted as follows: in prestimulus activity, the network is in a state with many unstable clusters. The network is thus in a 'preparatory state' such that many potential neuronal groups are available for encoding a given stimulus. During stimulus presentation, stable clusters emerge out of these 'potential' clusters, possibly representing the odor. This type of behavior is much less apparent using the C-distance, indicating that the effect of synchronization is of less importance than might be theoretically predicted for understanding population coding in the vertebrate olfactory system. This effect of stabilization of a pattern is compatible with the STOOP-hypothesis, because we expect an increased computation during a stimulus. This is expressed in more stable pattern that lead to more stable clusters when the LZ-distance is used for clustering. Further studies should focus on the question, whether clusters of spike trains can be related to specific stimuli. This would need more trials per stimulus condition to cope with the large inter-trial variability.

- Our pattern discovery results basically confirmed the existence of the three firing classes predicted by the STOOP-hypothesis. The classification obtained by analyzing the monkey data was, however, more difficult to achieve as we found more 'borderline' cases. This might result from the fact that the set of stimuli used for obtaining the monkey data was more uniform as compared to the cat data.

- The concept of pattern stability was only applied to files of two neurons. We consider this analysis as a preliminary test for the concept of pattern stability. It may be used in experiments, where we can separate between patterns emerging from temporal cues of the stimulus and pattern emerging as a result of a computational process. We had created an experimental protocol on this matter,[1] but were unable to find investigators that could perform the necessary experiments.

9.2 The Stoop-Hypothesis in the General Context

The STOOP-hypothesis unifies several seemingly contradictory concepts in neural coding: Although the generic scheme of neuronal locking is based on neuron-neuron interaction, the general framework emphasizes the role of groups of neurons. The hypothesis combines a rate coding perspective (which provides the driving of a neuron) with a temporal coding perspective (the perturbation of the limit cycle and the emergence of a phase-locked firing behavior). It attributes a precise role for the most important component of neuronal noise in the *in vivo* condition, the background activity, within neuronal information processing. It finally provides an alternative, dynamical systems based approach for understanding neural coding and neural computation. Based on a reconstruction of a system and a state space partition, the dynamics of neurons is encoded in state sequences that can be a further object of an analysis, using e.g. the STOOP-measure for computation. Results demonstrate that the computation improves when a neuron changes from a sub-optimal to an optimal stimulus [695]. Most of the discussed theoretical proposals concerning neural coding and neural computation – the synfire chain model, coincidence detection, and spike-timing dependent plasticity – can be related to the STOOP-hypothesis. Synfire chains, for example, might reliably transmit perturbation signals (synchronous spikes) to remote cortical areas. Coincidence detection provides us with a means to control the coupling parameter K by changing the number of coincident spikes and thus the perturbation strength. Also spike timing dependent plasticity might in principle be a way to change K. However, the chances in synaptic strength induced by LTP or LTD are rather small and we therefore suggest that spike timing dependent plasticity should be seen as a statistical effect changing many synapses and thus the driving of of a neuron – or, in other words, the parameter Ω. The temporal correlation hypothesis in the sense of SINGER et al., however, is not integrated in our framework, as we consider the binding of features as a result of locking expressed in the periodicity of a spike train. In this way, we avoid a discussion on the several controversial aspects of the temporal correlation hypothesis.

9.3 Outlook: Experimental Approaches

Several constraints limited the empirical investigation of the prediction of the STOOP-hypothesis. The funding of the project did not allow to perform own experiments and the

---

[1] Markus Christen: Spike patterns in neuronal information processing – an experimental setup. 2003 (working paper)

available data were not in all respect optimal for the questions we are interested in. Therefore, there is still a need for an empirical underpinning of central claims of the hypothesis. The main open questions are:

- To what physical limits extends locking? This includes the testing of biological perturbation profiles like different kinds of burst patterns, of the generic locking scheme (Fig. 6.7) with other neurons than pyramidal neurons *in vitro* and of the general locking scheme *in vivo*. Furthermore, the highest possible perturbation frequencies at which the locking phenomenon is still observable should be determined in *in vitro* experiments.

- Can locking persist under dynamic network activity? This includes the test *in vivo* that locking remains under homogeneously changed activity, the determination of the smallest input that triggers observable locking, and the determination of the dynamic range that is applicable without loss of locking. Finally, inhomogeneous changes should result in changing periodicities, which again should be observable in the *in vitro* condition, where the inhomogeneous change can be controlled by the experimentator. As already indicated in section 6.6, model studies demonstrate that locking persists also under changing driving conditions [698, 699]. This suggests that experiments in this respect lead to results in favor of the STOOP-hypothesis.

- What is the physiological background of the three classes of firing we postulated? One must find a clear reason how class I neurons are able to erase timing cues in the stimulus (for example provided by drifting gratings or the frame rate) and produce an output that only correlates with a general activity level.

- The integral framework presented by us can serve as a paradigm for large scale neural modelling using simple types of neurons, where the two basic types (class I and class III) are implemented and connected according Fig. 6.11. Within this scheme, also the effect of LTP as a means to increase the driving of limit cycle firing could be tested. Furthermore, recent work[2] focusses the connection between architecture and computation within neuronal networks. The results indicate that a fractal and recurrent network architecture is optimal for information speed and computation (measured according to Ref. [695]). These promising results are object of a follow-up PhD study by STEFAN MARTIGNOLI.

In this work, we have clarified the term 'spike pattern', discussed the problem of pattern detection and we have introduced novel methods for solving this problem. We furthermore have shown, how the STOOP-hypothesis is related to the current discussion on neural coding and neural computation, why the STOOP-hypothesis is interesting, and that the data available to us support prediction of this hypothesis. We believe that the tests outlined above would help to further understand and complete the hypothesis and exploit its potential for obtaining a deeper understanding of the information processing brain.

---

[2] A diploma thesis performed by Stefan Martignoli in 2005, as well as ongoing work that is currently submitted or in preparation (C Wagner & R Stoop: Neocortex's small world of fractal coupling / C Wagner & R Stoop: Neocortex's architecture optimizes synchronization and information propagation under a total connection length constraint. / S. Martignoli & R Stoop: Cortical recurrent connections maximize a measure of computation).

Part III

Appendix

List of Figures

1.1	Scheme of argumentation	12
2.1	SHANNON's scheme of a communication system	25
2.2	Scheme for historical analysis: before transition	30
2.3	The emergence of the spike	37
2.4	The emergence of the spike train	38
3.1	Scheme for historical analysis: after transition	40
3.2	Examples of network-structures from early neural modelling studies	62
4.1	The neuroscience boom since the Second World War	75
4.2	Journal analysis	76
4.3	Cluster analysis of conferences	82
4.4	Identifying important conferences	83
4.5	Citation analysis for MCCULLOCH	89
4.6	Citation analysis for GERARD	90
4.7	Citation analysis for BARLOW	91
4.8	Citation analysis for BULLOCK	92
4.9	Citation analysis for MACKAY	94
4.10	Citation analysis for REICHARDT	95
4.11	Overview historical developments	99
5.1	The dimensions of a code relation	107
5.2	Representations of spike trains	108
6.1	Types of spike patterns	119
6.2	The stability of an interval pattern	121
6.3	Methods of randomization	127
6.4	The degree of randomization introduced by shuffling procedures	128
6.5	Neuronal noise sources	130
6.6	Limit cycle firing	149
6.7	Locking and ARNOL'D coding	151
6.8	The coding problem in the framework of dynamical systems theory	153
6.9	ARNOL'D coding in the (Ω, K) parameter space	155
6.10	Classes of neuronal firing	158

6.11 STOOP-Hypothesis: Integral framework 160

7.1 Overview distance measures 167
7.2 Example of a 2D interval histogram 170
7.3 Pattern length estimation with 2D histogram 171
7.4 2D histograms for neuronal data 172
7.5 The TUCKWELL classification scheme 173
7.6 Exponential vs. power-law fits of distributions 174
7.7 Log-log plot steps from different ISI models 176
7.8 Test for log-log plot steps 177
7.9 Dependence of step appearance from noise level 179
7.10 Pattern-length indicator 180
7.11 Detection of sequences in random background 181
7.12 Log-log plot of cat V1 data 182
7.13 The basic template matching algorithm 184
7.14 The influence of the matching interval on the number of template matches . . 185
7.15 Relating patterns and clusters 189
7.16 Outline of the spike train clustering problem 190
7.17 Distances of pairs of spike trains 193
7.18 Clustering of simulated multi-train data 194
7.19 Comparison of distance measures 195
7.20 The pattern detection problem 197

8.1 Calculating the fit-coefficient for neurons in the visual system 202
8.2 Calculating the fit-coefficient for neurons in the olfactory system 203
8.3 Relation between the C/LZ-distances and the firing rate for Poisson spike trains 205
8.4 Firing reliability determination of visual neurons 206
8.5. Olfactory neural network clustering 208
8.6 Pattern discovery in visual neuron trains 210
8.7 Pattern stability 212

List of Tables

4.1	Conferences of the *Macy*-cluster	77
4.2	Work sessions and ISP of the NRP-cluster	79
4.3	Theory-Cluster	80
4.4	Biology cluster	80
4.5	Border crossers	85
4.6	Ambassadors	85
4.7	'Important people' of American origin	86
4.8	'Important people', emigrants to the US	86
4.9	'Important people', non-Americans	87
7.1	Overview histograms	169
7.2	Maximum number of steps $s(l,m)$	178
7.3	Slope ratio determination	182
8.1	Overview data set	200

Bibliography

[1] **Historical Sources (up to ~1980)**

[2] Abeles M, and MH Goldstein (1977): Multispike train analysis. *Proceedings of the IEEE* 65(5): 762-773.

[3] Abeles M, and Y Lass (1975): Transmission of information by the axon: II the channel capacity. *Biological Cybernetics* 19: 121-125.

[4] Adrian ED (1947): *The physical background of perception*. Clarendon Press, Oxford.

[5] Adrian ED, and G Moruzzi (1939): Impulses in the pyramidal tract. *Journal of Physiology* 97: 153-199.

[6] Adrian ED (1932): *The mechanism of nervous action*. University of Pennsylvania Press, Philadelphia.

[7] Adrian ED (1930): The effect of injury on mammalian nerve fibres. *Proceedings of the Royal Society of London B* 106: 596-618.

[8] Adrian ED (1928): *The basis of sensation*. Christophers, London.

[9] Adrian ED (1926): The impulses produced by sensory nerve endings. Part I. *Journal of Physiology* 61: 49-72.

[10] Adrian ED, and Y Zotterman (1926): The impulses produced by sensory nerve-endings. Part 2. The response of a single end-organ. *Journal of Physiology* 61: 151-171.

[11] Adrian ED, and Y Zotterman (1926): The impluse produced by sensory nerve endings. Part 3. Impulses set up by touch and pressure. *Journal of Physiology* 61: 465-483.

[12] Aitkin LM, CW Dunlop, and WR Webster (1966): Click-evoked response patterns of single units in the medial geniculate body of the cat. *Journal of Neurophysiology* 29(1): 109-123.

[13] Amassian VE, J Macy Jr., HJ Waller, HS Leader, and M Swift (1964): Transformation of afferent activity at the cuneate nucleus. In: RW Gerard, and JW Duyff: *Information processing in the nervous system*. Excerpta Medica Foundation, Amsterdam, New York: 235-254.

[14] Arbib MA (1972): *The metaphorical brain. An introduction to cybernetics as artificial intelligence and brain theory*. John Wiley & Sons Inc., New York.

[15] Arbib MA (1969): Automata theory as an abstract boundary condition for the study of information processing in the nervous system. In: KN Leibovic: *Information processing in the nervous system*. Springer Verlag, Berlin, Heidelberg, New York: 3-19.

[16] Arnol'd IV (1965): Small denominators I, Mappings of the circumference onto itself. *Transactions of the American Mathematical Society (Series 2)* 46: 213-284.

[17] Ashby WR (1966): Mathematical models and computer analysis of the function of the central nervous system. *Annual Review of Physiology* 28: 89-106.

[18] Ashby WR, H von Foerster, and CC Walker (1962): Instability of pulse activity in a net with threshold. *Nature* 196: 561-562.

[19] Ashby, WR (1952): *Design for a brain*. John Wiley & Sons Inc., New York.

[20] Ashby WR (1950): The stability of a randomly assembled nerve-network. *Electroencephalography and Clinical Neurophysiology* 2(4): 471-482.

[21] Atwood HL, and CAG Wiersma (1967): Command interneurons in the crayfish central nervous system. *Journal of Experimental Biology* 46: 249-261.

[22] Bar-Hillel Y, and R Carnap (1953): Semantic information. *British Journal of the Philosophy of Science* 4: 147-157.

[23] Barlow HB (1972): Single units and sensation: a neuron doctrine for perceptual psychology? *Perception* 1: 371-394. Reprinted in: JA Anderson, A Pellionisz, and E Rosenfeld (eds.): Neurocomputing 2. Directions for research. MIT Press, Cambridge: 218-234.

[24] Barlow HB (1969): Trigger features, adaptation and economy. In: KN Leibovic (ed.): *Information processing in the nervous system.* Springer Verlag, Berlin, Heidelberg, New York: 209-230.

[25] Barlow HB (1963): The information capacity of nervous transmission. *Kybernetik* 2(1): 1.

[26] Barlow HB (1961): Possible principles underlying the transformations of sensory messages. In: WA Rosenblith: *Sensory Communication,* MIT-Press, Cambridge: 217-234.

[27] Barlow HB (1961): The coding of sensory messages. In: WH Thorpe, and OL Zangwill (eds.): *Current problems in animal behaviour.* Cambridge University Press, Cambridge: 331-360.

[28] Barlow HB (1959): Sensory mechanisms, the reduction of redundancy, and intelligence. In: National Physical Laboratory: *Mechanisation of Thought-Processes,* volumes I/II. London, Her Majesty's Stationery Office: 537-574.

[29] Barlow HB (1953): Summation and inhibition in the frog's retina. *Journal of Physiology* 119: 69-88.

[30] Beurle RL (1956): Properties of a mass of cells capable of regenerating pulses. *Philosophical Transactions of the Royal Society of London B* 240(669): 55-94.

[31] Biscoe TJ, and A Taylor (1963): The discharge pattern recorded in chemoreceptor afferent fibres from the cat carotid body with normal circulation and during perfusion. *Journal of Physiology* 168: 332-344.

[32] Bishop PO, WR Levick, and WO Williams (1964): Statistical analysis of the dark discharge of lateral geniculate neurons. *Journal of Physiology* 170: 598-612.

[33] Bishop GH (1956): Natural history of the nerve impulse. *Physiological Review* 36: 376-399.

[34] Bishop GH, and M Clare (1953): Sequence of events in optic cortex response to volleys of impulses in the radiation. *Journal of Neurophysiology* 16(5): 490-498.

[35] Blair EA, and J Erlanger (1935): On the process of excitation by brief shocks in axons. *American Journal of Physiology* 114: 309-316.

[36] Blair EA, and J Erlanger (1933): A comparison of the characteristics of axons through their individual electrical responses. *American Journal of Physiology* 106: 524-564.

[37] Blair EA, and J Erlanger (1932): Responses of axons to brief shocks. *Proceedings of the Society of experimental Biology and Medicine* 29: 926-927.

[38] Blaschko H, M Cattell, and JL Kahn (1931): On the nature of the two types of response in the neuromuscular system of the crustacean claw. *Journal of Physiology* 73: 25-35.

[39] Blum M (1962): Properties of a neuron with many inputs. In: H von Foerster, and W George: *Principles of self-organization.* Pergamon Press, New York: 95-119.

[40] Braitenberg V (1961): Functional interpretation of cerebellar histology. *Nature* 190: 539-540.

[41] Braitenberg V, G Gambardella, and U Vota (1965): Observation on spike sequences from spontaneously active Purkinje cells in the frog. *Kybernetik* 2(5): 197-205.

[42] Brazier MA (1964): The biologist and the communication sciences. In: RW Gerard, and JW Duyff: *Information processing in the nervous system.* Excerpta Medica Foundation, Amsterdam: 87-94.

[43] Brazier MA (1963): How can models from information theory be used in neurophysiology? In: WS Fields, and W Abbott (eds.): *Information storage and neural control.* Charles C. Thomas Publisher, Springfield, Illinois: 230-242.

[44] Brazier MA (1957): Rise of neurophysiology in the 19th century. *Journal of Neurophysiology* 20(2): 212-226.

[45] Brink F, DW Bronk and MG Larrabee (1946): Chemical excitation of nerve. *Annals of the New York Academy of Sciences*, 47, 457-485, 1946.

[46] Bronk DW, and LK Ferguson (1934): The nervous control of intercostal respiration. *American Journal of Physiology* 110: 700-707.

[47] Brudno S, and TJ Marczynski (1977): Temporal patterns, their distribution and redundancy in trains of spontaneous neuronal spike intervals of the feline hippocampus studied with a non-parametric technique. *Brain Research* 125: 65-89.

[48] Bryant HL, and JP Segundo (1976): Spike initiation by transmembrane current: A white noise analysis. *Journal of Physiology* 260: 279-314.

[49] Bullock TH (1970): The reliability of neurons. *The Journal of General Physiology* 55: 565-584.

[50] Bullock TH (1968): The Representation of information in neurons and sites for molecular participation. *Proceedings of the National Academy of Science USA* 60(4): 1058-1068.

[51] Bullock TH (1967): Signals and neuronal coding. In: GC Quarton, T Melnechuk, and FO Schmitt (eds.): *The neurosciences. A study program*. The Rockefeller University Press, New York: 347-352.

[52] Bullock TH, and GA Horridge (1965): *Structure and function in the nervous systems of invertebrates*, volume I. W.H. Freeman and Company, San Francisco, London.

[53] Bullock TH (1964): Transfer functions at synaptic junctions. In: RW Gerard, and JW Duyff (eds.): *Information processing in the nervous system*. Excerpta Medica Foundation, Amsterdam: 98-108.

[54] Bullock TH (1961): The origins of patterned nervous discharge. *Behaviour* 17: 48-59.

[55] Bullock TH (1959): Neuron doctrine and electrophysiology. *Science* 129: 997-1002.

[56] Burke RE, P Rudomin, and FE Zajac (1970): Catch properties in single mammalian motor units. *Science* 168: 122-124.

[57] Burns BD (1968): *The uncertain nervous system*. Edward Arnold Ltd., London.

[58] Burns BD, and GK Smith (1962): Transmission of information in the unaenesthetized cat's isolated forebrain. *Journal of Physiology* 164: 238-251.

[59] Burns BD (1958): *The mammalian cerebral cortex*. Edward Arnold Ltd., London.

[60] Caianiello ER (ed.) (1968): *Neural networks. Proceedings of the school on neural networks*. Springer Verlag, Berlin, Heidelberg, New York.

[61] Caianiello ER (1964): Mathematical and physical problems in the study of brain models. In: RF Reiss: *Neural theory and modeling*. Stanford University Press, Stanford: 98-104.

[62] Caianiello ER (1961): Outline of a theory of thought-processes and thinking machines. *Journal of Theoretical Biology* 2: 204-235.

[63] Calvin WH (1968): Evaluating membrane potential and spike patterns by experimenter-controlled computer display. *Experimental Neurology* 21: 512-534.

[64] Calvin WH, and CF Stevens (1968): Synaptic noise and other sources of randomness in motorneuron interspike intervals. *Journal of Neurophysiology* 31(4): 574-587.

[65] Calvin WH, and CF Stevens (1967): Synaptic noise as a source of variability in the interval between action potentials. *Science* 155: 842-844.

[66] Cherry C (ed.) (1956): *Information Theory. Papers read at a symposium on 'information theory' held at the Royal Institution, London, Sept. 12-16 1955*. Butterworths Scientific Publications, London.

[67] Cherry C (ed.) (1961): *Information Theory. Papers read at a symposium on 'information theory' held at the Royal Institution, London, August 29th to September 2nd 1960*. Butterworths Scientific Publications, London.

[68] Cowan JD (1968): Statistical mechanics of nervous nets. In: ER Caianiello: *Neural networks*. Springer Verlag, Berlin, Heidelberg, New York: 181-188.

[69] Cox DR, and PAW Lewis (1966): *The statstical analysis of series of events*. Methuen & Co Ltd., London.

[70] Cox DR (1962): *Renewal theory*. Methuen & Co. Ltd., London.

[71] Derksen HE, and AA Verveen (1966): Fluctuations of resting neural membrane potential. *Science* 151: 1388-1389.

[72] Douglas WW, JM Ritchie, and W Schaumann (1956): A study of the effect of the pattern of electrical stimulation of the aortic nerve on the reflex depressor responses. *Journal of Physiology* 133: 232-242.

[73] Ead HW, JH Green and E Neil (1952): A comparison of the effects of pulsative and non-pulsative blood flow through the caroid sinus on the reflexogenic activity of the sinus baroreceptors in the cat. *Journal of Physiology* 118: 509-519.

[74] Eckhorn R, and B Pöpel (1974): Rigorous and extended application of information theory to the afferent visual system of the cat. I. Basic concepts. *Kybernetik* 16: 191-200.

[75] Emmers R (1976): Thalamic mechanisms that process a temporal pulse code for pain. *Brain Research* 103: 425-441.

[76] Evarts EV (1964): Temporal patterns of discharge of pyramidal tract neurons during sleep and waking in the monkey. *Journal of Neurophysiology* 27(2): 152-171.

[77] Färber G (1968): Berechnung und Messung des Informationsflusses der Nervenfaser. *Kybernetik* 5(1): 17-29.

[78] Fatt P, and B Katz (1952): Spontaneous subthreshold activity at motor nerve endings. *Journal of Physiology* 117: 109-128.

[79] Fatt P, and B Katz (1950): Some observations on biological noise. *Nature* 166: 597-598.

[80] Fehmi LG, and TH Bullock (1967): Discrimination among temporal patterns of stimulation in a coomputer model of a coelenterate nerve net. *Kybernetik* 3(5): 240-249.

[81] Fields WS, and W Abbott (eds.) (1963): *Information storage and neural control.* Charles C. Thomas Publisher, Springfield, Illinois.

[82] Fisher RA (1925): Theory of statistical estimation. *Proceedings of the Cambridge Philosophical Society* 22: 700-725.

[83] Fitch FB (1944): Review of: Warren S. McCulloch and Walter Pitts. A logical calculus of the ideas immanent in the nervous activity. *The Journal of Symbolic Logic* 9(2): 49-50.

[84] FitzHugh R (1958): A statistical analyzer for optic nerve messages. *The Journal of General Physiology* 41(4): 675-692.

[85] FitzHugh R (1956): The statistical detection of treshold signals in the retina. *The Journal of General Physiology* 40(6): 925-947.

[86] Forbes A, and C Thacher (1920): Amplification of action currents with the electron tube in recording with the string galvanometer. *The American Journal of Physiology* 52(3): 409-471.

[87] Forbes A, and A Gregg (1915): Electrical studies in mammalian reflexes. I. The flexion reflex. *American Journal of Physiology* 37: 118-176.

[88] Gasser HS, and J Erlanger (1922): A study of the action currents of nerve with the athode ray oscillograph. *American Journal of Physiology* 62: 496-524.

[89] George FH (1961): *The brain as a computer.* Pergamon Press, Oxford.

[90] Gerard LB, JG Miller, and A Rapoport (1975): Ralph Waldo Gerard. *Behavioral Science* 20(1): 1-8.

[91] Gerard RW, and JW Duyff (eds.) (1964): *Information processing in the nervous system. Proceedings of the International Union of Physiological Sciences*, volume III. Excerpta Medica Foundation, Amsterdam, New York.

[92] Gerard RW (1951): Some of the problems concerning digital notions in the central nervous system. In: H von Foerster, M Mead, and HL Teuber: *Cybernetics. Circular causal and feedback mechanisms in biological and social systems. Transactions of the seventh conference (March 23-24, 1950)*, New York: 11-57.

[93] Gerstein GL, DH Perkel, and KN Subramanian (1978): Identification of functionally related neural assemblies. *Brain Research* 140: 43-62.

[94] Gerstein GL, and DH Perkel (1972): Mutual temporal relationship among neuronal spike trains. *Biophysical Journal* 12: 453-473.

[95] Gerstein GL (1960): Analysis of firing patterns in single neurons. *Science*, 131, 1811-1812, 1960.

[96] Gerstein GL, and NY-S Kiang (1960): An approach to the quantitative analysis of electrophysiological data from single neurons. *Biophysical Journal* 1: 15-28.

[97] Gerstein GL, and B Mandelbrot (1964): Random walk models for the spike activity of a single neuron. *Biophysical Journal* 4: 41-68.

[98] Gerstein GL, and WA Clark (1964): Simultaneous studies of firing patterns in several neurons. *Science* 143: 1325-1327.

[99] Gerstein GL, and DH Perkel. Simultaneously recorded trains of action potentials: Analysis and functional interpretation. *Science* 164, 828-830, 1969.

[100] Gillary HL, and D Kennedy (1969): Neuromuscular effects of impulse pattern in a crustacean Motoneuron. *Journal of Neurophysiology* 32: 607-612.

[101] Gillary HL, and D Kennedy (1969): Pattern generation in a crustacean motoneuron. *Journal of Neurophysiology* 32: 595-606.

[102] Glaser EM, DS Ruchkin (1976): *Principles of neurobiological signal analysis*. New York, San Francisco, London. Academic Press.

[103] Goldstein K (1934): *Der Aufbau des Organismus*. Martinus Nijhoff, Den Haag.

[104] Goldstine HH (1961): Information theory. *Science* 133: 1395-1399.

[105] Gregory RL (1961): The brain as an engineering problem. In: WH Thorpe, and OL Zangwill (eds.): *Current problems in animal behaviour*. Cambridge University Press, Cambridge: 307-330.

[106] Griffith JS (1971): *Mathematical neurobiology*. Academic Press, London, New York.

[107] Griffith JS (1963): On the stability of brain-like structures. *Biophysical Journal* 3: 299-308.

[108] Grossman RG, and LJ Viernstein (1961): Discharge patterns of neurons in cochlear nucleus. *Science* 134: 99-101.

[109] Grüsser, O-J, KA Hellner, and U Grüsser-Cornehls (1962): Die Informationsübertragung im afferenten visuellen System. *Kybernetik* 1(5): 13-192.

[110] Grüsser O-J (1962): Die Informationskapazität einzelner Nervenzellen fr die Signalübermittlung im Zentralnervensystem. *Kybernetik* 1(5): 209-211.

[111] Grundfest H (1957): Excitations at synapses. *Journal of Neurophysiology* 20(3): 316-327.

[112] Hagiwara S, and H Morita (1963): Coding mechanisms of electroreceptor fibers in some electric fish. *Journal of Neurophysiology* 26: 551-567.

[113] Hagiwara S (1954): Analysis of interval fluctuation of the sensory nerve impulse. *Japanese Journal of Physiology* 4: 234-240.

[114] Harmon LD, and ER Lewis (1966): Neural Modeling. *Physiological Review* 46: 513-591.

[115] Harmon LD (1964): Problems in neural modeling. In: RF Reiss: *Neural theory and modeling*. Stanford University Press, Stanford: 9-30.

[116] Hartley RVL (1928): Transmission of information. *Bell System Technical Journal* 7: 535-563.

[117] Hartline HK, CH Graham (1932): Nerve impulses from single receptors in the eye. *Journal of Cellular and Comparative Physiology* 1(2): 277-295.

[118] Hassenstein B (1960): Die bisherige Rolle der Kybernetik in der biologischen Forschung. III. Die geschichtliche Entwicklung der biologischen Kybernetik bis 1948. *Naturwissenschaftliche Rundschau* 11: 419-424.

[119] Hebb DO (1949/1961): *The organization of behavior. A neuropsychological theory*. John Wiley & Sons Inc., New York.

[120] Herz AVM, O Creutzfeldt, and JM Fuster (1964): Statistische Eigenschaften der Neuronaktivität im ascendierenden visuellen System. *Kybernetik* 2(2): 61-71.

[121] Hind JE, JM Goldberg, DD Greenwood, and FE Rose (1963): Some discharge characteristics of single neurons in the inferior colliculus of the cat. II Timing of the discharges and observations on binaural stimulation. *Journal of Neurophysiology* 26(4): 321-341.

[122] Hintikka J (1968): The varieties of information and scientific explanation. In: B Rootselaar, JF van der Staal: *Logic, methodology and philosophy of science*, vol. 3. Amsterdam: 311-331.

[123] Hodgkin AL, and AF Huxley (1939): Action potentials recorded from inside a nerve fibre. *Nature* 144: 710-711.

[124] Householder AS, and HD Landahl (1945): *Mathematical biophysics of the central nervous system*. The Principia Press Inc., Bloomington, Indiana.

[125] Huygens C (1673): *Horologium Oscillatorium*. Muguet, Paris.

[126] Jacobson H (1951): The information capacity of the human eye. *Science* 113: 292-293.

[127] Jacobson H (1950): The information capacity of the huma ear. *Science* 112: 143-144.

[128] Jeffress LA (ed.) (1951): *Cerebral mechanisms in behavior. The Hixon symposium*. John Wiley & Sons, New York.

[129] Kalenich WA (ed.) (1966): *Information processing 1965. Proceedings of the IFIP congress 65*. Spartan Books, Washington D.C.

[130] Katsuki Y, S Yoshino, and J Chen (1950): Action currents of the single lateral-line nerve fiber of fish. 1. On the spontaneous discharge. *Japanese Journal of Physiology* 1: 87-99.

[131] Katz B, and R Miledi (1970): Membrane noise produced by acetylcholine. *Nature* 226: 962-963.

[132] Kenshalo DR (ed.) (1968): *The skin senses. Proceedings of the first international symposium on the skin senses*. Charles C. Thomas Publisher, Springfield, Illinois.

[133] Kolmogorov AN (1957): The theory of transmission of information. In: *The session of the Academy of Sciences of the USSR on Scientific Problems of Industrial Automation, October 15-20 1956*. Reprinted in: AN Shiryayev (ed.): Selected works of A.N. Kolmogorov, volume III: 6-30.

[134] Kolmogorov AN (1941): Interpolation und Extrapolation von stationären zufälligen Folgen. *Bulletin de l'académie des Sciences de l'Union des Républiques Soviétiques Socialistes* 5: 11-14.

[135] Kubie L (1930): A theoretical application to some neurological problems of the properties of excitation waves which move in closed circuits. *Brain* 53: 166-177.

[136] Kuffler SW, R FitzHugh, and HB Barlow (1957): Maintained activity on the cat's retina in light and darkness. *The Journal of General Physiology* 40(5): 683-702.

[137] Landahl HD, W McCulloch, and W Pitts (1943): A statistical consequence of the logical calculus of nervous nets. *Bulletin of Mathematical Biophysics* 5: 135-137.

[138] Landahl HD (1939): Contributions to the mathematical biophysics of the central nervous system. *Bulletin of Mathematical Biophysics* 1(2): 95-118.

[139] Lashley KS (1951): The problem of serial order in behavior. In: LA Jeffress (ed.): *Cerebral mechanisms in behavior*. John Wiley & Sons, New York: 112-146.

[140] Lashley KS (1933): Integrative functions of the cerebral cortex. *Physiological Reviews* 13(1): 1-42.

[141] Lashley KS (1931): Mass action in cerebral function. *Science* 73: 245-254.

[142] Leibovic KN (1969): *Information processing in the nervous system*. Springer Verlag, Berlin, Heidelberg, New York.

[143] Lempel A, and Ziv J (1976): On the complexity of finite sequences. *IEEE Transactions on Information Theory* IT-22: 75-81.

[144] Lewis RE (1968): The iron wire model of the neuron: A review. In: HL Oestreicher, DR Moore: *Cybernetic problems in bionics*. Gordon and Breach Science Publishers Inc., New York: 247-273.

[145] Ling G, RW Gerard (1949): The normal membrane potential of frog sartorius fibers. *Journal of Cellular and Comparative Physiology* 34: 383-394.

[146] Lucas K (1917): *The conduction of the nervous impulse*. Longmans, Green and Co., London.

[147] MacKay DM (1966): Information in brains and machines. In: WA Kalenich (ed.): *Information processing 65*. Spartan Books, Washington: 637-643.

[148] MacKay DM (1956): The place of 'meaning' in the theory of information. In: C Cherry (ed.): *Information theory*. Butterworths Scientific Publications, London: 215-225.

[149] MacKay DM (1954): On comparing the brain with machines. *American Scientist* 42: 261-268.

[150] MacKay DM, and W McCulloch (1952): The limiting information capacity of a neuronal link. *Bulletin of Mathematical Biophysics* 14: 127-135.

[151] Martin AR, and C Branch (1958): Spontaneous activity of betz cells in cats with midbrain lesions. *Journal of Neurophysiology* 21: 368-379.

[152] Marczynski TJ, and CJ Sherry (1971): A new analysis of trains of increasing or decreasing interspike intervals treated as self-adjusting sets of ratios. *Brain Research* 35: 533-538.

[153] McCulloch W (1965): *Embodiments of mind*. The MIT Press, Cambridge, Massachusetts.

[154] McCulloch W (1964): A historical introduction to the postulational foundations of experimental epistemology. In: FSC Northrop and HH Livingston (eds.): *Cross-cultural understanding: Epistemology in antrhopology*. Harper & Row Publishers Inc., New York. Reprinted in: W McCulloch: Embodiments of mind: 359-372.

[155] McCulloch W (1960): The reliability of biological systems. In: MC Yovits, and S Cameron: *Self-organizing systems*. Pergamon Press, Oxford: 264-281.

[156] McCulloch W (1957): Biological computers. *IRE Transactions on Electronic Computers*, September: 190-192.

[157] McCulloch W (1951): Why the mind is in the head. In: LA Jeffress (ed.): *Cerebral mechanisms in behavior*. John Wiley & Sons, New York: 42-57.

[158] McCulloch W (1949): The brain as a computing machine. *Electrical Engineering* 68: 492-497.

[159] McCulloch W, and W Pitts (1943): A logical calculus of the ideas immanent in nervous activity. *Bulletin of Mathematical Biophysics* 5: 115-133.

[160] Melzack R, and PD Wall (1962): On the nature of cutaneous sensory mechanisms. *Brain* 85: 331-356.

[161] Milner PM (1974): A model for visual shape recognition. *Psychological Review* 81(6): 521-535.

[162] Monnier A-M, HH Jasper (1932): Recherche de la relation entre les potentiels d'action élémentaires et la chronaxie subordination, nouvelle démonstration du fonctionnement par "tout ou rien" de la fibre nerveuse. *Comptes rendus des scéances de la société de biologie et ses filiales* 110: 547-549.

[163] Moore GP, DH Perkel and JP Segundo. Statistical analysis and functional interpretation of neuronal spike data. *Annual Review of Physiology*. 28, 493-522, 1966.

[164] Mountcastle VB (1975): The view from within: Pathways to the study of perception. *John Hopkins medical journal* (supplement) 136: 109-131.

[165] Mountcastle VB (1967): The problem of sensing and the neural code of sensory events. In: GC Quarton, T Melnechuk, and FO Schmitt (eds.): *The neurosciences. A study program*. The Rockefeller University Press, New York: 393-408.

[166] Nachmansohn D, H Merritt (1954): *Nerve Impulse*. Josiah Macy Jr. Foundation, New York.

[167] Nachrichtentechnische Gesellschaft im VDE Fachausschuss 'Informations- und Systemtheorie' (Hg.) (1961): *Aufnahme und Verarbeitung von Nachrichten durch Organismen*. S. Hirzel, Stuttgart.

[168] Nakahama H, S Nishioka, T Otsuka, and S Aikawa (1966): Statistical dependence between interspike intervals of spontaneous activity in thalamic lemniscal neurons. *Journal of Neurophysiology* 29(5): 921-941.

[169] National Physical Laboratory (1959): *Mechanisation of Thought-Processes*, volumes I/II. London, Her Majesty's Stationery Office.

[170] Oestreicher HL, and DR Moore (eds.) (1968): *Cybernetic problems in bionics*. Gordon and Breach Science Publishers Inc., New York, London, Paris.

[171] Pattee HH (1974): Discrete and continuous processes in computers and brains. In: M Conrad, W Güttinger, and M Dal Cin (eds.): *Physics and mathematics of the nervous system*. Springer-Verlag, Berlin, Heidelberg, New York: 128-148.

[172] Pecher C (1939): La fluctuation d'excitabilité de la fibre nerveuse. *Archives Internationales de Physiologie* XLIX(2): 129-152.

[173] Pecher C (1937): Fluctuations indépendantes de l'exitabilité de deux fibres d'un même nerf. *Comptes rendus des scéances de la société de biologie et ses filiales.* 124: 839-842.

[174] Peirce CS (1906/1960): Prolegomena to an apology for pragmaticism. In: C Hartshorne, and P Weiss: *Collected Papers of Charles Sanders Peirce*, volume III/IV. Cambridge. The Belknap Press of Harvard University Press: Paragraph 4.537.

[175] Perkel DH, GL Gerstein, MS Smith, and WG Tatton (1975): Nerve-impulse patterns: a quantitative display technique for three neurons. *Brain Research* 100: 271-296.

[176] Perkel DH (1970): Spike trains as carriers of information. In: FO Schmitt (ed.): *The Neurosciences. 2. study program.* The Rockefeller University Press, New York: 587-596.

[177] Perkel DH, and TH Bullock (1968): Neural Coding. *Neuroscience Research Progress Bulletin* 6(3): 221-348.

[178] Perkel DH, GL Gerstein, and GP Moore (1967): Neural spike trains and stochastic point processes. I. The single spike train. *Biophysical Journal* 7: 391-418

[179] Perkel DH, GL Gerstein and GP Moore. Neuronal spike trains and stochastic point processes. II Simultaneous spike trains. *Biophysical Journal*, 7, 419-440, 1967.

[180] Pfaffenhuber E (1974): Sensory coding and economy of nerve impulses. In: M Conrad, W Güttinger, and M Dal Cin (eds.): *Physics and mathematics of the nervous system.* Springer Verlag, Berlin, Heidelberg, New York: 469-483.

[181] Pierce WH (1965): *Failure-tolerant computer design.* Academic Press, New York, London.

[182] Pinneo LR (1966): On noise in the nervous system. *Psychological Review* 73(3): 242-247.

[183] Poggio GF, and LJ Viernstein (1964): Time series analysis of impulse sequences of thalamic somatic sensory neurons. *Journal of Neurophysiology* 27(4): 517-545.

[184] Pribram K (1960): A review of theory in physiological psychology. *Annual Review of Psychology* 11: 1-40,

[185] Pringle JWS, and VJ Wilson (1952): The response of a sense organ to a harmonic stimulus. *Journal of Experimental Biology* 29: 220-234.

[186] Province de Namur (1958): *1er congrès international de cybernétique.* Namur, Association Intèrnationale de Cybernétique.

[187] Quarton GC, T Melnechuk, and FO Schmitt (eds.) (1967): *The neurosciences. A study program.* The Rockefeller University Press, New York.

[188] Quastler H (1958): A primer of information theory. In: HP Yockey, RL Platzman, and H Quastler (eds.): *Symposium on information theory in biology.* Pergamon press, London: 3-49.

[189] Quastler H (1958c): The status of information theory in biology. In: HP Yockey, RL Platzman, and H Quastler (eds.): *Symposium on information theory in biology.* Pergamon Press, London: 399-402.

[190] Quastler H (1957): The complexity of biological computers. *IRE Transactions on Electronic Computers*, September: 192-194.

[191] H Quastler (ed.) (1953): *Essays on the use of information theory in biology.* University of Illinois Press, Urbana.

[192] Rapoport A (1964): Information processing in the nervous system. In: RW Gerard, JW Duyff (eds.): *Information processing in the nervous system.* Excerpta Medica Foundation, Amsterdam: 16-23.

[193] Rapoport A, and W Horvath (1960): The theoretical channel capacity of a single neuron as determined by various coding systems. *Information and Control* 3: 335-350.

[194] Rapoport A, and A Shimbel (1948): Steady states in random nets: I. *Bulletin of Mathematical Biophysics* 10: 211-220.

[195] Rapoport A (1948): Steady states in random nets: II. *Bulletin of Mathematical Biophysics* 10: 221-226.

[196] Rashevsky N (1948): *Mathematical Biophysics.* The University of Chicago Press, Chicago.

[197] Rashevsky N (1946): The neural mechanism of logical thinking. *Bulletin of Mathematical Biophysics* 8: 29-40.

[198] Reiss RF (ed.) (1964): *Neural theory and modeling. Proceedings of the 1962 Ojai Symposium.* Stanford University Press, Stanford.

[199] Reiss RF (1964): A theory of resonant networks. In: RF Reiss (ed.): *Neural theory and modeling.* Stanford University Press, Stanford: 105-137.

[200] Ripley SH, and CAG Wiersma (1953): The effect of spaced stimulation of excitatory and inhibitory axons of the crayfish. *Physiologia Comparata et Oecologia* 3: 1-17.

[201] Rodieck RW, NY-S Kiang, and GL Gerstein (1962): Some quantitative methods for the study of spontaneous activity of single neurons. *Biophysical Journal* 2: 351-368.

[202] Rose JE, JF Brugge, DJ Anderson, and JE Hind (1967): Phase-locked resonse to low-frequency tones in single auditory nerve fibers of the squirrel monkey. *Journal of Neurophysiology* 30: 769-793.

[203] Rose JE, DD Greenwood, JM Goldberg, JE Hind (1963): Some discharge characteristics of single neurons in the inferior colliculus of the cat. I. Tonotopical organization, relation of spike-count to tone intensity, and firing patters of single elements. *Journal of Neurophysiology* 26(2): 294-319.

[204] Rosenblatt F (1962): *Principles of neurodynamics.* Spartan Books, Washington D.C.

[205] Rosenblith WA (ed.) (1961): *Sensory Communication, Contributions to the symposium on principles of sensory communication.* MIT Press, Cambridge and John Wiley & Sons,New York, London.

[206] Rosenblith WA (1959): Some quantifiable aspects of the electrical activity of the nervous system (with emphasis upon responses to sensory stimuli). *Reviews of Modern Physics* 31(2): 532-545.

[207] Rosenblueth A, and N Wiener (1950): Purposeful and non-purposeful behavior. *Philosophy of Science* 17: 318-326.

[208] Rosenblueth A, N Wiener, and J Bigelow (1943): Behavior, Purpose and Teleology. *Philosophy of Science* 10: 18-24.

[209] Rudolfer SM, and HU May (1975): On the Markov properties of interspike times in the cat optic tract. *Biological Cybernetics* 19: 197-199.

[210] Saltzberg B (1963): What is information theory? In: WS Fields, and W Abbott(eds.): *Information storage and neural control.* Charles C. Thomas Publisher, Springfield, Illinois.

[211] Schiller PH, BL Finlay, and SF Volman (1976): Short-term resopnse variability of monkey striate neurons. *Brain Research* 105: 347-349.

[212] Schmitt FO, G Adelman, and FG Worden (eds.) (1977): *Neurosciences research symposium summaries,* volume 8. MIT Press, Cambridge, London.

[213] Schmitt FO, and F Worden (ed.) (1974): *The neurosciences. Third study program.* The MIT Press, Cambridge, London.

[214] Schmitt FO, G Adelman, and FG Worden (eds.) (1973): *Neurosciences research symposium summaries,* volume 7. MIT Press, Cambridge, London.

[215] Schmitt FO, G Adelman, T Melnechuk, and FG Worden (eds.) (1972): *Neurosciences research symposium summaries,* volume 6. MIT Press, Cambridge, London.

[216] Schmitt FO, G Adelman, T Melnechuk, and FG Worden (eds.) (1971): *Neurosciences research symposium summaries,* volume 5. MIT Press, Cambridge, London.

[217] Schmitt FO (ed.) (1970): *The neurosciences. Second study program.* The Rockefeller University Press, New York.

[218] Schmitt FO, T Melnechuk, GC Quarton, and G Adelman (eds.) (1970): *Neurosciences research symposium summaries,* volume 4. MIT Press, Cambridge, London.

[219] Schmitt FO, T Melnechuk, GC Quarton, and G Adelman (eds.) (1969): *Neurosciences research symposium summaries,* volume 3. MIT Press, Cambridge, London.

[220] Schmitt FO, T Melnechuk, and GC Quarton (eds.) (1967): *Neurosciences research symposium summaries,* volume 2. MIT Press, Cambridge, London.

[221] Schmitt FO, and T Melnechuk (eds.) (1966): *Neurosciences research symposium summaries*, volume 1. MIT Press, Cambridge, London.
[222] Segundo JP (1970): Communication and coding by nerve cells. In: FO Schmitt (ed.): *The neurosciences. Second study program.* The Rockefeller University Press, New York: 569-586.
[223] Segundo JP, DH Perkel, and GP Moore (1966): Spike probability in neurones: Influence of temporal structure in the train of synaptic events. *Kybernetik* 3(2): 67-82.
[224] Segundo JP, GP Moore, LJ Stensaas, and TH Bullock (1963): Sensitivity of neurones in Aplysia to temporal patternof arriving inpulses. *The Journal of Experimental Biology* 40: 643-667.
[225] Shannon CE, and J McCarthy (eds.) (1956): *Automata studies*. Princeton University Press, Princeton, New Jersey.
[226] Shannon CE, and W Weaver (1949): *The mathematial theory of communication*. University of Illinois Press, Urbana.
[227] Shannon CE (1948): A mathematical theory of communication. *The Bell Systems Technical Journal* 27: 379-423. Reprinted with corrections.
[228] Sherrington CS (1940/1951): *Man on his nature*. Cambridge University Press, Cambridge.
[229] Sherrington CS (1906/1909) *The integrative action of the nervous system*. Constable and Company Ldt., London.
[230] Sherry CJ, and TJ Marczynski (1972): A new analysis of neuronal interspike intervals based on inequality tests. *International Journal of Neuroscience* 3: 259-270.
[231] Shimbel A (1950): Contributions to the mathemathical biophysics of the central nervous system with special references to learning. *Bulletin of Mathematical Biophysics* 12: 241-275.
[232] Shimbel A, and A Rapoport (1948): A statistical approach to the theory of the central nervous system. *Bulletin of Mathematical Biophysics* 10: 41-55.
[233] Sholl DA, and AM Uttley (1953): Pattern discrimination and the visual cortex. *Nature* 171: 387-388.
[234] Smith DR, and GK Smith (1965): A statistical analysis of the continual activity of single cortical neurones in the cat unanaesthetized isolated forebrain. *Biophysical Journal* 5: 47-74.
[235] Snow CP (1959): *The two cultures and the scientific revolution*. Cambridge University Press, New York.
[236] Stark L, and JF Dickson (1966): Mathematical concepts of central nervous system function. In: FO Schmitt, and T Melnechuk (eds.): *Neuroscience Research Symposium Summaries* 1. MIT-Press, Cambridge: 109-178.
[237] Stein RB (1970): The role of spike trains in transmitting and distorting sensory signals. In: FO Schmitt (ed.): *The neurosciences. Second study program*. Rockefeller University Press, New York: 597-604.
[238] Stein RB (1967): The information capacity of nerve cells using a frequency code. *Biophysical Journal* 7: 797-826.
[239] Stein RB (1967): Some models of neuronal variability. *Biophysical Journal* 7: 37-68.
[240] Stein RB (1965): A theoretical analysis of neuronal variability. *Biophysical Journal* 5: 173-194.
[241] Steinbuch K (1961): *Automat und Mensch. Über menschliche und maschinelle Intelligenz*. Springer Verlag, Berlin, Göttingen, Heidelberg.
[242] Strehler BL (1977): *Time, cells, and aging*. New York, San Francisco, London. Academic Press.
[243] Strehler BL (1969): Information handling in the nervous system: An analogy to molecular-genetic coder-decoder mechanisms. *Perspectives in Biology and Medicine* 12: 584-612.
[244] Szilard L (1929): Über die Entropieverminderung in einem thermodynamischen System bei Eingriffen intelligenter Wesen. *Zeitschrift für Physik* 53: 840-856.
[245] Taylor R (1950): Comments on a mechanistic conception of purposefulness. *Philosophy of Science* 17: 310-317.
[246] Terzuolo CA (1970): Data transmission by spik trains. In: FO Schmitt (ed.): *The neurosciences. Second study program*. Rockefeller University Press, New York: 661-671.

[247] Tomko GJ, and DR Crapper (1974): Neuronal variability: non-stationary responses to identical visual stimuli. *Brain Research* 79: 405-418.

[248] Turing AM (1936): On computable numbers, with an application to the Entscheidungsproblem. *Proceedings of the London Mathematical Society* 42(2): 230-265.

[249] Uttal WR (1973): *The psychobiology of sensory coding.* Harper and Row, New York.

[250] Uttal WR (1969): Emerging principles of sensory coding. *Perspectives in Biology and Medicine* 12: 344-368.

[251] Uttal WR, and M Krissoff (1968): Response of the somesthetic system to patterned trains of electrical stimuli. In: DR Kenshalo (ed.): *The skin senses.* Charles C. Thomas Publisher, Springfield: 262-303.

[252] Uttal WR (1960): The three stimulus problem: A further comparison of neural and psychophysical responses in the somesthetic system. *Journal of Comparative and Physiological Psychology* 53(1): 42-46.

[253] Uttal WR (1959): A comparison of neural and psychophysical responses in the somesthetic system. *Journal of Comparative and Physiological Psychology* 52: 485-490.

[254] Uttley AM (1954): The classification of signals in the nervous system. *Electroencephalography and Clinical Neurophysiology* 6: 479-494.

[255] Verveen AA, and HE Derksen (1965): Fluctuations in membrane potential of axons and the problem of coding. *Kybernetik* 2(4): 152-160.

[256] Viernstein LJ, and RG Grossman (1961): Neural discharge patterns in the transmission of sensory information. In: C Cherry: *Information theory.* Butterworth, London: 252-269.

[257] Von Foerster H, and GW Zopf (eds.) (1962): *Principles of self-organization. International tracts in computer science and technology and their applications*, volume 9. Pergamon Press, New York.

[258] Von Foerster H (ed.) (1950): *Cybernetics. Circular causal and feedback mechanisms in biological and social systems, transactions of the sixth conference (march 24-25, 1949).* New York. Re-issued in C. Pias (2003).

[259] Von Foerster H, M Mead, and HL Teuber (eds.) (1951): *Cybernetics. Circular causal and feedback mechanisms in biological and social systems, transactions of the seventh conference (march 23-24, 1950).* New York. Re-issued in C. Pias (2003).

[260] Von Foerster H, M Mead, and HL Teuber (eds.): *Cybernetics. Circular causal and feedback mechanisms in biological and social systems, transactions of the eigth conference (march 15-16, 1951).* New York. Re-issued in C. Pias (2003).

[261] Von Foerster H, M Mead, and HL Teuber (1952): A note by the editors. In: H von Foerster, M Mead, and HL Teuber (eds.): *Cybernetics. Circular causal and feedback mechanisms in biological and social systems, transactions of the eigth conference (march 15-16, 1951).* New York: xi-xx.

[262] Von Foerster H, M Mead, and HL Teuber (eds.) (1953): *Cybernetics. Circular causal and feedback mechanisms in biological and social systems, transactions of the ninth conference (march 20-21, 1952).* New York. Re-issued in C. Pias (2003).

[263] Von Foerster H, M Mead, and HL Teuber (eds.) (1954): *Cybernetics. Circular causal and feedback mechanisms in biological and social systems, transactions of the tenth conference (april 22-24, 1953).* New York. Re-issued in C. Pias (2003).

[264] Von Monakow C (1914): *Die Lokalisation im Grosshirn und der Abbau der Funktion durch kortikale Herde.* Verlag von J.F. Bergman, Wiesbaden.

[265] Von Neumann J (1966): *The theory of self-reproducing automata.* Edited and completed by AW Burcks. University of Illinois Press, Urbana.

[266] Von Neumann J (1958): *The computer and the brain.* Yale University Press, New Haven, London.

[267] Von Neumann J (1956): Probabilistic logics and the synthesis of reliable organisms from unreliable components. In: CE Shannon, and J McCarthy (eds.): *Automata studies.* Princeton University Press, Princeton: 43-98.

[268] Von Neumann J (1951): The general and logical theory of automata. In: LA Jeffress (ed.): *Cerebral mechanisms in behavior.* John Wiley & Sons, New York: 1-41.

[269] Von Neumann J (1945/1993): First draft of a report on the EDVAC. Unpublished manuscript. Reprinted in: *IEEE Annals of the History of Computing* 15(4): 27-75.

[270] Wakabayashi T, and T Kuroda (1977): Effect of stimulation with impulse trains of various patterns, including adaptional type, on frog's nerve-muscle and spinal reflex preparations. *Tohoku Journal of experimental Medicine* 121: 219-229.

[271] Wakabayashi T, and T Kuroda (1977): Responses of crayfish muscle preparations to nerve stimulation with various patterns of impulse sequence. Effects of intermittent, intercalated and adaptational types of impulse sequence. *Tohoku Journal of experimental Medicine* 121: 207-218.

[272] Walker AE (1957): Stimulation and Ablation. Their role in the history of cerebral physiology. *Journal of Neurophysiology* 20(4): 435-449.

[273] Wall PD, JY Lettvin, W McCulloch, and W Pitts (1956): Factors limiting the maximum impulse transmitting ability of an afferent system of nerve fibres. In: C Cherry (ed.): *Information Theory*. Butterworths Scientific Publications, London: 329-344.

[274] Wayner MJ, and Y Oomura (1968): Meeting report: Neuronal Spike Trains. *Science* 160: 1025-1026.

[275] Weaver W (1948): Science and complexity. *American Scientist* 36: 536-544.

[276] Weddell G (1955): Somesthesis and the chemical senses. *Annual Review of Psychology* 6: 119-136.

[277] Whitsel BL, RC Schreiner, and GK Essick (1977): An analysis of variability in somatosensory cortical neuron discharge. *Journal of Neurophysiology* 40(3): 589-607.

[278] Wiener N (1949): *Extrapolation, interpolation and smoothing of stationary time series*. John Wiley & Sons Inc., New York.

[279] Wiener N (1949): Time, communication, and the nervous system. Annals of the New York Academy of Sciences 50(4): 197-220.

[280] Wiener N (1948): *Cybernetics, or control and communication in the animal and the machine*. MIT Press, Cambridge.

[281] Wiersma CA, and RT Adams (1950): The influence of nerve impulse sequence on the contractions of different crustacean muscles. *Physiologia Comparata et Oecologia* 2: 20-33.

[282] Wilson DM, and JL Larimer (1968): The catch property of ordinar muscle. *Proceedings of the National Academy of Science USA* 61: 909-916.

[283] Wilson DM, and WJ Davis (1965): Nerve impulse patterns and reflex control in the motor system of the crayfish claw. *Journal of Experimental Biology* 43: 193-210.

[284] Wilson DM, and RJ Wyman (1965): Motor output patterns during random and rhythmic stimulation of locust thoracic ganglia. *Biophysical Journal* 5: 121-143.

[285] Wilson DM (1964): The origin of the flight-motor command in grasshoppers. In: RF Reiss (ed.): *Neural theory and modeling*. Stanford University Press, Stanford: 331-345.

[286] Winograd S, and JD Cowan (1963): *Reliable computation in the presence of noise*. MIT Research Monograph, volume 2. MIT Press, Cambridge.

[287] Yockey HP, RL Platzman, and H Quastler (eds.) (1958): *Symposium on information theory in biology*. Pergamon Press, London.

[288] Young JZ (1964): *A model of the brain*. Clarendon Press, Oxford.

[289] Yovits MC, and S Cameron (1960): *Self-organizing systems. Proceedings of an interdisciplinary conference*. Pergamon Press, Oxford, London.

[290] Zangwill OL (1961): Lashley's concept of cerebral mass action. In: WH Thorpe, and OL Zangwill (eds.): *Current problems in animal behavior*. Cambridge University Press, Cambridge: 59-86.

[291] Ziv J, and A. Lempel (1978): Compression of individual sequences by variable rate coding. *IEEE Transactions on Information Theory* IT-24: 530-536.

Secundary Literature

[292] Abraham TH (2002): (Physio)logical circuits: The intellectual origins of the McCulloch-Pitts neural network. *Journal of the History of the Behavioral Sciences* 38(1): 3-25.

[293] Adelman G, and BH Smith (eds.) (2004): *Encyclopedia of neuroscience*, third edition. Elsevier, Amsterdam.

[294] Anderson JA, and E Rosenfeld (eds.) (1988): *Neurocomputing. Foundations of research*. MIT Press, Cambridge, London.

[295] Aspray W (1985): The scientific conceptualization of information: a survey. *Annals of the History of Computing* 7(2): 117-140.

[296] Bennett MR, and PMS Hacker (2003): *Philosophical foundations in neuroscience*. Blackwell Publishing, Malden MA.

[297] Borck C (2002): *Das elektrische Gehirn. Geschichte und Wissenskultur der Elektroenzephalographie*. Habilitationsschrift, Institut für Geschichte der Medizin, Freie Universität Berlin. Erschienen 2005 als: *Hirnströme. Eine Kulturgeschichte der Elektroenzephalographie*. Wallstein Verlag, Göttingen.

[298] Breidbach O (2001): The origin and development of the neurosciences. In: PK Machamer, R Grush, and P McLaughlin (eds.): *Theory and method in the neurosciences*. University of Pittsburgh Press, Pittsburgh: 7-29.

[299] Breidbach O (1997): *Die Materialisierung des Ichs: Eine Geschichte der Hirnforschung im 19. und 20. Jahrhundert*. Suhrkamp Verlag, Frankfurt.

[300] Breidbach O (1995): From associative psychology to neuronal networks: The historical roots of Hebb's concept of the memory trace. In: N Elsner and R Menzel (eds.): *Göttinger Neurobiologietagung 1995*. Thieme Verlag, Stuttgart: 58.

[301] Breidbach O (1993): Nervenzellen oder Nervennetze? Zur Entstehung des Neuronenkonzeptes. In: E Florey, and O Breidbach: *Das Gehirn - Organ der Seele?* City: Berlin Publisher: Akademie-Verlag Pages: 81-126

[302] Bullock TH (1996: Theodore H. Bullock. In: LR Squire: *The history of neuroscience in autobiography*, volume 1. Society of Neuroscience, Washington D.C.

[303] Christen M (2006): Varieties of publication patterns in neuroscience at the cognitive turn. Journal of the History of the Neurosciences, submitted.

[304] Clarke E, and LS Jacyna (1987): *Nineteenth century origins of neuroscientific concepts*. University of California Press, Berkeley.

[305] Cowan WM, DH Harter, and ER Kandel (2000): The emergence of modern neuroscience. Some implications for neurology and psychiatry. *Annual review of Neuroscience* 23: 343-391.

[306] Dennett DC (1991): Real patterns. *The Journal of Philosophy* 88(1): 27-51.

[307] Dretske FI (1981/1999): *Knowledge and the flow of information*. MIT Press, Cambridge.

[308] Eco U (1972/2002): *Einführung in die Semiotik*. Wilhelm Fink Verlag, München..

[309] Fox Keller E (2002): *Making sense of life. Explaining biological development with models, metaphors, and machines*. Harvard University Press, Cambridge.

[310] Galison P (1994): The ontology of the enemy: Norbert Wiener and the cybernetic vision. *Critical Inquiry* 21: 228-266.

[311] Gardner HE (1985): *The mind's new science: A history of the cognitive revolution*. Basic books, New York.

[312] Garson J (2003): The introduction of information into neurobiology. *Philosophy of Science* 70: 926-936.

[313] Glänzel W, and U Schoepflin (1994): A stochastic model for the ageing of scientific literature. Scientometrics 30(1): 49-64.

[314] Godfrey MD, and DF Hendry (1993): The computer as von Neumann planned it. *IEEE Annals of the History of Computing* 15(1): 11-21.

[315] Grush R (2001): The semantic challenge to computational neuroscience. In: PK Machamer, R Grush, and P McLaughlin: *Theory and method in the neurosciences*. University of Pittsburgh Press, Pittsburg: 155-172.

[316] Hagner M, and C Borck (2001): Mindful practices: on the neurosciences in the twentieth century. *Science in Context* 14(4): 507-510.

[317] Hagner M (1997/2000): *Homo cerebralis: Der Wandel vom Seelenorgan zum Gehirn.* Insel-Verlag, Frankfurt a.M.

[318] Hagner M (1993): Das Ende vom Seelenorgan. In: E Florey, and O Breidbach (eds.): *Das Gehirn. Organ der Seele.* Akademie Verlag, Berlin: 3-22.

[319] Hagner M (1999): *Ecce Cortex. Beiträge zur Geschichte des modernen Gehirns.* Wallstein Verlag, Göttingen.

[320] Hardcastle VG, C Stewart (2003): Neuroscience and the art of single cell recording. *Biology and Philosophy* 18: 195-208.

[321] Hoah H (2003): Remote control. *Nature* 423: 796-799.

[322] Kay LE (2001): From logical neurons to poetic embodiments of mind: Warren S. McCulloch's project in neuroscience. *Science in Context* 14(4): 591-614.

[323] Kay LE (2000/2001): *Who wrote the book if life? A history of the genetic code.* Stanford University Press, Stanford. German: *Das Buch des Lebens. Wer schrieb den genetischen Code.* Carl Hanser Verlag, München.

[324] Lettvin J (1999): McCulloch, Warren S. In: RA Wilson, and FC Keil: *The MIT encyclopedia of the cognitive sciences.* MIT press, Cambridge: 512-513.

[325] Lettvin J (1999): Pitts, Walter. In: RA Wilson, and FC Keil: *The MIT encyclopedia of the cognitive sciences.* MIT press, Cambridge: 651.

[326] Longuet-Higgins HC (1987): Donald MacKay (1922-1987). *Nature* 326: 446.

[327] Mendelsohn E, MR Smith, and P Weingart (eds.) (1988): *Science, technology, and the military.* Kluver, Dordrecht.

[328] Millman S (ed.) (1984): *A history of engineering and science in the Bell system. Communication Sciences (1925-1980).* AT&T Bell Laboratories, Indianapolis.

[329] Müller A (2000): Eine kurze Geschichte des BCL. *Österreichische Zeitschrift fr Geschichtswissenschaften* 11 (1): 9-30.

[330] Oyama S (1985/2002): *The ontogeny of information. Developmental systems and evolution.* Duke University Press, Durham.

[331] Penzlin H (2000): Die vergleichende Tierphysiologie In: I Jahn : *Geschichte der Biologie.* Spektrum akademischer Verlag, Heidelberg, Berlin: 461-498.

[332] Pias C (ed.) (2003): *Cybernetics - Kybernetik. The Macy conferences 1946-1953. Transactions/Protokolle.* Diaphanes Verlag, Zürich, Berlin.

[333] Pias C (ed.) (2004): *Cybernetics - Kybernetik. The Macy conferences 1946-1953. Essays & documents / Essays & Dokumente.* Diaphanes Verlag, Zürich, Berlin.

[334] Pias C (2004): Zeit der Kybernetik – eine Einstimmung. In: Pias C (ed.) (2004): *Cybernetics - Kybernetik. The Macy conferences 1946-1953. Essays & documents / Essays & Dokumente.* Diaphanes Verlag, Zürich, Berlin: 9-41.

[335] Piccinini G (2004): The first computational theroy of mind and brain: A close look at McCulloch and Pitt's "Logical calculus of ideas immanent in nervous activity'". *Synthese* 141.2: 175-215.

[336] Piccinini G (2004): Symbols, strings, and spikes. Unpublished manuscript, available under: http://artsci.wustl.edu/~gpiccini

[337] Revonsuo A (2001): On the nature of explanation in neuroscience. In: PK Machamer, R Grush, and P McLaughlin: *Theory and method in the neurosciences.* University of Pittsburgh Press, Pittsburg: 45-69.

[338] Schmiff FO (1992): The neurosciences research program: A brief history. In: F Samson, and G Adelman (eds.): *The neurosciences: Paths of discovery II.* Birkhäuser Verlag, Boston, Basel, Berlin.

[339] Schnelle H (1976): Information. In: *Historisches Wörterbuch der Philosophie.* Schwabe Verlag, Basel: 356-357.

[340] Shepherd GM (1991): *Foundations of the neuron doctrine*. Oxford University Press, Oxford.

[341] Swazey JP (1975/1992): Forging a Neuroscience Community: A Brief History of the Neurosciences Research Program. In: FG Worden, JP Swazey, and G Adelman (eds.): *The Neurosciences: Paths of Discovery*, 1. Birkhäuser, Boston, Basel, Berlin.

[342] Szentágothai J (1975): From the last skirmishes around the neuron theory to the functional anatomy of neuron networks. In: FG Worden, J Swazey, and G Adelman: The neurosciences – Paths of discovery. MIT Press, Cambridge: 103-120.

[343] Thagard P (2002): How molecules matter to mental computation. *Philosophy of Science* 69: 429-446.

[344] Williams MR (1993): The origins, uses and fate of the EDVAC. *IEEE Annals of the History of Computing* 15(1): 22-38.

[345] Young RM (1970): *Mind, brain, and adaption in the nineteenth century: Cerebral localization and its biological context from Gall to Ferrier*. Clarendon Press, Oxford.

Scientific Literature (~1980 – 2005)

[346] Abbott LF, WG Regehr (2004): Synaptic computation. *Nature* 431: 796-803.

[347] Abbott LF (1994): Decoding neuronal firing and modelling neural networks. *Quarterly Review of Biophysics* 27(3): 291-331.

[348] Abeles M, and I Gat (2001): Detecting precise firing sequences in experimental data. *Journal of Neuroscience Methods* 107: 141-154.

[349] Abeles M, H Bergman, I Gat, I Meilijson, E Seidemann, and N Tishby (1995): Cortical activity flips among quasi-stationary states. *Proceedings of the National Academy of Science USA* 92: 8616-8620.

[350] Abeles M, Y Prut, H Bergman, and E Vaadia (1994): Synchronization in neuronal transmission and its importance for information processing. *Progress in Brain Research* 102: 395-404.

[351] Abeles M, H Bergman, E Margalit, and E Vaadia (1993): Spatiotemporal firing patterns in the frontal cortex of behaving monkeys. *Journal of Neurophysiology* 70(4): 1629-1638.

[352] Abeles M (1991): *Corticonics. Neural circuits of the cerebral cortex*. Cambridge University Press, Cambridge.

[353] Abeles M, and GL Gerstein (1988): Detecting spatiotemporal firing patterns among simultaneously recorded single neurons. *Journal of Neurophysiology* 60(3): 909-924.

[354] Abeles M (1983): The quantification and graphic display of correlations among three spike trains. *IEEE Transactions on Biomedical Engineering* BME-30(4): 235-239.

[355] Abeles M, F de Ribaupierre, Y de Ribaupierre (1983): Detection of single unit responses which are loosely time-locked to a stimulus. *IEEE Transactions on Systems, Man, and Cybernetics* 13(5): 683-691.

[356] Abeles M (1982): Quantification, smoothing, and confidence limits for single-units' histograms. *Journal of Neuroscience Methods* 5: 317-325.

[357] Abeles M (1982): Role of the cortical neuron: Integrator or coincidence detector? *Israel Journal of Medical Sciences* 18: 83-92.

[358] Abeles M (1982): *Local cortical circuits. An electrophysiological study*. Berlin, Heidelberg, New York, Springer.

[359] Aertsen AMHJ, GL Gerstein, MK Habib, and G Palm (1989): Dynamics of neuronal firing correlation: Modulation of "effective connectivity". *Journal of Neurophysiology* 61(5): 900-917.

[360] Ahmed B, JC Anderson, KAC Martin, and JC Nelson (1997): Map of the synapses onto layer 4 basket cell of the primary visual cortex of the cat. *Journal of Computational Neuroscience* 380(2): 230-242.

[361] Allen C, and CF Stevens (1994): An evaluation for causes for unreliability of synaptic transmission. *Proceedings of the National Academy of Science USA* 91: 10380-10383.

[362] Alonso J-M, WM Usrey, and RC Reid (2001): Rules of connectivity between geniculate cells and simple cells in cat primary visual cortex. *The Journal of Neuroscience* 21(11): 4002-4015.

[363] Alonso J-M, WM Usrey, and RC Reid (1996): Precisely correlated firing in cells of the lateral geniculate nucleus. *Nature* 383: 815-819.
[364] Amigó JM, J Szczepański, E Wajnryb, and MV Sanchez-Vives (2004): Estimating the entropy rate of spike trains via Lempel-Ziv-complexity. *Neural Computation* 16: 717-736.
[365] Anderson JS, I Lampl, DC Gillespie, and D Ferster (2000): The contribution of noise to contrast invariance of orientation tuning in cat visual cortex. *Science* 290: 1968-1972.
[366] Arbib MA (ed.) (1998): *The Handbook of Brain Theory and Neural Networks.* MIT Press, Cambridge.
[367] Arbib MA (1987): *Brains, machines, and mathematics.* Springer Verlag, New York, Berlin, Heidelberg.
[368] Argyris J, G Faust, and M Haase (1995): *Die Erforschung des Chaos.* Vieweg, Braunschweig.
[369] Arieli A, A Sterkin, A Grinvald, and A Aertsen (1996): Dynamics of ongoing activity: Explanation of the large variability in the evoked cortical responses. *Science* 273: 1868-1871.
[370] Arthur B, and R Stoop (2004): *Drosophila m.* – where normal males tend to be female and the supermachos are fruitless. *Procceedings of NOLTA.*
[371] Atzori M, S Lei, DIP Evans, PO Kanold, E Phillips-Tanse, O McIntyre, and CJ McBain (2001): Differential synaptic processing separates stationary from transient inputs to the auditory cortex. *Nature Neuroscience* 4(12): 1230-1237.
[372] Azouz R, and CM Gray (2000): Dynamic spike treshold reveals a mechanism for synaptic coincidence detection in cortical neurons in vivo. *Proceedings of the National Academy of Science USA* 97(14): 8110-8115.
[373] Azouz R, and CM Gray (1999): Cellular mechanisms contributing to response variability of cortical neurons in vivo. *The Journal of Neuroscience* 19(6): 2209-2223.
[374] Bach M, and J Krüger (1986): Correlated neuronal variability in monkey visual cortex revealed by a multi-microelectrode. *Experimental Brain Research* 61: 451-456.
[375] Bair W (1999): Spike timing in the mammalian visual system. *Current Opinion in Neurobiology* 9: 447-453.
[376] Bair W, and C Koch (1996): Temporal precision of spike trains in extrastriate cortex of the behaving macaque monkey. *Neural Computation* 8: 1185-1202.
[377] Bair W, C Koch, W Newsome, and K Britten (1994): Power spektrum analysis of bursting cells in area MT in the behaving monkey. *The Journal of Neuroscience* 14(5): 2870-2892.
[378] Bair W, C Koch, W Newsome, and K Britten (1994): Relating temporal properties of spike trains from area MT neurons to the behavior of the monkey. In: G Buzsáki, R Llinas, W Singer, A Berthoz, and Y Christen: *Temporal coding in the brain.* Berlin, Heidelberg, New York. Springer Verlag: 221-250.
[379] Baker SN, and RN Lemon (2000): Precise spaciotemporal repeating patterns in monkey primary and supplementary motor areas occur at chance levels. *Journal of Neurophysiology* 84: 1770-1780.
[380] Barberini CL, GD Horwitz, and WT Newsome (2001): A comparison of spiking statistics in motion sensing neurones of flies and monkeys. In: JM Zanker, and J Zeil: *Motion vision – computational, neural and ecological constraints.* Springer Verlag, Berlin, Heidelberg, New York.
[381] Barlow HB (1996): Intraneuronal information processing, directional selectivity and memory for spatio-temporal sequences. *Network: Computation in Neural Systems* 7: 251-259.
[382] Barlow HB (1992): Single cells versus neuronal assemblies. In: A Aertsen, and V Braitenberg: *Information processing in the cortex.* Berlin, Heidelberg. Springer Verlag: 169-173.
[383] Barlow HB (1992): Neurons as computational elements. In: A Aertsen, and V Braitenberg: *Information processing in the cortex.* Berlin, Heidelberg. Springer Verlag: 175-178.
[384] Beierholm U, CD Nielsen, J Ryge, P Alstrøm, O Kiehn (2001): Characterization of reliability of spike timing in spinal interneurons during oscillating inputs. *Journal of Neurophysiology* 86: 1858-1868.
[385] Bekkers JM, GB Richerson, and CF Stevens (1990): Origin von variability in quantal size in cultured hippocampal neurons and hippocampal slices. *Proceedings of the National Academy of Science USA* 87: 5359-5362.

[386] Bennett CH, P Gács, M Li, PMB Vitányi, and WH Zurek (1998): Information distance. *IEEE Transactions on information theory* 44(4): 1407-1423.

[387] Bernander Ö, C Koch, and M Usher (1994): The effect of synchronized inputs at the single neuron level. *Neural Computation* 6: 622-641.

[388] Bernander Ö, RJ Douglas, KAC Martin, and C Koch (1991): Synaptic background activity influences spatiotemporal integration in single pyramidal cells. *Proceedings of the National Academy of Science USA* 88: 11569-11573.

[389] Berry MJ, and M Meister (1998): Refractoriness and neural precision. *The Journal of Neuroscience* 18(6): 2200-2211.

[390] Berry MJ, DK Warland, and M Meister (1997): The structure and precision of retinal spike trains. *Proceedings of the National Academy of Science USA* 94: 5411-5416.

[391] Bezrukov SM, and I Vodyanoy (1995): Noise-induced enhancement of signal transduction across voltage-dependent ion channels. *Nature* 378: 362-364.

[392] Bi G-q, and M-m Poo (2001): Synaptic modification by correlated activity: Hebb's postulate revisited. *Annual Review of Neuroscience* 24: 139-166.

[393] Bi G-q, and M-m Poo (1999): Distributed synaptic modification in neural networks induced by patterned stimulation. *Nature* 401: 793-796.

[394] Bialek W, F Rieke, R de Ruyter van Steveninck, and D Warland (1991): Reading a neural code. *Science* 252: 1854-1857.

[395] Blank DA (2001): *Firing rate amplification and collective bursting in models of recurrent neocortical networks*. PhD thesis, Swiss Federal Institute of Technology.

[396] Blank DA, and R Stoop (1999): Collective bursting in populations of intrinsically nonbursting neurons. *Zeitschrift für Naturforschung* A 54: 617-627.

[397] Bliss TVP, and GL Collingridge (1993): A synaptic model of memory: long-term potentiation in the hippocampus. *Nature* 361: 31-39.

[398] Borst A, and FE Theunissen (1999): Information theory and neural coding. *Nature Neuroscience* 2(11): 947-957.

[399] Bray D (1995): Protein molecules as computational elements in living cells. *Nature* 376: 307-312.

[400] Brenner N, O Agam, W Bialek, and R de Ruyter van Steveninck (2002): Statistical properties of spike trains: universal and stimulus-dependent aspects. *Physical Review E* 66: 031907-1-14.

[401] Brenner N, SP Strong, R Koberle, W Bialek, and R de Ruyter van Steveninck (2000): Synergy in a neural code. *Neural Computation* 12: 1531-1552.

[402] Brette R, and E Guigon (2003): Reliability of spike timing is a general property of spiking model neurons. *Neural Computation* 15: 279-308.

[403] Brezina V, PJ Church, and KR Weiss (2000): Temporal pattern dependence of neuronal peptide transmitter release: Models and experiments. *The Journal of Neuroscience* 20(18): 6760-6772.

[404] Brezina V, IV Orekhova, and KR Weiss (2000): The neuromuscular transform: The dynamic, nonlinear link between motor neuron firing patterns and muscle contraction in rhythmic behaviors. *Journal of Neurophysiology* 83: 207-231.

[405] Britten K, MN Shadlen, W Newsome, and JA Movshon (1993): Responses of neurons in macaque MT to stochastic motion signals. *Visual Neuroscience* 10: 1157-1169.

[406] Brivanlou IH, DK Warland, and M Meister (1998): Mechanisms of concerted firing among retinal ganglion cells. *Neuron* 20: 527-539.

[407] Brody CD (1998): Slow covariations in neuronal resting potentials can lead to artefactually fast cross-correlations in their spike trains. *Journal of Neurophysiology* 80: 3345-3351.

[408] Bullock TH (1997): Signals and signs in the nervous system: The dynamic anatomy of electrical signals is probably information-rich. *Proceedings of the National Academy of Science USA* 94: 1-6.

[409] Buračas GT, AM Zador, MR DeWeese, and TD Albright (1998): Efficient discrimination of temporal patterns by motion-sensitive neurons in Primate Visual Cortex. *Neuron* 20: 959-969.

[410] Burkitt AN, and GM Clark (1999): Analysis of integrate-and-fire neurons: synchronization of synaptic input and spike output. *Neural Computation* 11: 871-901.

[411] Butts DA (2003): How much information is associated with a particular stimulus. *Network: Computation in Neural Systems* 14: 177-187.

[412] Buzsáki G (2004): Large-scale recording of neuronal ensembles. *Nature Neuroscience* 7(5): 446-451.

[413] Buzsáki G, and A Draguhn (2004): Neuronal oscillations in cortical networks. *Science* 304: 1926-1929.

[414] Buzsáki G, and JJ Chrobak (1995): Temporal structure in spatially organized neuronal ensembles: a role for interneuronal networks. *Current Opinion in Neurobiology* 5: 504-510.

[415] Calabrese RL, and E De Schutter (1992): Motor-pattern-generating networks in invertebrates: modeling our way towards understanding. *Trends in Neuroscience* 15(11): 439-445.

[416] Calvin WH (1998): Cortical Columns, Modules, and Hebbian Cell Assemblies. In: MA Arbib (ed.). *The Handbook of Brain Theory and Neural Networks*. MIT Press, Cambridge Massachusetts: 269-272.

[417] Cariani P (1995): As if time really mattered: temporal strategies for neural coding of sensory information. *Communicaton and Cognition – Artificial Intelligence* 12(1-2): 157-219.

[418] Carr CE, and MA Friedman (1999): Evolution of time-coding systems. *Neural Computation* 11: 1-20.

[419] Carr CE (1993): Processing of temporal information in the brain. *Annual Review of Neuroscience* 16: 223-243.

[420] Cattaneo A, L Maffei, and C Morrone (1981): Patterns in the discharge of simple and complex visual cortical cells. *Proceedings of the Royal Society of London B* 212: 279-297.

[421] Chen Y-Q, and Y-H Ku (1992): Properties of favored patterns in spontaneous spike trains and responses of favored patterns to electroacupuncture in evoked trains. *Brain Research* 578: 297-304.

[422] Cecchi GA, M Sigman, J-M Alonso, L Martinez, DR Chialvo, MO Magnasco (2000): Noise in neurons is message dependent. *Proceedings of the National Academy of Science USA* 97(10): 5557-5561.

[423] Chi Z, PL Rauske and D Margoliash. Detection of spike patterns using pattern filtering, with application to sleep replay in birdsong. *Neurocomputing* 52-54, 19-24, 2003.

[424] Chow CC, and JA White (1996): Spontaneous action potentials due to channel fluctuations. *Biophysical Journal* 71: 3013-3021.

[425] Christen M, A Nicol, K Kendrick, T Ott, and R Stoop (2006): Odour encoding in olfactory neuronal networks beyond synchronisation. *NeuroReport*, submitted.

[426] Christen M, A Kohn, T Ott, and R Stoop (2006): A novel distance measure for classification of neuronal firing. *Journal of Neuroscience Methods*, in press.

[427] Christen M (2005): Proof that the LZ-distance satisfies the axioms of a metric. Working paper.

[428] Christen M, A Kern, J-J van der Vyver and R Stoop (2004): Pattern detection in noisy signals. *Proceedings of ISCAS 4*: 669-672.

[429] Christen M, A Kern, A Nikitchenko, W-H Steeb, and R Stoop (2004): Fast spike pattern detection using the correlation integral. *Physical Review E* 70: 011901-1-7.

[430] Christen M, T Ott, R Stoop (2004): Spike train clustering using a Lempel-Ziv distance measure. *Procceedings of NOLTA*: 379-382.

[431] Christen M, A Kern, and R Stoop (2003): A Correlation Integral Based Method for Pattern Recognition in Series of Interspike Intervals. In: *Proceedings of the IEEE Conference on Nonlinear Dynamics of Electronic Systems NDES 2003*: 49-52.

[432] Christodoulou C, and G Bugmann (2001): Coefficient of variation (CV) vs mean interspike interval (ISI) curves: What do they tell us about the brain? *Neurocomputing* 38-40: 1141-1149.

[433] Christodoulou C, and G Bugmann (2000): Near Poisson-firing produced by concurrent excitation and inhibition. *BioSystems* 58: 41-48.

[434] Churchland PS, and TJ Sejnowski (1992): *The computational brain*. MIT Press, Cambridge, England.

[435] Connors BW, and MJ Gutnick (1990): Intrinsic firing patterns of diverse neocortical neurons. *Trends in Neuroscience* 13(3): 99-104.

[436] Connors BW, MJ Gutnick, and DA Prince, (1982): Electrophysiological properties of neocortical neurons in vitro. *Journal of Neurophysiology* 488(6): 1302-1320.

[437] Cover TM, and JA Thomas (1991): *Elements of Information Theory*. John Wiley & Sons, New York.

[438] Croner LJ, KP Purpura, and E Kaplan (1993): Response variability in retinal ganglion cells of primates. *Proceedings of the National Academy of Science USA* 90: 8128-8130.

[439] Crutchfield JP (1994): The calculi of emergence: Computation, dynamics and induction. In: Proceedings of the Oji International Seminar 'Complex Systems – from Complex Dynamics to Artificial Reality', *Physica D* special issue.

[440] Dan Y, and M-m Poo (2004): Spike timing-dependent plasticity of neural circuits. Neuron 44: 23-30.

[441] Dan Y, J-M Alonso, WM Usrey, and RC Reid (1998): Coding of visual information by precisely correlated spikes in the lateral geniculate nucleus. *Nature Neuroscience* 1(6): 501-507.

[442] Date A, E Bienenstock, and S Geman (1998): *On the temporal resolution of neural activity*. Technical Report, Division of Applied Mathematics, Brown University, Providence.

[443] Dayan P, and LF Abbott (2001): *Theoretical neuroscience*. MIT Press, Cambridge, London.

[444] Dayhoff JE, and GL Gerstein. Favored Patterns in Spike Trains. I. Detection. *Journal of Neurophysiology*, 49(6), 1334-1348, 1983.

[445] Dayhoff JE, and GL Gerstein (1983). Favored Patterns in Spike Trains. II. Application. *Journal of Neurophysiology*, 49(6): 1349-1363.

[446] Deadwyler SA, and RE Hampson (1997): The significance of neural ensemble codes during behavior and cognition. *Annual Review of Neuroscience* 20: 217-244.

[447] Deadwyler SA, and RE Hampson (1995): Ensemble activity and behavior: What's the code? *Science* 270: 1316-1317.

[448] Dean AF (1981): The variability of discharge of simple cells in the cat striate cortex. *Experimental Brain Research* 44: 437-440.

[449] Debanne D, BH Gähwiler, and SM Thompson (1994): Synchronous pre- and postsynaptic activity induces associative long-term depression in area CA1 of the rat hippocampus in vitro. *Proceedings of the National Academy of Science USA* 91: 1148-1152.

[450] deCharms RC, and AM Zador (2000): Neural representation and the cortical code. *Annual Review of Neuroscience* 23: 613-647.

[451] deCharms RC (1998): Information coding in the cortex by independent or coordinated populations. *Proceedings of the National Academy of Science USA* 95(26): 15166-15168.

[452] deCharms RC, and MM Merzenich (1996): Primary cortical representation of sounds by the coordination of action-potential timing. *Nature* 381: 610-613.

[453] de Ruyter van Steveninck R, GD Lewen, SP Strong, R Koberle, and W Bialek (1997): Reproducibility and variability in neural spike trains. *Science* 275: 1805-1808.

[454] de Ruyter van Steveninck R, and W Bialek (1988): Real-time performance of a movement-sensitive neuron in the blowfly visual system: Coding and information transfer in short spike sequences. *Proceedings of the Royal Society of London B* 234: 379-414.

[455] Desmaisons D, J-D Vincent, P-M Lledo (1999): Control of action potential timing by intrinsic subtreshold oscillations in olfactory bulb output neurons. *The Journal of Neuroscience* 19(24): 10727-10737.

[456] DeWeese MR, and M Meister (1999): How to measure the information gained from one symbol. *Network: Computation in Neural Systems* 10: 325-339.

[457] Diesmann M, M-O Gewaltig, and A Aertsen (1999): Stable propagation of synchronous spiking in cortical networks. *Nature* 402: 529-533.

[458] Ding M, C Grebogi, E Ott, T Sauer, and JA Yorke (1993): Estimating correlation dimension fro a chaotic time series: when does plateau onset occur? *Physica D* 69: 404-424.

[459] Douglas RJ, and KAC Martin (2004): Neuronal circuits of the neocortex. *Annual Review of Neuroscience* 27: 419-451.

[460] Douglas RJ, and KAC Martin (1998): Neocortex. In: GM Sheperd: *The synaptic organization of the brain.* Oxford University Press: 459-509.

[461] Douglas RJ, and KAC Martin (1991): Opening the grey box. *Trends in Neuroscience* 14(7): 286-293.

[462] Douglas RJ, and KAC Martin (1991): A functional microcirquit for cat visual cortex. *Journal of Physiology* 440: 735-769.

[463] Douglass JK, L Wilkens, E Pantazelou, and F Moss (1993): Noise enhancement of information transfer in crayfish mechanoreceptors by stochastic resonance. *Nature* 365: 337-340.

[464] Eagan KP, and LD Partridge (1989): The sequential-interval state space: a means of displaying temporal information in neuron firing. *Journal of Neuroscience Methods* 27: 245-252.

[465] Eckhorn R, R Bauer, W Jordan, M Brosch, W Kruse, M Munk, and HJ Reitboek (1988): Coherent oscillations: a mechanism for feature linking in the visual cortex? *Biological cybernetics* 60: 121-130.

[466] Eckmann JP, OS Kamphorst and D Ruelle (1987): Recurrence plots of dynamical systems. *Europhysics Letters* 4: 973-977.

[467] Eggermont JJ (2001): Between sound and perception: reviewing the search for a neural code. *Hearing Research* 157: 1-42.

[468] Eggermont JJ (1998): Is there a neural code? *Neuroscience & Biobehavorial Reviews* 22(2): 355-370.

[469] Engel AK, P Fries, and W Singer (2001): Dynamic predictions: Oscillations and synchrony in top-down processing. *Nature Reviews Neuroscience* 2: 704-716.

[470] Engel AK, P König, AK Kreiter, and W Singer (1991): Interhemispheric synchronization of oscillatory neuronal responses in cat visual cortex. *Science* 252: 1177-1179.

[471] Engel AK, P König, and W Singer (1991): Direct physiological evidence for scene segmentation by temporal coding. *Proceedings of the National Academy of Science USA* 88: 9136-9140.

[472] Eysel UT, and U Burandt (1984): Fluorescent tube light evokes flicker responses in visual neurons. *Vision Research* 24(9): 943-948.

[473] Fee MS, PP Mitra, and D Kleinfeld (1996): Variability of extracellular spike waveforms of cortical neurons. *Journal of Neurophysiology* 76(6): 3823-3833.

[474] Feldmeyer D, and B Sakmann (2000): Synaptic efficacy and reliability of exitatory connections between the principal neurones of the input (layer 4) and output layer (layer 5) of the neocortex. *Journal of Physiology* 525: 31-39.

[475] Feldmeyer D, V Egger, J Lübke, and B Sakmann (1999): Reliable synaptic connections between pairs of exitatory layer 4 neurones within a single 'barrel' of developing rat somatosensory cortex. *Journal of Physiology* 521: 169-190.

[476] Fellin T, and G Carmignoto (2004): Neurone-to-astrocyte signalling in the brain represents a distinct multifunctional unit. Journal of Physiology 559(1): 3-15.

[477] Fellous J-M, PHE Tiesinga, PJ Thomas, and TJ Sejnowski (2004): Discovering spike patterns in neuronal responses. *Journal of Neuroscience* 24(12): 2989-3001.

[478] Fellous J-M, AR Houweling, RH Modi, RPN Rao, PHE Tiesinga, and TJ Sejnowski (2001): Frequency dependence of spike timing reliability in cortical pyramidal cells and interneurons. *Journal of Neurophysiology* 85: 1782-1787.

[479] Ferster D (1996): Is neural noise just a nuisance? *Science* 273: 1812.

[480] Ferster D, and N Spruton (1995): Cracking the neuronal code. *Science* 270: 756-757.

[481] Firestein S (2001): How the olfactory system makes sense of scents. *Nature* 413: 211-218.

[482] Fitzurka MA, and DC Tam (1999): A joint interspike interval difference stochastic spike train analysis: detecting local trends in the temporal firing pattern of single neurons. *Biological Cybernetics* 80: 309-326.

[483] Fleishman LJ, and JA Endler (2000): Some comments on visual perception and the use of video playback in animal behavior studies. *Acta Ethologia* 3: 15-27.

[484] Frégnac Y, V Bringuier, and A Baranyi (1994): Oscillatory activity in visual cortex: A critical re-evaluation. In: G Buzsáki, R Llinas, W Singer, A Berthoz, and Y Christen: *Temporal coding in the brain*. Berlin, Heidelberg, New York. Springer-Verlag: 81-102.

[485] Freeman TC, S Durand, DC Kiper, and M Carandini (2002): Suppression without inhibition in visual cortex. *Neuron* 35: 759-771.

[486] Frostig RD, Z Frostig, and RM Harper (1990): Recurring discharge patterns in multiple spike trains: I Detection. *Biological Cybernetics* 62: 487-493.

[487] Frostig RD, Z Frostig, and RM Harper (1984): Information trains. The technique and its uses in spike train and network analysis, with examples taken from the Nucleus Parabrachialis Medialis during sleep-wake states. *Brain Research* 322: 67-74.

[488] Furukawa S, and JC Middlebrooks (2002): Cortical representation of auditory space: information-bearing features of spike patterns. *Journal of Neurophysiology* 87: 1749-1762.

[489] Furukawa S, and JC Middlebrooks (2000): Coding of sound-source location by ensembles of cortical neurons. *The Journal of Neuroscience* 20(30): 1216-1228.

[490] Gat I, and N Tishby (2001): Spotting neural spike patterns using an adversary background model. *Neural Computation* 13: 2709-2741.

[491] Gautrais J, and S. Thorpe (1998): Rate coding versus temporal order coding: a theoretical approach. *BioSystems* 48: 57-65.

[492] Geisler WS, DG Albrecht, RJ Salvi, and SS Saunders (1991): Discrimination performance of single neurons: rate and temporal-pattern information. *Journal of Neurophysiology* 66: 334-362.

[493] Georgopoulos AP, AB Schwartz, and RE Kettner (1986): Neuronal population coding of movement direction. *Science* 233: 1416-1419.

[494] Gerstein GL, P Bedenbaugh and AMHJ Aertsen (1989): Neuronal Assemblies, *IEEE Transactions on Biomedical Engineering* 36(1): 4-14.

[495] Gerstein GL, and AMHJ Aertsen (1985): Representation of cooperative firing activity among simultaneously recorded neurons. *Journal of Neurophysiology* 54(6): 1513-1528.

[496] Gerstein GL, DH Perkel, and JE Dayhoff (1985): Cooperative firing activity in simultaneously recorded populations of neurons: detection and measurement. *The Journal of Neuroscience* 5(4): 881-889.

[497] Gerstein GL, MJ Bloom, IE Espinosa, S Evanczuk, and MR Turner (1983): Design of a laboratory for multineuron studies. *IEEE Transactions on Systems, Man, and Cybernetics* SMC-13(5): 668-676.

[498] Gerstner W, and W Kistler (2002): *Spiking Neuron Models: Single Neurons, Populations, Plasticity*. Cambridge University Press, Cambridge.

[499] Gerstner W, AK Kreiter, H Markram, and AVM Herz (1997): Neural codes: firing rates and beyond. *Proceedings of the National Academy of Science USA* 94: 12740-12741.

[500] Gerstner W, R Kempter, JL van Hemmen, and H Wagner (1996): A neuronal learing rule for submillisecond temporal coding. *Nature* 383: 76-78.

[501] Ghose GM, and RD Freeman (1992): Oscillatory discharge in the visual system: Does it have a functional role? *Journal of Neurophysiology* 68(5): 1558-1574.

[502] Gil Z, and BW Connors (1999): Efficacy of Thalamocortical and Intracortical Synaptic Connections: Quanta, Innervation and Reliability. *Neuron* 23: 385-397.

[503] Golledge HDR, CC Hilgetag, and MJ Tovée (1996): Information processing: A solution to the binding problem? *Current Biology* 6(9): 1092-1095.

[504] Grassberger P, and I Procaccia (1984): Dimensions and entropies of strange attractors from a fluctuating dynamics approach. *Physica D* 13: 34-54.

[505] Gray CM (1999): The temporal correlation hypothesis of visual feature integration: Still alive and well. *Neuron* 24: 31-47.

[506] Gray CM, and DA McCormick (1996): Chattering cells: superficial pyramidal neurons contributing to the generation of synchronous oscillations in the visual cortex. *Science* 274: 109-113.

[507] Gray CM (1994): Synchronous oscillations in neuronal systems: Mechanisms and functions. *Journal of Computational Neuroscience* 1: 11-38.

[508] Gray CM, P König, AK Engel, and W Singer (1989): Oscillatory responses in cat visual cortex exhibit inter-columnar synchronization with reflects global stimulus properties. *Nature* 338: 334-337.

[509] Grün S, M Diesmann, A Aertsen (2002a). Unitary events in multiple single-neuron spiking activity: I detection and significance. *Neural Computation*, 14: 43-80

[510] Grün S, M Diesmann, A Aertsen (2002b). Unitary events in multiple single-neuron spiking activity: II nonstationary data. *Neural Computation* 14: 81-119

[511] Grün S, M Diesmann, F Grammont, A Riehle, A Aertsen(1999). Detecting unitary events without discretization of time. *Journal of Neuroscience Methods* 93: 67-79.

[512] Grutzendler J, N Kasthuri, and W-B Gan (2002): Long-term dendritic spine stability in the adult cortex. *Nature* 420: 812-816.

[513] Gütig R, A Aertsen, S Rotter (2003). Analysis of higher-order neuronal interactions based on conditional inference. *Biological Cybernetics* 88: 352-359.

[514] Gütig R, A Aertsen, S Rotter (2001). Statistical significance of coincident spikes: Count-based versus rate-based statistics. *Neural Computation* 14: 121-153.

[515] Gur M, A Beylin, and DM Snodderly (1997): Response variability of neurons in primary visual cortex (V1) of alert monkeys. *The Journal of Neuroscience* 17(8): 2914-2920.

[516] Gutkin B, GB Ermentrout, and M Rudolph (2003): Spike generating dynamics and the conditions for spike-time precision in cortical neurons. *Journal of Computational Neuroscience* 15: 91-103.

[517] Halpern BP (2000): Sensory coding, decoding and representations: Unnecessary and troublesom constructs? *Physiology & Behavior* 69: 115-118.

[518] Harris KD, H Hirase, X Leinekugel, DA Henze, and G Buzsáki (2001): Temporal interaction between single spikes and complex spike bursts in hippocampal pyramidal cells. *Neuron* 32: 141-149.

[519] Hellwig B (2000): A quantitative analysis of the local connectivity between pyramidal neurons in layers 2/3 of the rat visual cortex. *Biological Cybernetics* 82: 111-121.

[520] Herrmann M, JA Hertz, and A Prügel-Benett (1995): Analysis of synfire chains. *Network: Computation in Neural Systems* 6: 403-414.

[521] Hessler NA, AM Shirke, and R Malinow (1993): The probability of transmitter release at a mammalian central synapse. *Nature* 366: 569-572.

[522] Ho N, and A Destexhe (2000): Synaptic background activity enhances the responsiveness of neocortical pyramidal neurons. *Journal of Neurophysiology* 84: 1488-1496.

[523] Holt GR, WR Softky, C Koch and R Douglas (1996): Comparison of discharge variability in vitro and in vivo in cat visual cortex. *Journal of Neurophysiology* 75(5): 1806-1814.

[524] Hopfield JJ, and CD Brody (2001): What is a moment? Transient synchrony as a collective mechanism for spatiotemporal integration. *Proceedings of the National Academy of Science USA* 98(3): 1282-1287.

[525] Hunter JD, and JG Milton (2003): Amplitude and frequency dependence of spike timing: Implications for dynamic regulations. *Journal of Neurophysiology* 90: 387-394.

[526] Hunter JD, JG Milton, PJ Thomas, and JD Cowan (1998): Resonance effect for neural spike time reliability. *Journal of Neurophysiology* 80: 1427-1438.

[527] Ikegaya Y, G Aaron, R Cossart, D Aronov, I Lampl, D Ferster, and R Yuste (2004): Synfire chains and cortical songs: temporal modules of cortical activity. *Science* 304: 559-564.

[528] Ip NY, and RE Zigmond (1984): Pattern of presynaptic nerve activity can determine the type of neurotransmitter regulating a postsynaptic event. *Nature* 311: 472-474.

[529] Izhikevich EM, NS Desai, EC Walcott, and FC Hoppensteadt (2003): Bursts as a unit of neural information: selective communication via resonance. *Trends in Neurosciences* 26(3): 161-167.

[530] Jin DZ (2004): Spiking neural network for recognizing spatiotemporal sequences of spikes. *Physical Review E* 69: 021905-1-13.

[531] Jin DZ (2002): Fast convergence of spike sequences to periodic patterns in recurrent networks. *Physical Review Letters* 89(20): 208102-1-4.

[532] Johnson DH, CM Gruner, K Baggerly, and C Seshagiri (2001): Information-theoretic analysis of neural coding. *Journal of computational neuroscience* 10: 47-69.

[533] Johnson KO (2000): Neural coding. *Neuron* 26: 563-566.

[534] Kaluźny P, and R Tarnecki (1993): Recurrence plots of neuronal spike trains. *Biological Cybernetics* 68: 527-534.

[535] Kandel ER, JH Schwartz, TM Jessell (2000): *Principles of neural science.* McGraw-Hill, New York.

[536] Kantz H, and T Schreiber (2000): *Nonlinear Time Series Analysis.* Cambridge University Press, Cambridge.

[537] Krolak-Salmon P, M-A Hénaff, C Tallon-Baudry, B Yvert, M Guénot, A Vighetto, F Mauguière and O Bertrand (2002): Human lateral geniculate nucleus and visual cortex respond to screen flicker. *Annals of Neurology* 53: 73-80.

[538] Kepecs A, and J Lisman (2003): Information encoding and computation with spikes and bursts. *Network: Computation in Neural Systems* 14: 103-118.

[539] Kisley MA, and GL Gerstein (1999): Trial-to-trial variability and state-dependent modulation of auditory-evoked responses in cortex. *The Journal of Neuroscience* 19(23): 10451-10460.

[540] Klemm WR, and CJ Sherry (1982): Do neurons process information by relative intervals in spike trains? *Neuroscience & Biobehavorial Reviews* 6: 429-437.

[541] Klemm WR, and CJ Sherry (1981): Serial order in spike trains: What's it "trying to tell us?" *International Journal of Neuroscience* 14: 15-33.

[542] Koch C, and I Segev (2000): The role of single neurons in information processing. *Nature Neuroscience* 3(supplement): 1171-1177.

[543] Koch C (1999): *Biophysics of Computation. Information Processing in Single Neurons.* Oxford University Press, New York, Oxford.

[544] Koch C, and G Laurent (1999): Complexity and the nervous system. *Science* 284: 96-98.

[545] Koch C, M Rapp, and I Segev (1996): A brief history if time (constants). *Cerebral Cortex* 6: 93-101.

[546] König P, AK Engel, and W Singer (1996): Integrator or coincidence detector? The role of the cortical neuron revisited. *Trends in Neuroscience* 19: 130-137.

[547] Kohn A, and MA Smith (2005): Stimulus dependence of neuronal correlation in primary visual cortex of the macaque. *The Journal of Neuroscience* 25(14): 3661-3673.

[548] Kretzberg J, M Egelhaaf, and A-K Warzecha (2001): Membrane potential fluctuations determine the precision of spike timing and synchronous activity: a model study. *Journal of Computational Neuroscience* 10: 79-97.

[549] Kupfermann I, and Weiss KR (1978): The command neuron concept. The Behavioral and Brain Sciences 1: 3-39.

[550] Ku Y-H, and X-Q Wang (1991): Favored patterns in spontaneous spike trains. *Brain Research* 559: 241-248.

[551] Langner G (1992): Periodicity coding in the auditory system. *Hearing Research* 60: 115-142.

[552] Larson J, D Wong, and G Lynch (1986): Patterned stimulation at the theta frequency is optimal for the induction of hippocampal long-term potentiation. *Brain Research* 368: 347-350.

[553] Laughlin SB, R de Ruyter van Steveninck, and JC Anderson (1998): The metabolic cost of neural information. *Nature Neuroscience* 1(1): 36-41.

[554] Laurent G, M Stopfer, RW Friedrich, MI Rabinovich, A Volkovskii, and HDI Abarbanel (2001): Odor encoding as an active, dynamical process: experiments, computation, and theory. *Annual Review of Neuroscience* 24: 263-297.

[555] Laurent G (1999): A systems perspective on early olfactory coding. *Science* 286: 723-728.

[556] Laurent G, M Wehr, and H Davidowitz (1996): Temporal representation of odors in an olfactory network. *The Journal of Neuroscience* 16(12): 3837-3847.

[557] Laurent G, and H Davidowitz (1994): Encoding of ofactory information with oscillating neural assemblies. *Science* 265: 1872-1875.

[558] Le Cun Y, and Y. Bengio (1998): Pattern recognition. In: MA Arbib: *The handbook of brain theory and neural networks*. The MIT Press, Cambridge: 711-715.

[559] Lee AK, and MA Wilson (2004): A combinatorical method for analyzing sequential firing patterns involving an arbitrary number of neurons based on relative time order. *Journal of Neurophysiology* 92: 2555-2573.

[560] Legéndy CR, and M Salcman (1985): Bursts and recurrence of bursts in the spike trains of spontaneously active striate cortex neurons. *Journal of Neurophysiology* 53(4): 926-939.

[561] Lestienne R (2001): Spike timing, synchronization and information processing on the sensory side of the central nervous system. *Progress in Neurobiology* 65: 545-591.

[562] Lestienne R, HC Tuckwell, M Chalansonnet, and M Chaput (1999): Repeating triplets of spikes and oscillations in the mitral cell discharge of freely breathing rats. *European Journal of Neuroscience* 11: 3185-3193.

[563] Lestienne R, and HC Tuckwell (1998): The significance of precisely replicating patterns in mammalian CNS spike trains. *Neuroscience* 82(2): 315-336.

[564] Lestienne R (1996): Determination of the precision of spike timing in the visual cortex of anesthetised cats. *Biological Cybernetics* 74: 55-61.

[565] Lestienne R (1988): Differences between monkey visual cortex cells in triplet and ghost doublet informational symbol relationships. *Biological Cybernetics* 59: 337-352.

[566] Lestienne R, and BL Strehler (1987): Time structure and stimulus dependence of precisely replicating patterns present in monkey cortical neuronal spike trains. *Brain Research* 437: 214-238.

[567] Letelier J-C, and PP Weber (2000): Spike sorting based on discrete wavelet transform coefficients. *Journal of Neuroscience Methods* 101: 93-106.

[568] Lewicki MS (1998): A review of methods for spike sorting: the detection and classification of neural action potentials. *Network: Computation in Neural Systems* 9: R53-R78.

[569] Lewis ER, KR Henry, and WM Yamada (2000): Essential roles of noise in neural coding and in studies of neural coding. *BioSystems* 58: 109-115.

[570] Li M, JH Badger, X Chen, S Kwong, P Kearney, and H Zhang (2001): An information-based sequence distance and its application to whole mitochondrial genome phylogeny. *Bioinformatics* 17(2): 149-154.

[571] Li M, and P Vitányi (1997): *An introduction to Kolmogorov complexity and its applications*. Springer Verlag, Berlin.

[572] Lindner JF, BK Meadows, TL Marsh, and WL Ditto (1998): Can neurons distinguish chaos from noise? *International Journal of Bifurcation and Chaos* 8(4): 767-781.

[573] Lindsey BG, KF Morris, R Shannon and GL Gerstein (1987): Repeated patterns of distributed synchrony in neuronal assemblies. *Journal of Neurophysiology* 78: 1714-1719.

[574] Lisman J (1997): Bursts as a unit of neural information: making unreliable synapses reliable. *Trends in Neuroscience* 20(1): 38-43.

[575] Liu RC, S Tzonev, S Rebrik, and KD Miller (2001): Variability and information in a neural code of the cat lateral geniculate nucleus. *Journal of Neurophysiology* 86: 2789-2806.

[576] Llinas R (1988): The intrinsic electrophysiological properties of mammalian neurons: Insights into central nervous system function. *Science* 242: 1654-1664.

[577] London M, A Schreibman, M Häusser, ME Larkum, and I Segev (2002): The information efficacy of a synapse. *Nature Neuroscience* 5(4): 332-340.

[578] London M, A Shcreibman, and I Segev (2002): Estimating information theoretic quantitites of spiketrains using the context three weighting algoithm. Supplementary material of London et al. (2002), *Nature Neuroscience* 5(4).

[579] London M, and I Segev (2001): Synaptic scaling in vitro and in vivo. *Nature Neuroscience* 4(9): 853-855.

[580] Lübke J, V Egger, B Sakmann, and D Feldmeyer (2000): Columnar organization of dendrites and axons of single and synaptically coupled excitatory spiny neurons in layer 4 of the rat barrel cortex. *Journal of Neuroscience* 20(14): 5300-5311.

[581] Lyskow E, V Ponomarev, M Sandström, KH Mild, and S Medvedev (1998): Steady-state visual evoked potentials to computer monitor flicker. *International Journal of Psychophysiology* 28: 285-290.

[582] MacLeod K, A Bäcker, and G Laurent (1998): Who reads temporal information contained across synchronized and oscillatory spike trains? *Nature* 395: 693-698.

[583] Magee JC (2003): A prominent role for intrinsic neuronal properties in temporal coding. *Trends in Neuroscience* 26(1): 14-16.

[584] Magee JC (2000): Dendritic integration of excitatory synaptic input. *Nature Reviews: Neuroscience* 1: 181-190.

[585] Magee JC, and EP Cook (2000): Somatic EPSP amplitude is independent of synapse location in hippocampal pyramidal neurons. *Nature Neuroscience* 3(9): 895-903.

[586] Magnusson MS (2000). Discovering hidden time patterns in behavior: T-patterns and their detection. *Behavior Research Methods, Instruments & Computers* 32(1): 93-110.

[587] Mainen ZF, and TJ Sejnowski (1996): Influence of dendritic structure on firing pattern in model neocortical neurons. *Nature* 382: 363-366.

[588] Mainen ZF, and TJ Sejnowski (1995): Reliability of spike timing in neocortical neurons. *Science* 268: 1503-1506.

[589] Makarov VA, VI Nekorkin, and MG Velarde (2001): Spiking behavior in a noise-driven system combining oscillatory and excitatory properties. *Physical Review Letters* 86(15): 3431-3434.

[590] Malenka RC, and MF Bear (2004): LTP and LTD: an embarrassment of riches. *Neuron* 44(1): 5-21.

[591] Manwani A, and C Koch (1999): Detecting and estimating signals in noisy cable structures I: neuronal noise sources. *Neural Computation* 11: 1797-1829.

[592] Mao B-Q, F Hamzei-Sichani, D Aronov, RC Froemke, and R Yuste (2001): Dynamics of spontaneous activity in neocortical slices. *Neuron* 32: 883-898.

[593] Markram H, J Lübke, M Frotscher, and B Sakmann (1997): Reglation of synaptic efficacy by coincidence of postsynaptic APs and EPSPs. *Science* 275: 213-215.

[594] Markram H (1997) A Network of Tufted Layer 5 Pyramidal Neurons. *Cerebral Cortex* 7: 523-533.

[595] Markram H, J Lübke, M Frotscher, A Roth, and B Sakmann (1997): Physiology and anatomy of synaptic connections between thick tufted pyramidal neurons in the developing rat neocortex. *Journal of Physiology* 500: 409-440.

[596] Marczynski TJ, LL Burns, and GT Marczynski (1980): Neuronal firing patterns in the feline hippocampus during sleep and wakefulness. *Brain Research* 185: 139-160.

[597] Marsalek P, C Koch, and J Maunsell (1997): On the relationship between synaptic input and spike output jitter in individual neurons. *Proceedings of the National Academy of Science USA* 94: 735-740.

[598] Martin KAC (1994): A brief history of the "feature detector". *Cerebral Cortex* 4: 1-7.

[599] Martin KAC, and D. Whitteridge (1984): Form, function and intracortical projection of spiny neurones in the striate visual cortex of the cat. *Journal of Physiology* 353: 463504.

[600] Martignon L, G Deco, K Laskey, M Diamond, W Freiwald, E Vaadia (2000). Neural coding: Higher-order temporal patterns in the neurostatistics of cell assemblies. *Neural Computation* 12: 2621-2653.

[601] Maynard EM, NG Hatsopoulos, CL Ojakangas, BD Acuna, JR Sanes, RA Normann, and JP Donoghue (1999): Neuronal interactions improve cortical population coding of movement direction. *The Journal of Neuroscience* 19(18): 8038-8093.

[602] McCormick DA, BW Connors, JW Lighthall, and DA Prince (1985): Comparative electrophysiology of pyramidal and sparsely spiny stellate neurons of the neocortex. *Journal of Neurophysiology* 54(4): 782-806.

[603] McCormick DA (1999) Spontaneous activity: signal or noise? *Science* 285: 541-543.
[604] Mechler F, RM Shapley, MJ Hawken, and DL Ringach (1996): Video refresh entrains neurons in monkey V1 cortex. *Society of Neuroscience Abstracts* 22: 1704.
[605] Meister M, and MJI Berry (1999): The neural code of the retina. *Neuron* 22: 435-450.
[606] Meister M (1996): Multineuronal codes in retinal signalling. *Proceedings of the National Academy of Science USA* 93: 609-614.
[607] Meister M, L Lagnado, and DA Baylor (1995): Concerted signalling by retinal ganglion cells. *Science* 270: 1207-1210.
[608] Melssen WJ, and WJM Epping (1987): Detection and estimation of neural connectivity based on crosscorrelation analysis. *Biological Cybernetics* 57: 403-414.
[609] Middlebrooks JC, AE Clock, L Xu, and DM Green (1994): A panoramic code for sound location by cortical neurons. *Science* 264: 842-844.
[610] Montgomery JM, and DV Madison (2004): Discrete synaptic states define a major mechanism of synaptic plasticity. Trends in Neurosciences 27(12): 744-750.
[611] Mountcastle VB (1997): The columnar organization of the neocortex. *Brain* 120: 701-722.
[612] Mountcastle VB (1993): Temporal order determinants in a somesthetic frequency discrimination: Sequential order coding. *Annals of the New York Academy of Sciences* 682: 150-170.
[613] Nádasdy Z (2000): Spike sequences and their consequences. *Journal of Physiology (Paris)* 94: 505-524.
[614] Nádasdy Z, H Hirase, A Czurkó, J Csicsvari, and G Buzsáki (1999): Replay and time compressio of recurring spike sequences in the Hippocampus. *The Journal of Neuroscience* 19(21): 9497-9507.
[615] Nakahara H, and S Amari (2002). Information-geometric measure for neural spikes. *Neural Computation* 14: 2269-2316.
[616] Nawrot MP, A Aertsen, and S Rotter (2003): Elimination of response latency variability in neuronal spike trains. *Biological Cybernetics* 88: 321-334.
[617] Nelson SB (2002): Cortical microcircuits: Diverse or canonical? Neuron 36: 19-27.
[618] Nicolelis MAL, AA Ghanzanfar, BM Faggin, S Votaw, and LMO Oliveira (1997): Reconstructing the engram: Simultaneous, multisite, many single neuron recordings. *Neuron* 18: 529-537.
[619] Nicoll A, and C Blakemore (1993) Patterns of local connectivity in the neocortex. *Neural Computation* 5: 665-680.
[620] Nikitchenko A, M Christen, and R Stoop(2004): Correlation integral properties of embedded periodic patterns. Unpublished manuscript.
[621] Nirenberg S, and PE Latham (2003): Decoding neuronal spike trains: How important are correlations? *Proceedings of the National Academy of Science USA* 100(12): 7348-7353.
[622] Novak K, de Camargo AB, Neuwirth M, Kothbauer K, Amassian VE, and Deletis V (2004): The refractory period of fast conducting corticospinal tract axons in man and its implications for intraoperative monitoring of motor evoked potentials. *Clinical Neurophysiology* 115(8): 1931-41.
[623] Nowak LG, R Azouz, MV Sanchez-Vives, CM Gray, and DA McCormick (2003): Electrophysiological classes of cat primary visual cortical neurons in vivo as revealed by quantitative analysis. *Journal of Neurophysiology* 89: 1541-1566.
[624] Nowak LG, MV Sanchez-Vives, and DA McCormick (1997): Influence of low and high frequency inputs on spike timing in visual cortical neurons. *Cerebral Cortex* 7: 487-501.
[625] Optican LM, and BJ Richmond (1987): Temporal encoding of two-dimensional patterns by single units in primate inferior temporal cortex. III information theoretic analysis. *Journal of Neurophysiology* 57(1): 162-178.
[626] Oram MW, D Xiao, B Dritschel, and KR Payne (2002): The temporal resolution of neural codes: dies response latency have a unique role? *Philosophical Transactions of the Royal Society of London B* 357: 987-1001.

[627] Oram MW, MC Wiener, R Lestienne, and BJ Richmond (1999): Stochastic nature of precisely timed spike patterns in visual system neuronal responses. *Journal of Neurophysiology* 81: 3021-3033.

[628] Oshio K-i, S Yamada, and M Nakashima (2003): Neuron classification based on temporal firing patterns by the dynamical analysis with changing time resolution. *Biological Cybernetics* 88: 438-449.

[629] Ott T, Kern A, Steeb WH, and Stoop R. (2005): Sequential clustering: Tracking down the most natural clusters. *Journal of Statistical Mechanics: Theory and Experiment*: P 11014

[630] Ott T, A Kern, A Schuffenhauer, M Popov, P Acklin, E Jacoby, and R Stoop (2004): Sequential superparamagnetic clustering for unbiased classification of high-dimensional chemical data. *Journal of Chemical Information and Computer Sciences* 44(4): 1358-1364.

[631] Ottersen OP, and JP Helm (2002): How hardwired is the brain? *Nature* 420: 751-752.

[632] Palanca BJA, and GC DeAngelis (2005): Does neuronal synchrony underlie visual feature grouping? *Neuron* 46: 333-346.

[633] Pareti G, and A. De Palma (2004): Does the brain oscillate? The dispute on neuronal synchronization. *Neurological Sciences* 25: 41-47.

[634] Pei X, L Wilkens, and F Moss (1996): Noise-mediated spike timing precision from aperiodic stimuli in an array of Hodgekin-Huxley-type neurons. *Physical Review Letters* 77(22): 4679-4682.

[635] Peinke J, J Parisi, OE Roessler and R Stoop (1992): *Encounter with Chaos.* Springer Verlag, Berlin.

[636] Petersen CCH, and B Sakmann (2000): The Exitatory Neuronal Network of Rat Layer 4 Barrel Cortex. *Journal of Neuroscience* 20(20): 7579-7586.

[637] Phillips WA, and W Singer (1997): In search of common foundations for cortical computation. *Behavioral and brain Sciences* 20: 657-722.

[638] Pouzat C, M Delescluse, P Viot, and J Diebolt (2004): Improved spike-sorting by modeling firing statistics and burst-dependent spike amplitude attenuation: A Markov chain Monte Carlo approach. *Journal of Neurophysiology* 91: 2910-2928.

[639] Prut Y, E Vaadia, H Bergman, I Haalman, H Slovin, and M Abeles (1998): Spatiotemporal structure of cortical activity: properties and behavioral relevance. *Journal of Neurophysiology* 79: 2857-2874.

[640] Quian RQ, Z Nadasdy, and Y Ben-Shaul (2004): Unsupervised spike detection and sorting with wavelets and superparamagnetic clustering. *Neural Computation* 16: 1661-1687.

[641] Quirk MC, and MA Wilson (1999): Interaction between spike waveform classification and temporal sequence detection. *Journal of Neuroscience Methods* 94: 41-52.

[642] Rabinovich MI, and HDI Abarbanel (1998): The role of chaos in neural systems. *Neuroscience* 87(1): 5-14.

[643] Reich DS, JD Victor, and BW Knight (1998): The power ratio and the interval map: spiking models and extracellular recordings. *The Journal of Neuroscience* 18(23): 10090-10104.

[644] Reich DS, JD Victor, BW Knight, T Ozaki, and E Kaplan (1997): Response variability and timing precision of neuronal spike trains *in vivo*. *Journal of Neurophysiology* 77: 2836-2841.

[645] Reinagel P, and RC Reid (2002): Precise firing events are conserved across neurons. *The Journal of Neuroscience* 22(16): 6837-6841.

[646] Reinagel P (2002): Information theory in the brain. *Current Biology* 10(15): R542-544.

[647] Richmond BJ, LM Optican, and H Spitzer (1990): Temporal cording of two-dimensional patterns by single units in primate primary visual cortex. I. Stimulus-response relations. *Journal of Neurophysiology* 64(2): 351-369.

[648] Richmond BJ, and LM Optican (1990): Temporal encoding of two-dimensional patterns by single units in primate primary visual cortex. II. Information transmission. *Journal of Neurophysiology* 64(2): 370-380.

[649] Richmond BJ, LM Optican, M Podell, and H Spitzer (1987): Temporal encoding of two-dimensional patterns by single units in primate inferior temporal cortex. I. Response characteristics. *Journal of Neurophysiology* 57(1): 132-146.

[650] Riehle A, S Grün, M Diesmann, and A Aertsen (1997): Spike synchronization and rate modulation differentially involved in motor cortical function. *Science* 278: 1950-1953.

[651] Rieke F, D Warland, R de Ruyter van Steveninck, and W Bialek (1999): *Spikes. Exploring the Neural Code.* MIT Press, Cambridge.

[652] Ritter H, and F Schwenker (1996): Kodierung. In: G Strube (ed.): *Wörterbuch der Kognitionswissenschaft.* Klett-Cotta, Stuttgart: 302.

[653] Rose GM, and TV Dunwiddie (1986): Induction of hippocampal long-term potentiation using physiologically patterned stimulation. *Neuroscience Letters* 69: 244-248.

[654] Rose D (1979): An analysis of the variability of unit activity in the cat's visual cortex. *Experimental Brain Research* 37: 595-604.

[655] Rosenmund C, JD Clements, and GL Westbrook (1993): Nonuniform probability of glutamate release at a hippocampal synapse. *Science* 262: 754-757.

[656] Sakurai Y (1996): Population coding by cell assemblies – what it really is in the brain. *Neuroscience Research* 26: 1-16.

[657] Sauer T (1994): Reconstruction of dynamical systems from interspike intervals. *Physical Review Letters* 72: 3811-3814.

[658] Schindler KA, CA Bernasconi, R Stoop, PH Goodman, and RJ Douglas (1997): Chaotic spike patterns evoked by periodic inhibition of rat cortical neurons. *Zeitschrift für Naturforschung A* 52: 509-512.

[659] Schneidman E, B Freedman, and I Segev (1998): Ion channel stochasticity may be critical in determining the reliability and precision of spike timing. *Neural Computation* 10: 1679-1703.

[660] Schnitzer MJ, and M Meister (2003): Multineuronal firing patterns in the signal from eye to brain. *Neuron* 37: 499-511.

[661] Schöner G, and JAS Kelso (1988): Dynamic pattern generation in behavioral and neural systems. *Science* 239: 1513-1520.

[662] Schreiber S, J-M Fellous, PHE Tiesinga, and TJ Sejnowski (2004): Influence of ionic conductances on spike timing reliability of cortical neurons for suprathreshold rhythmic inputs. *Journal of Neurophysiology* 91: 194-205.

[663] Schreiber S, J-M Fellous, D Whitmer, PHE Tiesinga, and TJ Sejnowski (2004): A new correlation-based measure of spike timing reliability. *Neurocomputing* 52-54: 925-931.

[664] Schubert D, JF Staiger, N Cho, R Kötter, K Zilles, and HJ Luhmann (2001): Layer-Specific intracolumnar and transcolumnar functional connectivity of layer V pyramidal cells in rat barrel cortex. *Journal of Neuroscience* 21(10): 3580-3592.

[665] Sejnowski TJ, C Koch, PS Churchland (1988): Computational neuroscience. *Science* 241: 1299-1306.

[666] Shadlen MN, and JA Movshon (1999): Synchrony unbound: A critical evaluation of the temporal binding hypothesis. *Neuron* 24: 67-77.

[667] MN Shadlen and WT Newsome (1998): The variable discharge of cortical neurons: Implications for connectivity, computation and information coding. *Journal of Neuroscience* 18(10): 3870-3896.

[668] Shadlen MN, and TW Newsome (1995): Is there a signal in the noise? *Current Opinion in Neurobiology* 5: 248-250.

[669] Shadlen MN, and WT Newsome (1994): Noise, neural codes and cortical organization. *Current Opinion in Neurobiology* 4: 569-579.

[670] Shalizi CR, LK Shalizi and JP Crutchfield (2002): An algorithm for pattern discovery in time series. *Santa Fe Insitute* Working Paper 02-10-060.

[671] Shalizi CR, and JP Crutchfield (2001): Computational mechanics: Pattern and Prediction, Structure and Simplicity. *Journal of Statistical Physics* 104(3/4): 817-879.

[672] Sherry CJ, DL Barrow, and WR Klemm (1982): Serial dependencies and Markov properties of neuronal interfspike intervals from rat cerebellum. *Brain Research Bulletin* 8: 163-169.

[673] Sherry CJ, and WR Klemm (1982): Failure of large doses of ethanol to eliminate Markovian serial dependencies in neuronal spike train intervals. *International Journal of Neuroscience* 17: 109-117.

[674] Siebler M, H Köller, G Rose HW Müller (1991): An improved graphical method for pattern recognition from spike trains of spontaneously active neurons. *Experimental Brain Research* 90: 141-146.

[675] Siegel RM, and HL Read (1993): Temporal processing in the visual brain. *Annals of the New York Academy of Sciences* 682: 171-178.

[676] Singer W (1999): Time as coding space? *Current Opinion in Neurobiology* 9: 189-194.

[677] Singer W (1998): Synchronization of neuronal responses as a putative binding mechanism. In: MA Arbib: *The handbook of brain theory and neural networks*. The MIT Press, Cambridge: 960-964.

[678] Singer W, and CM Gray (1995): Visual feature integration and the temporal correlation hypothesis. *Annual Review of Neuroscience* 18: 555-586.

[679] Singer W (1994): Time as coding space in neocortical processing: A hypothesis. In: G Buzsáki, R Llinas, W Singer, A Berthoz, and Y Christen: *Temporal coding in the brain*. Berlin, Heidelberg, New York. Springer-Verlag: 51-79.

[680] Sipser M (1997): *Introduction to the Theory of Computation*. PWS Publishing Company, Boston.

[681] Skaggs WE, and BL McNaughton (1996): Replay of neuronal firing sequences in rat hippocampus during sleep following spatial experience. *Science* 271: 1870-1873.

[682] Skottun BC, RL de Valois, DH Grosof, JA Movshon, DG Albrecht and AB Bonds (1991): Classifying simple and complex cells on the basis of response modulation. *Vision Research* 31(7/8): 1079-1086.

[683] Sneddon R (2001): Neural computation as an extension of information theory. *Proceedings of the 8th Joint Symposium on Neural Computation*. The Salk Institute, La Jolla, California.

[684] Snider RK, JF Kabara, BR Roig, and AB Bonds (1998): Burst firing and modulation of functional connectivity in cat striate cortex. *Journal of Neurophysiology* 80: 730-744.

[685] Softky WR, and C Koch (1993): The higly irregular firing of cortical cells is inconsistent with temporal integration of random EPSPs. *The Journal of Neuroscience* 13(1): 334-350.

[686] Softky WR (1995): Simple codes versus efficient codes. *Current Opinion in Neurobiology* 5: 239-247.

[687] Softky WR (1996): Fine analog coding minimizes information transmission. *Neural Networks* 9(1): 15-24.

[688] Sporns O, G Tononi, and GM Edelman (2002): Theoretical neuroanatomy and the connectivity of the celebral cortex. *Behavioral Brain Research* 135: 69-74.

[689] Spors H, and A Grinvald (2002): Spatio-temporal dynamics of odor representations in the mammalian olfactory bulb. Neuron 34: 301-315.

[690] Stahel WA (2002): *Statistische Datenanalyse*. Vieweg, Braunschweig, Wiesbaden.

[691] Staiger JF, R Kötter, K Zilles, and HJ Luhmann (2000): Laminar characteristics of functional connectivity in rat barrel cortex revealed by stimulation with caged-glutamate. *Neuroscience Research* 37: 49-58.

[692] Stevens CF, and AM Zador (1998): Input synchrony and the irregular firing of cortical neurons. *Nature Neuroscience* 1(3): 210-217.

[693] Stevens CF, and AM Zador (1995): The enigma of the brain. *Current Biology* 5(12): 1370-1371.

[694] Stoop R, and Steeb W-H (2005): *Berechenbares Chaos*. Birkhäuser, Basel.

[695] Stoop R, and N Stoop (2004): Natural computation measured as a reduction of complexity. *Chaos* 14(3): 675-679.

[696] Stoop R, and N Stoop (2004): Computation by natural systems defined. *Proceedings of ISCAS* 5: 664-667.

[697] Stoop R, N Stoop, and LA Bunimovich (2004): Complexity of dynamics as variability of predictability. *Journal of Statistical Physics* 114(3): 1127-1137.

[698] Stoop R, J Buchli, and M Christen (2004): Phase and frequency locking in detailed neuron models. In: *Proceedings of the International Symposium on Nonlinear Theory and its Applications NOLTA 2004*: 43-46.

[699] Stoop R, J-J van der Vyver, M Christen, and A Kern (2004): Where noise and precision join forces: Coding of neural information via limit cycles. In: *Proceedings of the International Symposium on Nonlinear Theory and its Applications* NOLTA 2004: 375-378.

[700] Stoop R, and BI Arthur (2004): Drosophila.m – where normal males tend to be female, and the supermachos are fruitless. In: *Proceedings of the International Symposium on Nonlinear Theory and its Applications* NOLTA 2004: 387-390.

[701] Stoop R, DA Blank, J-J van der Vyver, M Christen, and A Kern (2003): Synchronization, chaos and spike patterns in neocortical computation. *Journal of Electrical & Electronic Engineering* 3(1): 693-698.

[702] Stoop R, J Buchli, G Keller, and W-H Steeb (2003): Stochastic resonance in pattern recognition by a holographic neuron model. *Physical Review E* 67: 061918-1-6.

[703] Stoop R, and C Wagner (2003): Scaling properties of simple limiter control *Physical Review Letters* 90: 154101-1-4.

[704] Stoop R, DA Blank, A Kern, J-J van der Vyver, M Christen, S Lecchini, and C Wagner (2002): Collective bursting in layer IV: Synchronization by small thalamic inputs and recurrent connections. *Cognitive Brain Research* 13: 293-304.

[705] Stoop R, J-J van der Vyver, and A Kern (2001): Detection of noisy and pattern responses in complex systems. In: FL Neerhoff, A van Staveren, and CJM Verhoeven (eds.): *Proceedings of the IEEE Conference on Nonlinear Dynamics of Electronic Systems NDES*: 113-116.

[706] Stoop R, DA Blank, J-J van der Vyver, and A Kern (2001): Synchronization-based computation, chaos and spike patterns in neocortical neural networks. In: V Porra, M Valtonen, I Hartimo, M Ilmonen, O Simula, and T Veijola (eds.): *Circuit paradigm in the 21st century* 1: 221-224.

[707] Stoop R (2000): Efficient Coding and Control in Canonical Neocortical Microcircuits. *Proceedings of the IEEE Conference on Nonlinear Dynamics of Electronic Systems NDES*: 278-282.

[708] Stoop, K Schindler, and L Bunimovich (2000): Generic origins of irregular spiking in neocortical networks. *Biological Cybernetics* 83: 481-489.

[709] Stoop R, KA Schindler, and L Bunimovich (2000): When pyramidal neurons lock, when they respond chaotically, and when they like to synchronize. *Neuroscience Research* 36: 81-91.

[710] Stoop R, K Schindler, and LA Bunimovich (2000): Neocortical networks of pyramidal neurons: from local locking and chaos to macroscopic chaos and synchronization. *Nonlinearity* 13(5): 1515-1529.

[711] Stoop R, K Schindler, and LA Bunimovich (2000): Noise-driven neocortical interaction: A simple generation mechanism for complex neuron spiking. *Acta Biotheoretica* 48: 149-171.

[712] Stoop R, KA Schindler, and LA Bunimovich (1999): Inhibitory connections enhance pattern recurrence in networks of neocortical pyramidal cells. *Physics Letters A* 258: 115-122.

[713] Stopfer M, S Bhagavan, BH Smith, and G Laurent (1997): Impaired odour discrimination on desynchronization of odour-encoding neural assemblies. *Nature* 390: 70-74.

[714] Strehler BL, and R Lestienne (1986): Evidence on precise time-coded symbols and memory of patterns in monkey cortical neuronal spike trains. *Proceedings of the National Academy of Science USA* 83: 9812-9816.

[715] Strong SP, R Koberle, R de Ruyter van Steveninck, and W Bialek (1998): Entropy and information in neural spike trains. *Physical Review Letters* 80(1): 197-200.

[716] Takens F (1981): Detecting strange attractors in turbulence. In: DA Rand, and LS Young (eds.): *Dynamical systems and turbulence*. Lecture Notes in Mathematics 898, Springer, Berlin: 366-381.

[717] Tank DW, A Gelperin, and D Kleinfeld (1994): Odors, oscillations, and waves: Does it all compute? *Science* 265: 1819-1820.

[718] Tank DW, and JJ Hopfield (1987): Neural computation by concentrating information in time. *Proceedings of the National Academy of Science USA* 84: 1896-1900.

[719] Tarczy-Hornoch K, KAC Martin, KJ Stratford, and JJB Jack (1999): Intracortical excitation of spiny neurons in layer 4 of cat striate cortex in vitro. *Cerebral Cortex* 9: 833-843.

[720] Tetko IV AEP Villa (2001): A pattern grouping algorithm for analysis of spatiotemporal patterns in neuronal spike trains. 1. Detection of repeated patterns. *Journal of Neuroscience Methods* 105: 1-14.

[721] Tetko IV, and AEP Villa (2001): A pattern grouping algorithm for analysis of spatiotemporal patterns in neuronal spike trains. 2. Application to simultaneous single unit recordings. *Journal of Neuroscience Methods* 105: 15-24.

[722] Tetko IV, and AEP Villa (1997): Fast combinatorial methods to estimate the probability of complex temporal patterns of spikes. *Biological Cybernetics* 76: 397-407.

[723] Theunissen FE, and JP Miller (1995): Temporal encoding in nervous systems: A rigorous definition. *Journal of Computational Neuroscience* 2: 149-162.

[724] Thorpe S, D Fize, and C Marlot (1996): Speed of processing in the human visual system. *Nature* 381: 520-522.

[725] Tiesinga PHE, J-M Fellous, and TJ Sejnowski (2002): Attractor reliability reveals deterministic structure in neuronal spike trains. *Neural Computation* 14: 1629-1650.

[726] Tovée MJ, and ET Rolls (1992): The functional nature of neuronal oscillations. *Trends in Neuroscience* 15(10): 387.

[727] Trachtenberg JT, BE Chen, GW Knott, G Feng, JR Sanes, E Welker, and K Svoboda (2002): Long-term in vivo imaging of experience-dependent synaptic plasticity in adult cortex. *Nature* 420: 788-794.

[728] Traynelis SF, and F Jaramillo (1998): Getting the most out of noise in the central nervous system. *Trends in Neuroscience* 21: 137-145.

[729] Troyer TW, and KD Miller (1997): Physiological gain leads to high ISI variability in a simple model of cortical regular spiking cells. *Neural Computation* 9: 971-983.

[730] Tsuda I (2001): Toward an interpretation of dynamic neural activity in terms of chaotic dynamical systems. *Behavioral and Brain Sciences* 24: 793-847.

[731] Tsukada M, and N Sugano (1978): Effect of correlated adjacent interspike interval sequences of the input on the output response in the computer-simulated neuromuscular system. *Biological Cybernetics* 29: 69-73.

[732] Tuckwell HC (1988): *Introduction to theoretical neuroniology: Volume 2. Nonlinear and stochastic theories.* Cambridge University Press, Cambridge.

[733] Usrey WM (2002): The role of spike timing for thalamocortical processing. Current Opinion in Neurobiology 12: 411-417.

[734] Usrey WM, and RC Reid (1999): Synchronous activity in the visual system. *Annual Review of Physiology* 61: 435-456.

[735] Vaadia E, and A Aertsen (1992): Coding and computation in the cortex: single-neuron activity and cooperative phenomena. In: A Aertsen, and V Braitenberg: *Information processing in the cortex.* Berlin, Heidelberg. Springer Verlag: 81-121.

[736] van Rossum MCW (2001): The transient precision of integrate and fire neurons: effect of background activity and noise. *Journal of Computational Neuroscience* 10: 303-311.

[737] van Rossum MCW (2001): A novel spike distance. *Neural Computation* 13: 751-763.

[738] VanRullen R, R Guyonneau, and S Thorpe (2005): Spike times make sense. Trends in Neurosciences 28(1): 1-4.

[739] Victor JD (2000): How the brain uses time to represent and process visual information. *Brain Research* 886: 33-46.

[740] Victor JD (1999): Temporal aspects of neural coding in the retina and lateral geniculate. *Network: Computation in Neural Systems* 10: R1-R66.

[741] Victor JD and KP Purpura (1997): Metric-space analysis of spike trains: theory, algorithms and application. *Network: Computation in Neural Systems* 8: 127-164.

[742] Victor JD and KP Purpura (1996): Nature and precision of temporal coding in visual cortex: A metric space Analysis. *Journal of Neurophysiology* 76(2): 1310-1326.

[743] Vilim FS, EC Cropper, DA Price, I Kupfermann, and KR Weiss (1996): Release of peptide cotransmitters in Aplysia: regulation and functional implicance. *The Journal of Neuroscience* 16(24): 8105-8114.

[744] Villa AEP, and M Abeles (1990): Evidence for spatiotemporal firing patterns within the auditory thalamus of the cat. *Brain Research* 509: 325-327.

[745] Volgushev M, and UT Eysel (2000): Noise makes sense in neuronal computation. *Science* 290: 1908-1909.

[746] Volgushev M, M Chistiakova, and W Singer (1998): Modification of discharge patterns of neocortical neurons by induced oscillations of the membrane potential. *Neuroscience* 83(1): 15-25.

[747] von der Malsburg C (1981): *The correlation theory of brain function*. Max-Planck-Institute for Biophysical Chemistry, Göttingen. Internal Report 81-2.

[748] Walter MA, LJ Stuart, and R Borisyuk (2003): Spike Train Correlation Visualization. In: *Proceedings of the Seventh International Conference on Information Visualization*: 555-560.

[749] Warzecha A-K, J Kretzberg, and M Egelhaaf (2000): Reliability of a fly motion-sensitive neuron depends on stimulus parameters. *The Journal of Neuroscience* 20(23): 8886-8896.

[750] Warzecha A-K, and M Egelhaaf (1999): Variability in spike trains during constant and dynamic stimulation. *Science* 283: 1927-1930.

[751] Warzecha A-K, J Kretzberg, and M Egelhaaf (1998): Temporal precision of the encoding of motion information by visual interneurons. *Current Biology* 8: 359-368.

[752] Wehr M, and G Laurent (1996): Odour encoding by temporal sequences of firing in oscillating neural assemblies. *Nature* 384: 162-166.

[753] Wessberg J, CR Stambaugh, JD Kralik, PD Beck, M Laubach, JK Chapin, J Kim, SJ Biggs, MA Srinivasan, and MAL Nicolelis (2000): Real-time prediction of hand trajectory by ensembles of cortical neurons in primates. *Nature* 408: 361-365.

[754] Wiesenfeld K, and F Jaramillo (1998): Minireview of stochastic resonance. *Chaos* 8(3): 539-548.

[755] White JA, JT Rubinstein, and AR Kay (2000): Channel noise in neurons. *Trends in Neuroscience* 23: 131-137.

[756] White JA, R Klink, AA Alonso, and AR Kay (1998): Noise from voltage-gated ion channels may influence neuronal dynamics in the entorhinal cortex. *Journal of Neurophysiology* 80: 262-269.

[757] Wollman DE, and LA Palmer (1995): Phase locking of neuronal responses to the vertical refresh of computer display monitors in cat lateral geniculate nucleus and striate cortex. *Journal of Neuroscience Methods* 60: 107-113.

[758] Wu J-Y, Y Tsau, H-P Hopp, LB Cohen, and AC Tang, (1994): Consistency in nervous systems: Trial-to-trial and animal-to-animal variations in the responses to repeated applications of a sensory stimulus in Aplysia. *The Journal of Neuroscience* 14(3): 1366-1384.

[759] Young MP, K Tanaka, and S Yamane (1992): On oscillating neuronal responses in the visual cortex of the monkey. *Journal of Neurophysiology* 67(6): 1464-1474.

[760] Zador AM (2001): Synaptic connectivity and computation. *Nature Neuroscience* 4(12): 1157-1158.

[761] Zador AM (2000): The basic unit of computation. *Nature Neuroscience* 3 (supplement): 1167.

[762] Zador AM (1998): Impact of synaptic unreliability on the information transmitted by spiking neurons. *Journal of Neurophysiology* 79: 1219-1229.

[763] Zhou C, J Kurths, and B Hu (2001): Array-enhanced coherence resonance: Nontrivial effects of heterogeneity and spatia independence of noise. *Physical Review Letters* 87(9): 098101-1-4.

[764] Zhu JJ, and BW Connors (1999): Intrinsic firing patterns and whisker-evoked synaptic responses of neurons in the rat barrel cortex. *Journal of Neurophysiology* 81: 1171-1183.

Index

A

Abeles, Moshe .... 47, 144 f
Adaptation 32, 34
Adrian, Edgar..................... 32 ff, 36
Alphabet 104
Arnol'd coding 155
Ashby, Ross 28, 63, 68
Auditory system 45, 49

B

Background................. 116, 123 – 129
    activity 130
Barlow, Horace .... 42, 45, 47, 85, 87 f, 91
Bibliometrics 73
Binning 168
Bitstring 108
Brain-computer analogy.............. 64, 66
Brazier, Mary .... 50, 55, 57, 85 f
Bullock, Theodore .... 42, 50 f, 88, 92
Burns, Delisle .... 41, 64 f
Burst .... 51, 118, 120 f, 137, 143, 174

C

Channel 24, 47
    capacity................. 25, 43, 45 f, 48
    neural 27, 43
Channel noise .... 131 f
Cherry, Colin 26, 85, 87
Circle map 150
Clustering .... 78, 176, 186 f
    of spike trains .... 206 – 209
Code
    definition 104
    neural 19
    neural, decoding 108
    neural, definition 103
    neural, history 47
    neural, NRP-session 51
    population 113
    rate .... 34, 110 f
    sensory 110
    temporal .... 111 f
Code relation 104
Code transformation 106
Codeword 104
Coefficient of variation 124
Coincidence detection .... 143 f
Computation .... 113 f
Correlation integral 175
Cortical layers 137
Cybernetics .... 23, 27, 77, 90 f

D

Distance measures .... 79, 165 ff, 178
    LZ-distance .... 190 – 196

E

Embedding 169
Encoding a dynamical system 154
Erlanger, Joseph 35, 37
Event .... 106, 116, 118 – 121
    unitary 118

F

Fano factor 124
Feature detector 112
Fit-coefficient 175

G

Gasser, Herbert Spencer 37

Gerard, Ralph Waldo .... 20, 38, 74, 87 f
Gerstein, George 59, 145

H

Histogram .... 59, 164, 167 – 175
Hixon symposium .... 62, 67, 80 f
Hodgkin, Alan 37
Huxley, Andrew 37

I

Information 20
   and meaning 26
   concept 24
   conceptualization 23
   distance 166
   history of neural 56
   in molecular biology 56
   introduction in neuroscience 34
   measure of 24
   train 168
   vocabulary 27
Information age 23
Information capacity 43
Information theory 43, 57
   conferences 80
Interval pattern 117

J

Jitter 117

K

Kay, Lily 22, 57
Kolmogorov complexity 115

L

Lashley, Karl Spencer 30, 64
Limit cycle 149
Localization 30

M

MacKay, Donald 27, 44, 67, 88, 93
Macy conferences .... 26, 77 f

McCulloch, Warren ..... 23, 43 f, 71, 88 f, 96
Measurement, definition 152
Message .... 24, 34 f, 104

N

Neocortex 136
Network .... 31, 61 f, 70, 136 – 139
Neural modelling 69
   conferences 81
Neuron doctrine 31, 42
   single 42
Neuronal firing classes 120, 158
Neuroscience
   computational 77
   definition 20
   introduction of the term 20
   journals 74
   Society of 74
Neurosciences Research Program 22
   conferences 77
Noise 24, 27, 54, 65, 135
   additive 129
   definition 129
   external 129
   functional value .... 42, 55, 135 f, 156
   internal 129
   level 117
   membrane 131
   neural .... 27, 53 f, 98, 129 – 132
   synaptic .... 131 f
Noise sources 130

O

Order pattern 117
Oscillation .... 139 f
   definition 139

P

Parsing 191
Pattern discovery .... 164, 175 – 182
   experimental .... 209 f
Pattern group 122
Pattern quantification .... 164, 186 – 189
Pattern recognition .... 164, 183 – 186

Pattern stability .... 121 ff
    experimental determination .... 211 f
Pecher, Charles 35
Perceptron 69
Period 149
Periodicity 139, 149
Phase locking 151
Phrase 191
Pitts .... 23, 62, 70 f, 85 f
Poisson hypothesis .... 123 – 126
    experimental testing .... 201 ff
Poisson model 60, 110, 121, 126
    definition 124
Precision 117

Q

Quastler, Henry 26, 56, 66

R

Randomization .... 126 – 129
Randomness 54, 127
    connectivity 61
Rapoport, Anatol .... 45, 48, 62, 85 f
Rashevsky, Nicolas 70
Reichardt, Werner 85, 90, 95
Reliability
    experimental determination .. 203 – 206
Reliability of firing .... 132 – 135

S

Scaled pattern 117
Schmitt, Francis 22
Shannon, Claude .... 23 – 26, 85 f
Sherrington, Charles Scott 31, 34
Singer, Wolf 141
Spike .... 33, 37, 106, 118 f
    as symbol 106
Spike count 133, 166
Spike pattern 48, 119
    definition 116
    examples 144
    in the auditory system 146
    in the hippocampus 147
    in the olfactory system 146

in the visual system 145
    protagonists 145
Spike sorting 119
Spike timing 133
Spike train .... 37 f
    definition 108
Spike-time dependent plasticity 143
Spontaneous activity .... 65 f
State 153
State series 154
Stationarity 124
Stochastic resonance 135
Stoop, Ruedi 114, 148
Symbol 104, 106
Synchronization .... 140 f
Synfire hypothesis 138, 145

T

T-pattern 118, 165
Template .... 183 – 186
Temporal correlation hypothesis 141
Time series, definition 153
Timing pattern 118
Token 106
Transmitter release probability 142
Tuckwell classification scheme 172
Type 106

U

Uttal, William 49, 53

V

Variability .... 35, 53, 66, 98, 129, 131 f
Variation 117
Vocabulary 191
von Neumann, John .... 23, 47 f, 66 f

W

Weaver, Warren 21, 26
Wiener, Norbert 23, 27, 67
Wiersma, Cornelius .... 48, 85 f
Word 104

i want morebooks!

Buy your books fast and straightforward online - at one of world's fastest growing online book stores! Environmentally sound due to Print-on-Demand technologies.

Buy your books online at
www.get-morebooks.com

Kaufen Sie Ihre Bücher schnell und unkompliziert online – auf einer der am schnellsten wachsenden Buchhandelsplattformen weltweit! Dank Print-On-Demand umwelt- und ressourcenschonend produziert.

Bücher schneller online kaufen
www.morebooks.de

VDM Verlagsservicegesellschaft mbH
Heinrich-Böcking-Str. 6-8   Telefon: +49 681 3720 174   info@vdm-vsg.de
D - 66121 Saarbrücken   Telefax: +49 681 3720 1749   www.vdm-vsg.de

Printed by Books on Demand GmbH, Norderstedt / Germany